Rachel Armstrong

Vibrant Architecture: Matter as a CoDesigner of Living Structures

Rachel Armstrong

Vibrant Architecture: Matter as a CoDesigner of Living Structures

Managing Editor: Monika Michałowicz

Associate Editor: Davina Jackson

DE GRUYTER
OPEN

Published by De Gruyter Open Ltd, Warsaw/Berlin.

ISBN 978-3-11-040372-5
e- ISBN 978-3-11-040373-2

Bibliographic information published by the Deutsche Nationalbibliothek
The Deutsche Nationalbibliothek lists this publication in the Deutsche Nationalbibliografie; detailed bibliographic data are available in the Internet at http://dnb.dnb.de.

Managing Editor: Monika Michałowicz
Associate Editor: Davina Jackson

www.degruyteropen.com

Cover illustration: © Rachel Armstrong

Contents

Preface

This book sets out the conditions in which the need for a new approach to the production of architecture in the 21st century is established, where our homes and cities are facing increasing pressures from environmental challenges that are compromising our wellbeing and our lives. Vibrant architecture embodies a new kind of architectural design practice that explores how lively materials, or 'vibrant matter', may be incorporated into our buildings to confer on them some of the properties of living things, such as movement, growth, sensitivity and self-repair.

My research examines the theoretical and practical implications of how this may occur through the application of a new group of materials in the production of our living spaces, collectively referred to as 'vibrant matter'. Characteristically, these substances possess some of the properties of living systems but may not have the full status of being truly 'alive' and include forms of chemical 'artificial' life, such as 'dynamic droplets' or synthetically produced soils. These complex systems are able to directly communicate with the natural world using a shared language of chemistry and so negotiate their continued survival in restless contexts.

These chemical conversations may become a design strategy by applying the principles of an emerging scientific field called natural computing, which is evolving Alan Turing's interest in the computational powers of Nature. Natural computing shapes the outcomes of vibrant matter and offers a range of new tools for design through a new technical operating system identified as an 'assemblage'. Assemblages provide a unique set of associated concepts, operating principles and qualitatively distinctive outcomes from machines. A range of design projects that demonstrate the principles of a new approach to the choreography of space are explored through the construction of spatial programs and formulating design tactics in projects such as the 'Hylozoic Ground' installation, a collaboration with architect Philip Beesley for the 2010 Venice Architecture Biennale. Further experimental and speculative development of the assemblage operating system is explored through further design work in 'Vibrant Venice', which proposes to grow an artificial limestone reef underneath the foundations of the city. Urban-scale outcomes are also explored in 'Vibrant Cities', which applies synthetic soils as a material and technological strategy that optimizes environmental performance in underused and poorly imagined sites within urban environments.

Collectively, these technical and design studies suggest that the theory and practice of vibrant matter may give rise to new kinds of material solutions within the practice of the built environment, which could be applied to architectural design as 'vibrant architecture', which is stochastic and life-promoting – and stands in stark contrast to the prevalent view of sustainable practices, which are centred on an industrial approach to resource conservation.

Vibrant architecture may create new opportunities for architectural design practice that venture beyond top-down form-finding programs, by enabling architects to codesign in partnership with human and non-human collectives, which result in the production of post-natural landscapes. Ultimately, vibrant architecture may operate as an ecological platform for human development; one that augments, not diminishes, the liveliness of our planet.

Acknowledgements

There are many people to whom I owe a great deal of gratitude for their support in undertaking this research. Firstly, I should like to thank Professor Neil Spiller, Professor Stephen Gage and Associate Professor Martin Hanczyc for their continuing advice, experience and patience during its completion.

The research itself could not have been conducted without the support from the Center for Fundamental Living Technology, the European Centre for Living Technology, Philip Beesley Architect and the many staff and researchers within these institutions that offered their experience and advice along the way.

Certain individuals also deserve special mention, including Dan Slavinsky, Simone Ferracina, Phil Watson, Steve Fuller, Mike Phillips, Murray Fraser, Jim Clark, Joshua Galloway, Mark Morris, Davina Jackson, Carlos Olguin, Peter Feltham, Davide De Lucrezia, Robert Styblo, Ross Lysinger, Steen Rasmussen, Andrew Adamatzky, Michael Simon Toon, Jelena Guga, Monika Michałowicz and Stephen Murray.

Finally, I should like to acknowledge the support I received more generally from the staff at University College London in compiling this publication.

1 Introduction

The city, great cemetery of the animal kingdom, was closed, aseptic, over the final buried corpses with their last fleas and their last germs. Man had finally re-established the order of the world which he had himself upset: no other living species existed to cast any doubts. (Calvino, 1997, p.160)

1.1 Overview

At the start of the third millennium, the material conditions of the planet appear to have changed in ways that are causing global-scale problems (Jha, 2011). While our planet has always been a complex, turbulent system, we have recently been shielded from its tempestuous nature by global, technological advances that have created the impression that we can control Nature. Although industrialization has set incredible new standards, its disposable consumer culture and fossil fuel-based metabolisms have contaminated and poisoned our environment on a geological scale (Crutzen and Stoermer, 2000). Consequently, the material exchanges that spontaneously occur in natural systems are resulting in a new kind of planetary condition that defines the 21st century where global warming ravages our terrestrial landscapes (US National Research Council, 2010, p.3) and our oceans and abysses are contaminated by continent-sized garbage patches (McLendon, 2010) (see Fig. 1.1).

Figure 1.1: Marine organisms in the Venice lagoon produce solid materials through biological assemblages, which function as a Nature-based technology. Here, mussels, barnacles and oysters act as an accretion platform that may be used as an architectural material. Photograph, Rachel Armstrong, August 2012..

These physical transformations are feeding back negatively on those natural systems that support us. We are witnessing these impacts on our quality of life in devastating ways such as increasingly severe, changeable weather patterns (Lee, 2013) and a dramatic loss of biodiversity (Braun, 2010). Indeed, our environment appears to be changing so dramatically and irreversibly that Timothy Morton (2007) and Slavoj Žižek (Williams, 2009; Žižek, 2011) petition us to abandon our nostalgic ideas about Nature and re-engage with the material conditions in which we are *actually* immersed. Over the last few decades, environmentalists have successfully lobbied governments and industries in tackling some of the anthropogenic causes of these seismic shifts in materiality and have set in motion a worldwide desire for more 'sustainable' forms of human development.

> *They talk instead about 'sustainable development' – but there is no such thing. Not only can development not be sustained; even the existing fabric cannot be sustained any longer.* (Beer, 1992)

Sustainability has been the buzzword of the new millennium, whose colours have been pinned to an aspiration for a 'better' kind of human development (World Commission on Environment and Development, 1987). The term is widely applied to a whole range of activities, such as business (Sustainable Business, 2011), social networks (Imperial College London, 2013) and sustainable living (*Guardian*, 2013), where we try to minimize our environmental impacts. There are many different lobby groups and vested interests in how these changes are implemented, which stretch beyond the original concerns of environmentalists (Næss, 1989; Benson and Roe, 2000) owing to the economic implications in changing social, commercial and regulatory practices across a whole range of activities. Perhaps unsurprisingly, then, the term 'sustainability' embodies many contradictions that arise from conflicting paradigms. In its current form, sustainable practice is an industrial process, inspired by the Brundtland Report (World Commission on Environment and Development, 1987), which strives to meet the needs of the present generation without compromising the ability of future populations to also satisfy their own requirements. Sustainable approaches also propose to inflict less harm on ecological systems than previous forms of industrialization (Carson, 1962), while maintaining the basic principles of an industrialized society. For example, sustainable growth proposes that 'steady state' growth coupled to progressive, economically productive activity is achievable (Mourmoras, 1991), despite long standing concerns that large-scale resource consumption will inevitably lead to poverty and societal collapse (Malthus and Gilbert, 2008; Meadows et al, 1972). Yet, even mature ecosystems are in flux, where populations and resources wax and wane, with periodic regularity (MacLulich, 1937; Turchin, 2009). Indeed, in Nature it appears that only prehistoric oddities such as the coelacanth may live 'sustainably' – or perhaps more precisely, remain unaltered and even 'preserved' by evolutionary processes (Connor, 2013).

Despite humanity's mastery of technology, and our industrial liberation from the 'natural order' of the world, the Brundtland principle has already been breached many times. This implies that a purely conservational approach, based on incremental changes in our practices and lifestyles and characterized by acts of austerity, will simply not be sufficient to ensure future prosperity. Fossil fuel supplies will not provide for current generations, let alone for those yet to come (US Department of Energy, 2011). Worldwide water shortages have prompted civil conflict, such as the current hostilities between Pakistan and India (Malik, 2010) and famines are precipitated by social upheaval and natural disasters, such as those in North Korea (Yani, 2011). Moreover, advocates of sustainable approaches cannot clearly demonstrate the environmental benefits of their policies, since validation of changes today may not be evident for generations to come (Asheim, Buchholz and Tungodden, 2001). Yet, while environmental policies promote the idea of qualitatively different outcomes than industrial systems, they fundamentally rely on the same economic systems to validate their cause (Swartz, 2010). Qualitatively different paradigms for human development must, therefore, go beyond the established modes of practice and seek Buckminster Fuller's goal of building new models that make existing ones obsolete (Buckminster Fuller Institute, not dated). Yet, this does not mean that industry is obsolete or could be completely ousted by another dominant paradigm. Rather, it implies that the machine paradigm on which modern industry rests is an incomplete approach to making – where there is much room for technological improvement and opportunities to hybridize with alternative modes of production. Indeed, my research proposes to increase the range of available technological species by positioning Nature itself as an operating system that develops connections between unlike bodies. Nature-based technologies, therefore, may not only find ways of integrating with machines but in doing so, may transform their environmental impacts into more diverse and even environmentally positive outcomes. In this way, it may be possible to extend the range of technical systems available to us (see Fig. 1.2). Ultimately, such diversity catalyses the possibility of technological convergence and increases our options for identifying alternative platforms for underpinning human development (Armstrong, 2010b).

Contemporary architecture is at the heart of human development, and is therefore a cornerstone in initiating and developing different kinds of practices which could transform the way we live. Architectural practice may be viewed as a complex human activity that accommodates concepts such as form, structure, process, spatiality, time and materiality. It also embodies human interests in the ecological realm. As such, the built environment becomes a site for 'heterogeneous discourses, overlapping spatial programs and the constant interaction between movement, sensual experience and conceptual acrobatics' (Tschumi, 2012, p.193). These are not exclusively human concerns but are also relevant to the many other actants and systems that shape and enable the production of architecture. In this sense, the urban environment offers a rich fabric for potential change, which does not take place in steady, linear sequences but

Figure 1.2: Moss flourishing on a wall in Monte Vertia, Switzerland. The 'bioprocesses' that make up the metabolic functions of vibrant matter may be harnessed in a technological capacity. Photograph, Rachel Armstrong, July 2013.

may be provoked by interactions between different bodies, within a potent landscape of non-linear probabilities. The diffuse nature of architecture, therefore, may respond to many simultaneous events and accommodate the needs of heterogeneous societies and house our rapidly expanding global populations (Parker, 2011) while aiming to consume resources more mindfully and equitably in the process. The relationship between human consumption, materiality and quality of experience are key to the success of 21st century human development. Yet contemporary 'sustainable' architecture is reactively shaped by bucolic infatuations and formal discourses that reinforce, rather than extend, our current practices by prioritizing industrial, technological and political initiatives over enhancing lively urban fabrics, such as the incidental blooms of lichens and algae on the walls of buildings (see Fig. 1.3).

Buildings are not branded ecological for the quality of events and relationships they facilitate, but because they (literally) appear 'green', owing to the amount of vegetation overlaid on 'grey' concrete (Dean, 2011, p.230). Other approaches efficiently apply formal principles of material conservation to achieve their environmental goals, i.e. they consume less energy, use fewer resources and produce lower carbon emissions compared with those that do not follow sustainable criteria, such as the Leadership in Energy and Environmental Design (LEED) rating systems (US Green Building Council, 2013) and the UK's Building Research Establishment Environmental

Figure 1.3: Algae thrive in Venetian canals owing to the abundance of minerals, sunlight and carbon dioxide. They produce a range of pigments (green, red and brown) to differentially absorb the ultraviolet radiation spectrum. Photograph, Rachel Armstrong, August 2012.

Assessment Methodology (BREEAM, Building Research Establishment, 2010–2013). While sustainable architecture is firmly wedded to a particular kind of industrial thinking and practice, the potential role that architects could play in a much bigger environmental picture is being obscured – namely, the opportunity to identify a different paradigm for the production of architecture (Armstrong, 2012b), with qualitatively different kinds of outcomes for humans and non-humans, which may better orchestrate the material exchanges that flow through and are orchestrated by our cities (Armstrong, 2013e).

My research does more than suppose this possibility but also gathers evidence from scientific experiments, and applies this through the construction of architectural prototypes and models, to establish a toolset of conceptual and practical approaches for producing new kinds of architectures. I do not propose to construct a particular architecture to embody these ideas, but to identify a new technological platform based on the interactions of lively, material assemblages that may increase the range of possible species of new architectures. These options are outlined in a manifesto, so that spatial programs may be regarded as transformers of matter, such as green leaves converting sunlight and carbon dioxide into sugar and water, rather than simply positioning buildings as geometric obstacles in a landscape around which living things are compelled to move. To develop a new approach that acknowledges the agency of the material world, we must realize that cities are sites for only a tiny fraction of the planet's exchanges of matter. Other exchanges constantly occur through 'unbuilt' living systems such as seas, soils and rainforests. Natural networks draw on a much larger reserve of creativity than exists within our man-made urban spaces, and ally with the potency of robust, self-replenishing environmental cycles – our ecologies (see Fig. 1.4). Indeed, Jane Bennett has observed that matter possesses differing degrees of agency that can shape human events – characteristics that are not appreciated by industrial modes of practice (Bennett, 2010). My research is intended to develop a design approach for more equitable architectural practices that support both human and non-human interests by thinking much more broadly about the performance and the innate creativity of materials. I will also consider how these qualities and relationships may be developed through infrastructures that enable overlapping programs to be developed and mutually coexist in elemental media, such as water, air and earth.

These challenges are more than simply about developing a single high-performance, efficient materials such as ethylene tetrafluoroethylene (ETFE), or new technologies like aerosol sealing (UC Davis, 2013). They instead require architects to engage with ecologies of lively interactions between diverse architectural agents that mutually (re)inform each other. It may be possible to develop design tactics and spatial programs that deal with the complex exchanges between assemblages, whose effects may be shaped to exert influences at the urban scale. For example, chemical assemblages that can metabolize or 'fix' pollutants (*Treehugger*, 2005; *Raconteur*, 2012; Fraunhofer, 2012) could improve air quality, raise the quality of life of residents and increase the prosperity of the area. Ultimately these collaborative exchanges may shape the gargantuan streams of matter around our cities so they work synergistically with us, rather than in opposition.

By recognizing our connection with the material world (Sagan, 2007; Bennett, 2010), my research builds a case for an approach to architectural design that treats matter as possessing technological potential, so that it is no longer a passive subject of design interventions, but becomes an active codesigner of spatial systems. The parallel processing abilities of chemistry enable lively materials to function simultaneously as substrates, infrastructures and technologies, which respond directly to environmental changes without the need for interpretation, such as through software. Such directness and lack of abstraction enriches the networks of exchange that exist between local ecosystems, human lifestyles and the biosphere – rather than weakening or destroying them (Armstrong, 2011c).

My research tests the hypothesis that by reframing our expectations of matter, it may be possible to establish principles for an architectural design practice that is not simply environmentally benign, but active, participatory and subversive (Spiller and Armstrong, 2011, p.18). Potentially, the convergence of vibrant matter with a range of technological platforms such as digital, nanotechnological and biological (Roco and Bainbridge 2003), may even revitalize the ecological networks that entangle the living and non-living world (see Fig. 1.5).

Figure 1.4: The potential synthesis between natural and human-made systems is captured in this photograph where abundant resources, such as sunlight and carbon dioxide, are transformed by algae into a gelatinous material. Photograph, Rachel Armstrong, August 2012.

The scope and design of my research is ambitious and has been produced by working in concert with international, collaborative, multidisciplinary teams. I have employed various approaches that span the fields of architectural design and chemistry, including experimental work, model building, probabilistic approaches and speculative design. The terminology and concepts which are used throughout my work are outlined in a chapter on 'definitions', since they are drawn from diverse sources that span the Two Cultures (Snow, 1959) and based on experimental findings that are not already widely reported in the literature. By theoretically and experimentally investigating the principles of vibrant matter I aim to identify ways of working that may provide new materials, technologies, tactics and spatial programs, which enable and promote the incredible material vitality in which we are immersed. Potentially, such explorations

Figure 1.5: Dynamic droplets can transform simple mineral solutions into complex crystalline formations. Photograph, Rachel Armstrong, August 2012.

may expand the range of materials and technologies available to architectural design practice and suggest new platforms for human development which do not deplete but enliven our ecological relationships.

1.2 Vibrant Matter in Practice

The first chapter proposes that material agency is more than a theory but a real and testable proposal. My particular view of vibrant matter differs from Bennett's in its experimental approach, which is directed towards building prototypes and architectural models. Moreover, Bennett invokes non-material essences to account for the lively nature of matter such as vitalistic forces (Bergson, 1922; Driesch, 1929; Bennett, 2010; Spuybroek, 2011), or by inviting viewers to participate in metaphors[1] (Deleuze and Guattari, 1979)[21] and provoking psychological responses (Bennett, 2010, p.4), while my research takes a materialist perspective that attributes the lively

1 Emergence is a term that proposes an alternative roadmap of organization between a mechanistic view of the world and a vitalistic one. It originates from John Stuart Mill's work in the mid 19th century that sought to describe the nature of vital substances as being composed of inanimate materials (O'Connor and Wong, 2013).

properties of vibrant matter to the quantum world. I propose that it may be possible to accept the codesignership of materials (see Fig. 1.6) as a reality by recognizing how quantum physics shapes everyday experiences, such as colour and gravity. Understanding the physical systems that shape our sensory experiences may also help us develop spatial programs and design tactics to perhaps work with these qualities differently in the construction of our living spaces.

1.3 Dynamic Droplets

This chapter establishes a model system for examining the principles of vibrant matter as a series of experiments. Bütschli droplets, which emerge at the interface between strong aqueous alkali and olive oil, possess striking lifelike properties that can be observed at the human scale. They embody the principles of vibrant matter and offer a robust platform that can be applied as material, infrastructure and technology (see Fig. 1.7). Bütschli droplets are programmable, modifiable and capable of producing environmentally responsive events that can be practically explored within an architectural design context.

Figure 1.6: Chemical self-assembly in dynamic droplets may produce a range of living phenomena from movement to microstructures. Photograph, Rachel Armstrong, February 2011.

Figure 1.7: Dynamic droplets may be practically explored within an architectural design context. Photograph, Rachel Armstrong, August 2012.

1.4 Hylozoic Ground

Applications of a range of lively chemistries are explored in this chapter as part of an architectural design context for the 'Hylozoic Ground' installation. This was a collaboration with architect Philip Beesley that was exhibited at the 2010 Venice Architecture Biennale (Armstrong and Beesley, 2011) (see Fig. 1.8). Far from equilibrium chemical systems were identified as models for vibrant matter that could respond morphologically, chemically and poetically to the installation's themes and programs that explored notions of 'life', ecology, the quality of spatial experience and dynamic systems. The question of scale was implicit in this design challenge (Haldane, 1926; Thompson, 1917, pp.22–77; Morris, 2011, p.49) and, for example, the millimetre scale field of action of the Bütschli system could be further increased to the metre scale. This was achieved by slowing down the metabolism of the droplets and entangling distributed populations of the chemistry in flasks within the cybernetic matrix. By working with a range of materials and infrastructures, an initial set of design principles for working with vibrant matter in an architectural context was developed. This enabled further speculation and reflection on the range of architectural design contexts in which vibrant matter could be applied.

Figure 1.8: Modified Bütschli droplets within the cybernetic matrix of the Hylozoic Ground installation are processing carbon dioxide and turning it into a solid. Photograph, Rachel Armstrong, July 2010.

1.5 Vibrant Venice

An urban-scale application of vibrant matter was proposed as an architectural project to develop the design principles established during the Hylozoic Ground installation. Programmable chemistries, or 'protopearls', were envisaged to transform Venice's foundations, so the city could fight back against the ravages of the elements in a struggle for survival by growing an artificial limestone reef underneath the foundations of the city. The expansion of the solid base of the city, which historically rests on narrow woodpiles, was imagined, to attenuate the city from sinking into the soft delta soils on which it was founded. This proposal was tested in the laboratory by developing a model system for protopearls that were capable of producing shell-like structures, and was tested in the field using samples of local water. Indeed, Venice was chosen because of its unique situation in a lagoon, where water provided the site for the action of droplets as well as a resource-rich medium for chemical ingredients needed by the droplets, such as minerals and carbon dioxide. Concurrently, a series of collaborative drawings by architect/artist Christian Kerrigan and GMJ were developed to explore how dynamic droplets may perform within a design context. Additionally, a photographic survey was conducted to characterize the behaviour of the marine ecology around the canals, which revealed that the marine organisms were already forming an accretion technology similar to the one proposed, by strategically introducing dynamic droplets into the waterways (see Fig. 1.9). This project proposed substantial, transformative change within the city's foundations and raised ethical and moral questions that would be expected to inform a possible architectural design practice where non-human actants work alongside humans as codesigners of their environment. The complex spatial and material tactics proposed for 'Vibrant Venice' were envisaged as robust synthetic ecologies around the shoreline as a post-natural, coastal landscape that seamlessly integrates with the city. These design principles may not only be relevant to Venice but also to other sites that are challenged by periods of flooding, intercoastal zones and reclaimed land, such as Songdo (Cisco, not dated).

Figure 1.9: In this Venetian canal,hrombolite-like mineral accretions are produced by the native ecology. Photograph, Rachel Armstrong, August 2012.

1.6 Vibrant Cities

This chapter considers applications of vibrant matter within a non-coastal setting by proposing that soil can be used as a chemical technology and infrastructure to strategically process and shape the flow of matter. Soils offer an architectural medium by enabling the simultaneous existence of overlapping architectural programs that can be read through chemical software (see Fig. 1.10). These tactics enable the material systems to change with time and produce a new kind of experience for communities by increasing the material fertility of their surroundings. Enabled by these new production platforms alternative social groupings may form around oases of productive potency that may influence the cultural or economic value of soils within our cities.

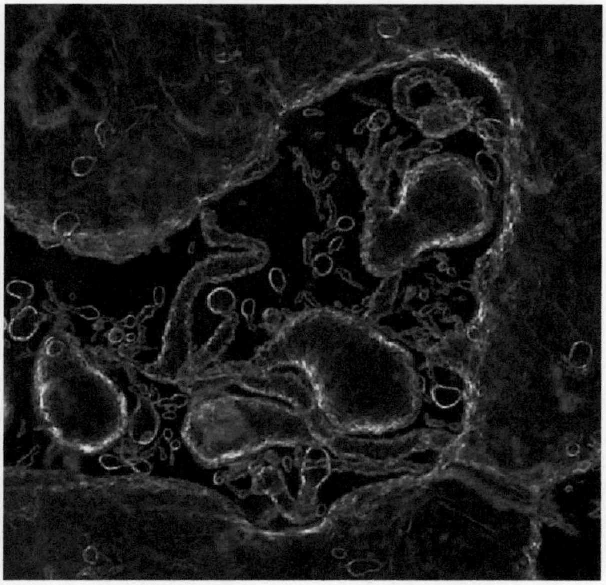

Figure 1.10: These soft, crystalline casts are soil-like in their appearance and are spun from bodies of dynamic droplets that function like 'chemical worms'.Micrograph, magnification 4×, Rachel Armstrong, August 2012.

1.7 Manifesto for Vibrant Architecture

As a way to integrate my research findings, I developed a manifesto for 'vibrant architecture' that advocates the production of post-natural fabrics (see Fig. 1.11). It proposes a vision of what a new kind of architectural design practice might be (Woods, 2011). The manifesto was inspired by writings such as the 'Protocell Manifesto' that I coauthored with Neil Spiller and Martin Hanczyc, which was based on a Dadaist text that countered biological formalism (Spiller, Armstrong and Hanczyc, 2011) and also Hundertwasser's manifestos which advocate for harmony with Nature as well as individual self-expression (Hundertwasser, not dated; Hundertwasser, 1976).

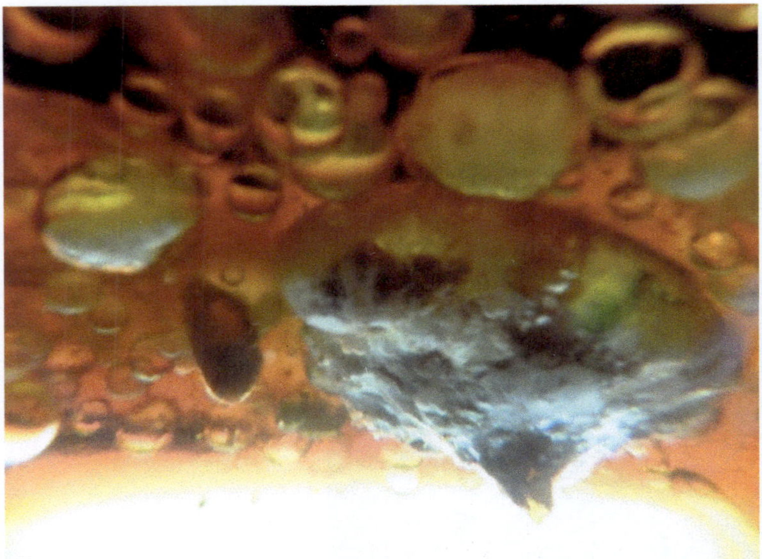

Figure 1.11: Dynamic droplet experiments inform the principles of practice for a manifesto for vibrant architecture. Photograph, Rachel Armstrong, February 2010.

1.8 Speculative Narratives

In keeping with the exploratory character of the materials and technologies that inform my research, I wrote five short speculative fictional narratives which relate to the subject matter under discussion. These compositions serve as design experiments, used in illustrative contexts. They have also been used to reflect on key issues during different stages of my research and include:
— 'A Short Story of a Short Life'
— 'Biolime: Mock Rock'
— 'Post-natural Venice'
— 'Japan: 2060'
— 'The Greatest Alluvian Poet that Ever Lived'

1.9 Summary

As a design theory, vibrant architecture advocates a way of imagining and working equally with assemblages of human and non-human, living and non-living matter. It proposes new architectural substrates, tactics and programs across many scales which operate within a probabilistic worldview and may build and enhance the ecosystems of actors that inhabit and shape our living spaces.

2 Definitions

These are interesting times for architecture. As existing paradigms are being eroded by external factors – be it the economic crisis or the impressive rise of mobile media – the profession seems wandering in search of a solid ground from which to rethink itself and its production. (Botazzi, 2012)

2.1 Overview

My research brings together the humanities and sciences whose intellectual systems are culturally distinct (Snow, 1959).[2] Early definition of the various terms of reference is therefore required to familiarize readers with the broad range of subjects addressed in my work and to establish their critical context.

2.2 Aim of Research

Practical steps are established to enable a transition from industrial to ecological architectural practices. Using the parallel processing powers of the material world, a Nature-based production platform is informed, which evades traditional binary divisions that characterize modern architecture, such as Nature/machine, humanism/environmentalism and matter/information. In dissolving these divisions that influence design decisions, I aim to increase the connectivity of architectural substrates and enable designers to explore new ways of thinking and making. In establishing the philosophical and cultural context for involving material properties in design solutions, I have drawn on Bennett's notion of vibrant matter (Bennett, 2010), which invites us to recognize that we share a common ontology with matter through stardust (*Science Daily*, 1999; Sagan, 2007), which is intrinsically lively. The fabric of our existence can therefore respond directly to spatial programs, codesign events and ultimately exert collective effects through a new technological platform based on assemblages. The properties of this platform are experimentally demonstrated using chemistry at non-equilibrium states and in different contexts, which are further developed and critiqued in project work. Research findings are used to develop a manifesto that establishes a series of principles for the practice of vibrant architecture. The conceptual and practical transitions in my research demonstrate how matter may:

---

2 The first popular airing of the growing 20th century rift between the humanities and science is usually attributed to C.P. Snow's 1959 Rede lecture 'The Two Cultures and the Scientific Revolution' (Snow, 1959), which proposed that the cultural differences between the arts and sciences presented a 'gulf of mutual incomprehension' that could not be reconciled.

- Be lively
- Perform computational tasks when specific infrastructural and environmental conditions are met
- Provide self-supporting infrastructures
- Become a technology, or an architectural production platform

The themes that run through my research have been derived in a particular way and are influenced by contingent thoughts from a range of disciplines. It is therefore essential to outline the terminology that informs these ideas so that multidisciplinary influences may converge to clearly invoke specific sets of concepts from the outset and bring clarity to the proposal.

2.3 Process Philosophy

Although my research spans disciplines, it is united through a common set of ideas, which I refer to as 'process philosophy' (Seibt, 2012). While classical western philosophy engages with a world whose matter is effectively static where movement and process are secondary or derivative, process philosophy is concerned with the idea of 'becoming' as a primary quality of experience both for humans and the material world. Process philosophers share a set of principles that describe reality as a combination of physical, organic, social and cognitive processes, which are entangled across many levels of organization (see Fig. 2.1). Yet, within this broad framework, debates vary on how a world forged by process is construed. Examples of process philosophers include Heraclitus, Gottfried Wilhelm Leibniz, Georg Hegel, Friedrich Nietzsche, Martin Heidegger, Jacques Derrida, Alfred North Whitehead, Henri Bergson, William James and John Dewey, although this list is not exhaustive (Seibt, 2012).

2.4 Complexity

While the language and metaphysics of process philosophy provide the semantics that inform the design principles that I apply in my research, complexity is the science and engineering framework through which these ideas may be experimentally tested and constructed. Although complexity is a relatively new science, it is a very old idea and its principles have been explored and expressed across disciplines throughout the ages, in many ways, to deal with notions of an ever-changing reality. Works could include, for example, William Blake's notion of evanescence, the poetry and prose of the Romantics,[4] Henri Lefebvre's Rhythmanalysis (Lefebvre, 2004), Jan Christiaan Smuts' idea of 'holism' (Smuts, 1998), Norbert Wiener's 'cybernetics' (Wiener, 1948;

Figure 2.1: Dynamic interactions between non-object systems such as sunlight and water movement are best conveyed in terms of their processes and captured for a moment here on the surface of the Venetian lagoon. Photograph, Rachel Armstrong, August 2012.

Wiener, 1998), James Lovelock's Gaia Theory (Lovelock, 1979), Ilya Prigogine's 'dissipative' structures (Prigogine, 1997, p.27), Jane Bennett's notion of vibrant matter (Bennett, 2010), Giles Deleuze and Felix Guattari's engagement with 'agentised' matter (Deleuze and Guattari, 1979), the mathematics of chaos (Gleick, 2008) and the science of non-linear systems (Schuster, 2007). In short, these many definitions and explorations denote an intellectual field that spans disciplines and, therefore, engages with a variety of methods and perspectives for working with complexity (Armstrong, 2013b).

Complexity was scientifically formalized by Ludwig von Bertalanffy, who proposed 'general systems theory' as a theoretical framework (Von Bertalanffy, 1950). This was further developed in collaboration with Ross Ashby in the emerging field of cybernetics (Ashby, 1947; Ashby, 1956) and enabled researchers across many fields to apply its principles. For example, Gregory Bateson linked together ideas in a new multidisciplinary research field (Bateson, 1972). The rise of computers enabled a graphical display of the networks and flows that underpin its organizational principles. Today, complexity is most commonly represented by endlessly branching graphics that appear like fractal trees on computer screens. Yet complexity is not an abstraction. It can be deduced from the relationships inherent in many kinds of everyday systems, which may be as diverse as the metabolism of cells (Warr, 2013), air-traffic flight patterns (National Aeronautics and Space Administration, not dated) or the movement of people around cities (Portugali, 2000). Complexity is more than speculative connections between things but encompasses embodied relationships. Theoretical physicist Albert-László Barabási characterized the behaviour of complex systems as being surprisingly stable, conservative, robust and resilient (Liu, Slotine and Barabási, 2011). Yet they also have the capacity to be ceaselessly creative and unpredictable. Indeed, the science of complexity is still an emerging[3] discipline, so its governing principles and applications are still in development (Armstrong, 2013b).

Such perspectives are not simply intellectual fashions; they are responses to infrastructural changes in the way we live. We find ourselves at the event horizon of a generation of globally connected, digital natives whose day-to-day understanding of reality is complex, strange, disobedient and full of paradoxes. This is much more in keeping with the turbulent perspectives of process philosophy than the ordered, bounded hierarchies of the classical world and provides a way of imagining the world anew. Simultaneously, we also live at a time of quantum entanglement where it is possible for something to be in two places at once, or where we may inhabit different characters without confusion. In cyberspace, contradictions seamlessly coexist and are extensions of our natural selves – and although it does not matter if you are a dog, a tin man, or a chatbot – it does matter how well we are connected. Yet the significance of complexity has only burst into the public domain with the rise of modern computing, which enables us to describe, diagram and imagine how this reality may be constructed. With advances in processing power and speed, it has been possible to observe recognizable structures, such as veined mushroom clouds of connections that explode upwards to the megascale and downwards to the nanoscale. My research aims to explore some of these characteristics as functions – not of virtuality but through material empowerment. From an experimental perspective, complexity provides the practical frameworks that link the material realm through chemistry and architectural design, in ways that do not need instruction by external agencies – yet they may be shaped and coerced into new configurations to provide an alternative practice of 'making'.

Terms derived from process philosophy that describe the new possibilities afforded by complexity are key to designing within it. Of particular relevance to my research are the units of reality. In classical western philosophy these units are objects, the process-led equivalents being 'actants'. The operational unit of classical western philosophy is 'machine' and the process-led equivalent is 'assemblage'. These concepts are key to developing spatial programs and architectural design tactics that are not only imagined, but also designed and implemented (see Fig. 2.2).

3 Emergence is a term that proposes an alternative roadmap of organization between a mechanistic view of the world and a vitalistic one. It originates from John Stuart Mill's work in the mid 19th century that sought to describe the nature of vital substances as being composed of inanimate materials (O'Connor and Wong, 2013).

Figure 2.2: Complex interactions need not be abstracted from their environment as in this film of bubbles on the sandy shore of the Lido in Venice. Photograph, Rachel Armstrong, August 2012.

2.5 Actant

Actant is a literary term that is used to describe an agent that denotes the locus of action within actor network theory (ANT).[4] Actants are nouns that operate as verbs and exert their diverse effects through collaboration and emergence (Latour, 1996). They may arise from a diverse assemblage of bodies including structural traits, corporate bodies, individuals or loose aggregates of individuals. Yet, none of these configurations are regarded as being more or less realist, concrete, abstract or artificial than the others in their vicinity (Latour, 2005, p.54). Bennett uses the term to denote the empowerment of matter, which renders obsolete the notion of passive objects. Actants increase their force by forming assemblages with other actants, so that non-human actors can exert autonomous forces that shape events (Bennett, 2010, p.vii). For example, power grids (Bennett, 2005) and garbage possess the capacity

4 Latour comments that actor network theory is 'a name that is so awkward, so confusing, so meaningless that it deserves to be kept ... which is ... perfectly fit for a blind, myopic, workaholic, trail-sniffing and collective traveler. An ant writing for other ants, this fits my project very well!' (Latour, 2005, p.9).

to perform work, which is conferred by many synergistic chemical processes that transform one set of substances into another (Bennett, 2010, p.6). Latour does not constrain the idea of actant to material forms (Latour, 2005, p.54) but observes that non-material entities such as ideas may produce effects. Yet, throughout my research, I use the term actant in an entirely material context, whose causal chains of action may be described through the laws of classical physics and as quantum phenomena. In this sense, actants may be regarded as embodied agents of potentiality rather than determined states of matter, such as objects. Actants may coherently exist and behave within both complex and classical scientific frameworks and, therefore, concurrently and coherently act as objects and systems. The existential paradox of matter was first observed and experimentally tested for the electromagnetic spectrum, in which light was demonstrated to simultaneously behave as a stream of particles and also as waves (Moskowitz, 2012). My research also proposes that, in keeping with the Correspondence Principle (Nielsen, 1976, pp.241–282), the paradoxical nature of the electromagnetic spectrum may be mirrored at the human and architectural scales, where linking the quantum world to the macroscale, through vibrant matter, could potentially offer new opportunities for architectural design practice. By achieving the status of a real material phenomenon, actants potentially offer an important conceptual and practical bridge between the subatomic and macroscopic world, where they may coherently operate in the worlds of objects as well as participating in the flows of material that characterize systems (see Fig. 2.3).

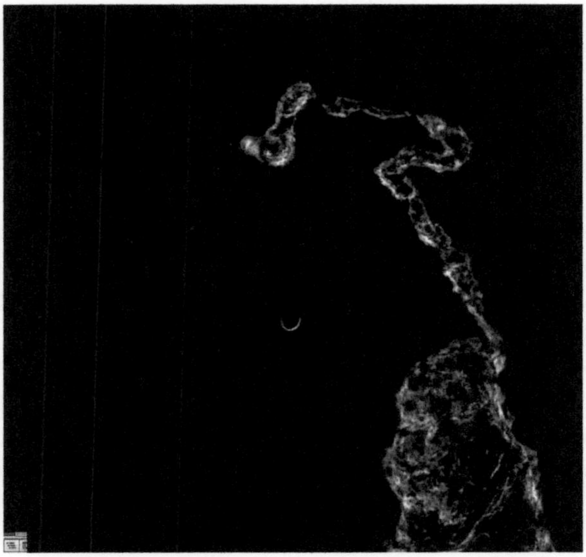

Figure 2.3: Actants are far from equilibrium systems that have the potential to form molecular bonds with other bodies. Here, a dynamic droplet is depositing a complex 'osmotic crystal skin' in the environment, while moving in a strikingly lifelike manner. Micrograph, magnification 4×, Rachel Armstrong, November 2009.

2.6 Assemblage

The concept of assemblage is from the French word *agencement*, used by Gilles Deleuze and Felix Guattari to denote specific connections between groupings of actants[5] that form loose, reversible associations with each other. Their combined actions produce complex agents composed of many different, interacting bodies such as soil. Bennett notes, 'Bodies enhance their power in or as a heterogeneous assemblage ... Assemblages are ad hoc groupings of diverse elements, of vibrant materials of all sorts' (Bennett, 2010, p.23). Through their interactions, assemblages behave according to the laws of complexity and may give rise to new meanings or effects in creative and often unpredictable ways (Deleuze and Guattari, 1986, p.22) (see Fig. 2.4). Since changing their internal, or external, conditions may alter the behaviour and effects of assemblages, they may also be considered as a form of technology that can be shaped by morphological computing techniques and may be considered as a Nature-based production platform. For example, the Dutch generative artists Erwin Driessens and Maria Verstappen use scientific processes to evolve images and sculptures. Driessens and Verstappen explore chemical systems such as wax and sand, which undergo transformations caused by physical changes in the local environment controlled by heat or vibration (Whitelaw, 2003; Siggraph, not dated) that take these substances away from an equilibrium state. The agency that is giving rise to these transformations is an assemblage, since it does not reside in any one component of their installation, as the potency of process is dependent on the artificial environments set up by the artists and also resides within the chemistry of the transmuting material itself.

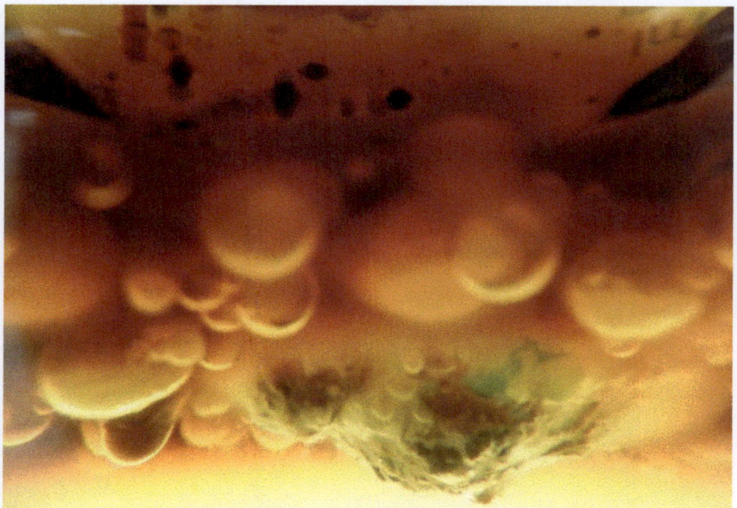

Figure 2.4: Modified Bütschli droplets form assemblages that interact and transform one another to produce novel events such as crystalline structures. Photograph, Rachel Armstrong, February 2010.

5 Actant, agent and body are considered synonyms and used interchangeably.

2.7 Nature

> *Nature is really more of a dialectical formation where we endlessly remake Nature, and Nature remakes us.* (Gissen, 2011)

My research suggests that architectural design practice may explore a more Nature-based form of making, by proposing that matter can spontaneously act through a new technological platform enabled by assemblage formation (see Fig. 2.5). In this context, Nature can be thought of as being composed of terrestrial, heterogeneous (material)[6] agents that are not commanded by a single providence, or (meta)agency, but are free to differentiate (Latour, 2013). Indeed, my research proposes that not only is Nature an agency that can be materially moulded by physical interventions such as gardening, but may be further shaped according to cultural and technological milieux (Van Mensvoort and Grievink, 2012).

Figure 2.5: Nature is shaped by assemblage formations that shape the natural world, such as resource exchanges between flowers and butterflies. Photograph, Rachel Armstrong, July 2013.

6 In his Gifford lecture series, held at the University of Edinburgh, 2013, Latour specifies these active material agents as 'chemical instabilities', or metabolisms. The role of these agents is to act so they prevent material systems from 'rushing towards equilibrium' (Latour, 2013).

Latour proposes that western cultures have 'prematurely unified' the idea of Nature (Latour, 2013), leaving us conditioned by a 'closed' set of relationships. He observes that James Lovelock and Lynn Margulis's notion of Gaia serves as a secular model of Nature that creates the condition by which our living world may possess its own agency and, therefore, effectively constitutes a living body (Lovelock, 1979). Indeed, Latour proposes that, based on the potentially infinite possibilities that may influence our experience of Nature, our planet is a discontinuous multiverse[7] from which we may begin to 'compose our cosmos' (Latour, 2013). Whitehead notes the importance of humans in shaping Nature and proposes that without human experience and subjectivity the natural world is little more than 'a dull affair, soundless, scentless, colourless; merely the hurrying of material, endlessly, meaninglessly' (Whitehead, 1970, p.54). Since humans and all of these factors that influence our encounters with reality are constantly changing, Nature invariably refers to different models of experience. It therefore requires a specific definition, which is referred to as 'Millennial Nature'. Like Morton, I capitalize the word to highlight its 'unnatural' qualities (Morton, 2012, p.3) that may be described using the laws of complexity and metaphysics of process philosophy.

2.8 Millennial Nature

In my research, the term Nature refers to a very specific set of cultural, technological and ecological conditions (Van Mensvoort and Grievink, 2012) that pertain to 21st century challenges. This 'Millennial Nature' is not the bucolic, untouched wilderness that the Romantics swooned over but a version of Nature that has been deconstructed and stripped of its aestheticisms to reveal its raw, relentlessly material character (Morton, 2007). Millennial Nature is not an Enlightenment 'standing reserve' (Heidegger, 1993) that awaits mechanical instruction, nor is it an anti-modern, vengeful force that seeks to usurp humankind (Koolhaas, not dated). Rather, it is forged through the horizontal coupling[8] of different species of lively material agents, which negotiate many difficult relationships through the production of assemblages. Yet these are not utopian fabrics devoid of struggle or contradiction. They are wilful and must be managed – not repressed – so their constituents may respond favourably to human requirements. Millennial Nature is not anti-human but implies a new relationship with humanity that demands to be engaged and nurtured, not tamed (see Fig. 2.6). It embraces many

7 Latour's term of reference for a non-prematurely-unified notion of the cosmos is 'multiverse', or 'pluriverse' (Latour, 2013).

8 By 'horizontal coupling' I mean the propensity for all matter at all scales to form relationships through subatomic interactions and the forging of molecular bonds so that there is effectively no top-down, or bottom-up hierarchical ordering but continual negotiation between assemblages that occur at all scales.

different substrates such as inorganic agents (Woods, 2012a), biological systems, weather, geological forces, soils, oceans, atmosphere, gravity, light, star systems, black holes and humans. Indeed, Millennial Nature is the fabric of reality, at all scales, which can only be perceived in relationship to human activity – but which acts entirely independently of us. Millennial Nature is not just an alternative organizing system but possesses technological characteristics that construe an alternative production platform to machines for the synthesis of new systems and fabrics.

Figure 2.6: A dinghy afloat in Lake Maggiore, Ascona, Switzerland encapsulates the character of Millennial Nature as a powerful set of forces that are neither 'for' or 'against' humans. Photograph, Rachel Armstrong, July 2013.

2.9 Post-natural

> *... the new biotechnological paradigm in architecture, which marries mixed (for old and new) technologies and biology, creating the means by which as logic of life can be mapped and engaged in architectural design processes ...* (Frichot, 2011)

If machines are the embodiment of an object-centred culture and Nature is a cultural expression of processes, then the post-natural is where these worldviews converge through new material syntheses and transformations. The post-natural indulges the complex interplay between cultures, Nature, ecology and biotechnology, which results in the seamless integration of artificially designed and engineered agents into the environment, such as genetically modified organisms or cybernetic systems (Gilster, 2014).

Post-natural landscapes are cultural readings of the environment where human and non-human agents are materially integrated through fluctuating chemical exchanges and are invisible to inhabitants, who consider these fabrics as everyday experiences, such as the 'new normal' (Luebkeman, 2012), or 'Next Nature' (Van Mensvoort and Grievink, 2012.). In the same way that the post-digital age has witnessed a generation of children that poke at paper expecting it to jump into interaction with them (YouTube, 2011), post-natural natives expect their environments to be lively, not inert, and respond to them at the human scale. The convergent terrains harbour a spectrum of conditions and experiences that are as varied as the configuration of life itself. Examples of emerging post-natural landscapes include the transformation of native biomes by industrial toxins (Paul, 2011) and more strategic proposals to entangle artificial and natural systems (YouTube, 2009; Meier, 2013; Sterling, 2013) (see Fig. 2.7).

Figure 2.7: Reflections of a boat hull in Lake Maggiore, Ascona, Switzerland produce maps of connections from which post-natural fabrics may be formed. Photograph, Rachel Armstrong, July 2013.

2.10 Ecology

The term 'ecology' is a scientific view of Nature lacking subjective interpretation (Whitehead, 1970, p.54). It has been widely adopted as a way of talking about relationships between actants that exist in the natural world that have not been culturally aestheticized or defined by their materiality. Yet the idea of ecology itself has evolved since Ernst Haeckel first coined the term 'Oecologie' in 1886 (Wimberley, 2009, p.6; Haeckel, 1866). His organism-centred perspective was inspired by Charles Darwin's theory of modification by descent (Darwin, 1999) and was concerned with the nature of the relationships between organisms and their environment, which influenced the development of new species. By the middle of the 20th century, Herbert Andrewartha and Charles Birch recognized that a complex web of non-biotic factors influenced the abundance and distribution of animals (Andrewartha and Birch, 1984). However, Howard and Eugene Odum emphasized the importance of the mineral world in shaping evolutionary events and gave it equal status to the animal world. They championed Arthur Tansley's ecosystem view of the world (Willis, 1997), which was centred on how creatures thrive through the continual exchange of vital materials (Stuart, Matson and Mooney, 2002, pp.7–11). In a contemporary ecological context, these networks not only relate to the development of biotic communities over time but may also refer to the flow of matter and energy through elemental cycles such as carbon, nitrogen, phosphorus and water.

As the scientific study of complex systems has developed (Capra, 1996), its material embodiment has become equated with ecology and is practised through a range of multidisciplinary collaborations between scientific disciplines such as biology, physics, chemistry, permaculture, soil studies, bacteriology, atmospheric science, geology, oceanography and geography and non-scientific fields such as ethics, panarchy, economics, art, humanities, politics, design, philosophy, law and resource management. A broad range of commercial practitioners also currently qualifies the environmentally beneficial nature of their services using the term ecology. For example, the construction engineering firm WSP Group defines 'urban' ecology as 'bringing the power of ecology to buildings ... to create solutions that integrate Life, Energy, Water, Nutrients and Future potential into interlinked cycles' (WSP, not dated). An ecological approach to architecture, therefore, not only deals with the scientific ideas embodied in complexity but also raises questions about the production of space (Lefebvre, 1991). Therefore, ecology inevitably deals with the cultural ideas that shape our relationship with the natural world but through the lens of science.

Accordingly, architects have embraced a diverse set of terms to refer to specific aspects of these relationships, such as sustainable (Turrent, 2008), 'green' design (Edwards, 2010), eco-architecture (McDonough and Braungart, 2002), bioclimactic architecture (Dahl, 2009), biomimicry (Beynus, 1997), zero (carbon) footprint buildings (Godfrey-Cook, 2011), passive houses (Cotterby and Dadeby, 2012) and architectures made from recycled materials (Laylin, 2012). In the context of my research, ecology refers to a scientific view of a multidisciplinary and practical engagement with complex embodied systems. These relationships shape and build functional networks that are culturally and aesthetically entangled with diverse communities of actors (Armstrong, 2012b; Armstrong, 2013b). An '(eco)systems' view of reality challenges the conceptual frameworks that depict the natural world as composed of discrete objects, bodies or forces such as biology, geology, climate and the oceans (see Fig. 2.8) and links these heterogeneous actants through internal metabolic processes and external, global material flows.

Indeed, Ecologist Fern Wickson wonders whether – if humans are intertwined in a complex web of biological systems and therefore cannot be included within a definition of Nature, where 'an atom bomb becomes as "natural" as an anthill' – an alternative understanding of Nature is needed (Wickson, 2008). However, changing the definition of Nature is not the solution to Wickson's conundrum – the paradigm in which it is imagined requires adjusting. The classical scientific method, based on Cartesian thinking – which Kauffman calls the 'Galilean spell' (Kauffman, 2008, pp.129–149) – is actually responsible for the paradox Wickson seeks to contest. The Galilean spell is founded on irreducible binaries such as mind/body, human/machine and Nature/artifice, so if the problem of human connectedness to the natural world is to be resolved, then the scientific framework used to describe it needs to change.

Figure 2.8: Reflections within a Venetian canal encapsulate complex, ecological interactions between humans and the natural world. Photograph, Rachel Armstrong, August 2012.

Robert Rosen observed that when physics is used to describe biology, a generalization occurs that distorts reality (Rosen, 1996, pp.199–214). He proposed that formal definitions of 'life' set limits that are caused by an excess of 'rigour' in mathematics, and 'objectivity' or 'context independence' in science (Rosen, 1999, p.2). Similarly, Turing noted in his essay on morphogenesis that mathematical abstraction could not capture the richness of the natural world (Turing, 1952), since life is a complex system that is governed by a variety of unique processes that machines simply do not possess. Living things continually respond and adapt to their environment and are made up of networks of assemblages which enable them to self-regulate and express increasing degrees of agential autonomy (Maturana and Varela, 1980). Yet Deutsch notes that reductionism is not the only form of science and proposes that scientific inquiry is not about making predictions based on observations, but serves a broader function in finding explanations (Deutsch, 1997, p.71).

This is exactly what complexity does, but rather than being based on inductivism,[9] it is informed by process philosophy (Seibt, 2012). Since a complex worldview is based on concepts that relate to networks, relationship and flows that cannot be reduced into their parts, it challenges the ontological and epistemological basis on which technologies that have been championed by modern science and industry, are founded. So what does 'systems' science (Von Bertalanffy, 1950), or complexity, mean for our relationship with Nature? Are we separate from or intrinsically connected to the natural world? In a complex system, we are both, and experience reality as being formed by both processes and objects and also from subjective and objective perspectives. For example, a building may be thought of as an object but it is also a site for many processes that may have brought about its construction, enable site activities and ultimately contribute to its decay. We may also observe and measure these material changes and consider them beautiful and encourage them, or disagreeable and seek to prevent or repair them. However, our understanding of which processes are important are culturally determined and are not Platonic truths. Indeed, Bruce Sterling proposes a play on Arthur C. Clarke's dictum[10] and wryly observes, 'Any sufficiently advanced technology is indistinguishable from its garbage' (Sterling, 2012). Sterling's reference to garbage suggests how we can be connected to Nature – but not in an unlimited way. Garbage-making is a subjective exercise in which we distinguish ourselves from the natural world, and our technologies, by 'editing' our material networks to reflect our cultural conditioning. We choose what is important to us by applying cultural and aesthetic, rather than material criteria, to make the appropriate selections. Turing had already grasped the importance of personal bias in dealing with complex systems

---

9 According to Deutsch, inductivism produces scientific theories by extrapolating the results of observations, which are justified when corroborating evidence is obtained (Deutsch, 1997, p.71).

10 Clarke's dictum from his Third Law, in the essay 'Hazards of prophesy: The failure of imagination' states 'any sufficiently advanced civilization is indistinguishable from its technology' (Clarke, 1973, p.21).

and devised the ‘imitation game’ to address the conundrum of intelligence, which evaded an easy empirical solution. This is now more popularly known as the ‘Turing Test’ and is now being used more widely to fathom complex systems and to identify ‘life’ (Cronin et al, 2006). Ultimately, the flow and structure of information within our ecological systems are important in establishing what technology, architecture, garbage and ‘life’ is – and indeed, what design is.

2.11 Technology

Throughout my research, I use Martin Heidegger’s notion of technology, which does not restrict the term to any specific material form or instrumental process but considers how the idea of technology reveals truths about the world (Heidegger, 1993). Heidegger engages the concept of technology as a process and resists collapsing the concept into descriptions of objects and their hierarchies. Yet, Heidegger equates technology with industrial machines, which he views as being destructive in the thought processes they embody and the kind of influence this exerts on the material world. At the heart of Heidegger’s quest is an ecological project where he seeks a free relationship between technology and humanity, to establish a new relationship with Nature, which is not ultimately self-destructive. Heidegger’s idea of technology may be extended to non-mechanical systems, such as process-led ecological living technologies (ELT: see 2.13) that harness the properties of lifelike systems, (see Fig. 2.9). My research develops the notion that a non-object-centred view of the world allows us to conceive of more Nature-based forms of technology and design solutions that have qualitatively different environmental impacts than mechanical technologies.

2.12 Machines

Machines may be regarded as an expression of a human culture of objects through the design and construction of functional object hierarchies. These chains of order require centralized, hard control systems that are designed to operate the objects when they are at equilibrium – the lowest possible energy state within an environment. They also need an external energy source to tip the system away from equilibrium, so they may conduct useful work. Machines embody an atomistic worldview and are based on Cartesian principles characterized by binaries, objects, hierarchies of order and geometry (see Fig. 2.10). While we have increasingly come to equate technology with machines, or gadgets, my research makes reference to other kinds of technology, whose operational principles are more in keeping with a Nature-based form of practice. Alternative technological systems to machines may, for example, possess innate energy by existing at far from equilibrium states, engage parallel forms of

Figure 2.9: This reflection in a Venetian canal notionally suggests that buildings may be grown from an elemental recipe, rather than constructed using prefabricated components. Photograph, Rachel Armstrong, July 2013.

Figure 2.10: This Venetian mask design depicts the human body as a machine. Photograph, Rachel Armstrong, August 2012.

multidimensional processing and invite soft forms of control, since their agency is distributed throughout their constituent assemblages.

2.13 Ecological Living Technology

Ecological living technology (ELT) is a process-led platform which is composed of non-hierarchical groupings of material assemblages that can perform useful work and whose outcomes may be shaped by altering the internal and external conditions of the system. This is different to the expectations of classical machines whose performance is controlled by designing programs that control the internal state of the machine. ELT is an amalgamation of two concepts, 'living technology' and 'ecological technology', which deal with uniquely 21st century technological imperatives, namely:

— The desire to produce more 'ecological' forms of technology with operational processes that benefit, rather than harm the environment.
— The increasing lifelike nature of technological developments.

Bütschli droplets are an example of ELT. They are programmable, chemical agents capable of growing microstructures (see Fig. 2.11). ELT also includes modified or natural biological systems, such as synthetic biology, which harness metabolic qualities to produce their lifelike outputs, like growth, repair, reproduction and locomotion. From a design perspective, ELT aims to reveal and harness new opportunities in architectural design using different species of vibrant matter as a form of technology. These produce work through the formation and destruction of chemical bonds that can be instructed by embodied computational methods.

2.13.1 Living Technology

'Living technology' was coined by a consortium of interdisciplinary researchers[11] at the Initiative for Science, Society and Policy (ISSP) at the University of Southern Denmark, which refers to technology that is based on the core features of life (Bedau et al, 2010). Living technology has a broad definition that refers to technologies with lifelike qualities, which may be machines or organisms, and is subdivided into 'primary' and 'secondary' living technologies (Bedau, 2009).

— 'Primary' living technologies exhibit lifelike behaviours yet do not possess any biological parts.
— 'Secondary' living technologies harness organisms to perform their tasks, such as genetically modified bacteria, or in vitro fertilization.

11 I was a member of this group (University of Southern Denmark, 2012).

Figure 2.11: Modified Bütschli droplets process spatial chemical programs and form dark mineral deposits within the body of a droplet cluster. Photograph, Rachel Armstrong, February 2010.

Living technology is also connected with other scientific fields including artificial life (soft, hard, and wet), synthetic biology[12] (top-down and bottom-up: see 2.14), and the convergence of nano- and bioinformation, and cognitive (NBIC) technologies (Bedau et al, 2010).[13] However, what may or may not count as living technology is debatable. For example, a horse may be considered a living technology dedicated to transport, while a dog is a living technology bred for companionship and security. Additionally, simple tools, such as a spanner or hammer, could be considered as secondary living technologies if they are regarded as a system of work-performing assemblages, where human and instrument are reversibly coupled together to achieve a task. Living

12 Synthetic biology has been popularized as an exclusively biological form of technology but originated as an offshoot of synthetic chemistry (Leduc, 1911) and is further discussed by A.G. Cairns-Smith (Cairns-Smith, 1971; Cairns-Smith, 1987; Cairns-Smith, 1990).

13 The so-called NBIC (Nano, Bio, Info, Cogno) convergence is an innovation initiative that followed an NSF (National Science Foundation) sponsored report, which has been particularly influential in precipitating a new kind of scientific approach suggesting unification of the sciences as a common goal through converging advanced technologies to provide the practical basis for the retranslation of humanity, in keeping with the pursuit of Julian Huxley's vision of transhumanism where 'the human species will be on the threshold of a new kind of existence' (Huxley, 1957, pp.13–17).

technology also prioritizes the status of mechanical systems (primary) over that of modified, or technologized biological ones (secondary). Indeed, the all-embracing nature of the terminology led to criticism during the initial ISSP workshops (University of Southern Denmark, 2012) that the term may be applied so generally that it is almost meaningless to use it. However, from an applied design perspective, living technology conjures up different expectations of technology and conditions associated with their use (Armstrong, 2010c), which is relevant to my research ambitions in identifying more environmentally enriching technological approaches.

2.13.2 Ecological Technology

William Mitsch used the term 'ecological technology' in reference to systems with properties that distinguished them from conventional, industrial technologies (Mitsch, 1993). These technologies could be designed to perform particular kinds of work by choosing initial species and establishing the starting conditions of the system (see Fig. 2.12). Ecological technologies are partly created by natural forces such as molecular self-assembly, but also shape the output of the technical system by virtue of their own agency.

Figure 2.12: Could trees be regarded as ecological technology? A tree captures sunlight, carbon dioxide and uses water to produce outputs that may be shaped by human agencies to perform useful work (Living Root Bridges, 2009). Photograph, Rachel Armstrong, July 2013.

2.14 Synthetic Biology

Synthetic biology is a relatively new scientific field that involves the rational design and engineering of living systems, which provides the tools and methods that underpin novel problem-solving tools and materials. Synthetic biology may be regarded as a specific example of ELT which provides access to the assemblage-based technology implicit in the natural world. With the rise of biotechnology, synthetic biology has become equated with genetically modified agents. Yet, the term was coined by Stephane Leduc (Leduc, 1911) and refers to the transition from chemical to biological systems (Cairns-Smith, 1971; Cairns-Smith 1987; Cairns-Smith 1990) (see Fig. 2.13).

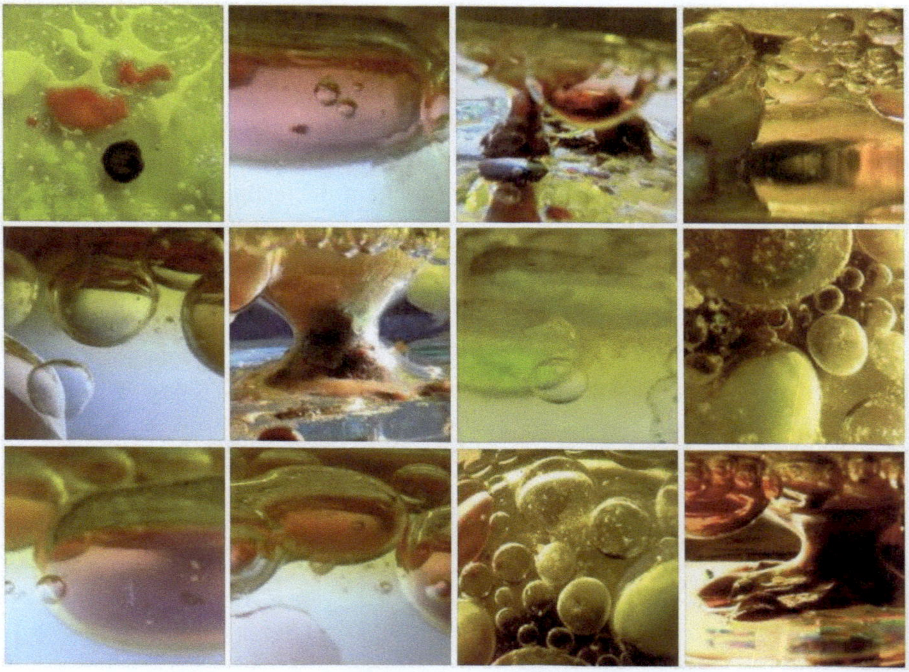

Figure 2.13: In this photograph complex chemical species are behaving in lifelike ways (Leduc, 1911). Photographs and collage, Rachel Armstrong, July 2013.

2.15 Synthetic Ecology

Synthetic ecology refers to the rational design and engineering of environmental networks and can be thought of as the 'systems' version of synthetic biology.

2.16 Protocell

Non-biological materials may also exert lively effects. My research examines how chemical assemblages that are not technically 'alive' may be considered and applied as ELT. My experimental work explores a range of vibrant materials that take the form of a series of non-equilibrium chemistries ranging from Liesegang rings (Liesegang, 1869) to Traube cells (Traube, 1867) and dynamic droplets (Bütschli, 1892; Hanczyc et al, 2007; Toyota et al, 2009). Most of my technological research is conducted using a specific species of dynamic droplet called the Bütschli system, whose products are generally referred to as 'dynamic droplets' rather than 'protocells', which has been used in other publications. My choice of terminology in this particular instance is based on the controversial and ambiguous nature of the term protocell, which invites a broad range of definitions. Sometimes it has been used interchangeably with 'vesicle' (Szostak, Bartel and Luisi, 2001), while at other times it may indicate fully artificial chemical cells capable of replication (Rasmussen et al, 2003; Rasmussen et al, 2008). My preferred definition, however, is an etymological one, where protocells are simple chemical systems with lifelike properties that serve as models for the earliest form of natural living cells.[14] I have therefore used the term dynamic droplet, or Bütschli droplet, to refer to a specific example of a broader portfolio of agents that may also be described as protocells. Since I have already written on the subject of dynamic chemistries using the term protocell (Armstrong, 2011d; Armstrong, 2012g), the term appears in the text where it references a specific piece of work, or where it is used in a popular context, such as in the speculative narratives like 'Post-Natural Venice'.

2.17 Natural Computing

The term 'natural computing' was inspired by Alan Turing's interest in the computational powers of Nature (Turing, 1952; Denning, 2007), which operates at the level of molecular interaction. Natural computing processes inform the behaviour of ELT and ultimately inform the networks of chemical exchange in the living world, therefore enabling a more Nature-based form of technology. The possibilities of natural computing have been further developed by David Deutsch to embrace the quantum realm. Deutsch proposes that Turing's abstract computer must be substituted by an actual, physical, universal quantum computer derived from the Church–Turing–Deutsch principle, which could reveal the computational possibilities of the natural world (Deutsch, 1985; Copeland, 1997) (see Fig. 2.14).

---

14 Protocells are 'primordial molecular globules, situated in the environment through the laws of physics and connected through the language of chemistry' (Spiller and Armstrong, 2011).

Figure 2.14: A leaf may be regarded in a technological context as a water collecting system. Photograph, Rachel Armstrong, July 2013.

The field of natural computing is inspired by the capabilities of natural organisms and embraces broad, overlapping and multidisciplinary practices such as digital modelling of biological systems and unconventional computing, as well as some aspects of robotics.[15] The main goal is to develop programmable, lifelike systems using a spectrum of platforms to better understand and reflect the properties of living things, such as adaptation, learning, evolution, growth, development and robustness. Natural computing informs the technological character of vibrant matter and therefore guides the process of assemblage production that underpins ELT. However, as is the case with living technology, the term natural computing is very broad and relatively recently established, so its application has been developed and interpreted according to the aims of the various participating research groups. Research practices include the study of biomimicry in digital computing that notionally engages with material processes, mainly through representations in 'genetic algorithms'. However, my work relates to the direct manipulation of the chemical agency of matter to produce events which can be orchestrated by computational processes. With this focus in mind, two separately evolving yet overlapping practices that are grouped within the field of natural computing are of particular interest, namely, unconventional and morphological computing, which examine the direct impacts of material processes on computational tasks and may therefore assist in developing the technological potential of vibrant matter so that it can be applied within architectural design contexts.

15 Notably, the field of soft robotics is concerned with the computational properties of materials (Whitesides Research Group, 2011).

2.17.1 Unconventional Computing

Unconventional computation has been defined from a range of perspectives, including technical, logical, scientific-theoretical and philosophical (MacLennan, 2011; Paun, 2005). According to Mihai Oltean, unconventional computers 'have been only recently invented, operate with some exotic principles, and ... have not been yet introduced on the market' (Oltean, 2009). Andrew Adamatzsky's research group at the University of South West England has built unconventional computing into a unique interdisciplinary research area and practice of analogue computing, which is concerned with the applications of biological and chemical systems to solve a range of challenges. For example, slime mould computing (see Fig. 2.15) has been used to identify the shortest route in compound pathways (Adamatzky et al, 2013), and dynamic chemistries can process complex information (Adamatzky, Costello and Ratcliffe, 2002; Adamatzky and De Lacy Costello, 2003). Adamatzky's aim is to enrich, or go beyond, the standard models of computing such as the Turing machine and von Neumann architectures, which have dominated computer science for more than half a century (Adamatzky et al, 2007; Armstrong, 2011b).

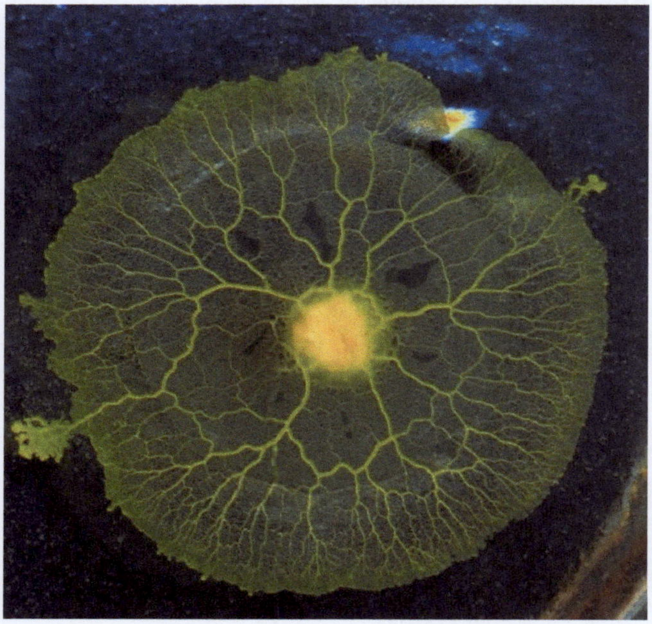

Figure 2.15: Slime mould Physarum polycephalum is a network of biochemical oscillators (Matsumoto et al, 1988) that can solve challenges such as, identifying transport routes within models of cities (Adamatzky et al, 2012). Photograph, courtesy Andrew Adamatzky, July 2013.

2.17.2 Morphological Computing

Morphological computing incorporates the properties of matter to program the behaviour of a system. The term originated from the field of robotics, where the physical properties of robot bodies are harnessed to spontaneously produce and control the robot. Robots may exist as kinetic devices that act synergistically with the system's mechanics to automate it (Theo Jansen's Strandbeest, not dated), or may use entirely chemical approaches, such as Shingo Maeda's chemical inchworm (Gyr, 2009), which responds physically to changes in its environment. The idea is that the physical properties and morphology of agents such as living organisms, complex chemistries or machines constrain robotic interactions with the environment and therefore play a role in their development, growth and reconfiguration (Pfeifer and Iida, 2005) (see Fig. 2.16). Helmut Hauser observes that morphological computing is a way of outsourcing some of the computing processes performed by a central digital processing system to the physical body.

This is achieved by using non-classical engineering principles that embrace the qualities of high dimensionality, non-linearity, compliance and noise (Hauser, 2013). Technological assemblages may be regarded as morphological computers whose control systems are distributed throughout the operational body rather than being centralized. Morphological computing therefore needs soft control strategies,

Figure 2.16: Jellyfish are able to swim without central control over this movement by virtue of the physical properties of their gel matrix.Photographs and collage, Rachel Armstrong, February 2011.

which require orchestration rather than hierarchical command, and are already used in practices such as gardening and cooking. Additionally, control methods are coupled to the natural computing body and are able to respond variably to a range of different conditions within definable limits of probability. For example, the nervous and musculoskeletal system in the body have different levels and scales of control that are tightly coupled and variably engaged, depending on the sets of tasks they are required to address – from interactions at the cellular level to reflex movements and 'higher' centralized motor control. Varied outcomes exist within a possible set of circumstances, which are context sensitive and constrained by the physics and chemistry of the system. Morphological computing refers to a very specific research interest in the computational powers of dynamic chemistries (Armstrong and Hancyzc, 2013; Tangen and McCaskill, 2004–2008; Simonite, 2009) and material assemblages (Chu, 2012), which are regarded as 'soft robots' (Whitesides Research Group, 2011). The fundamental design units of morphological computing are not objects but leaky systems which spontaneously form as chemical and biological assemblages. These groupings enable information flow (droplets, proteins, DNA) and, ideally, can remain open (e.g. cells) throughout the computational process to transform substrates, rather than consume them. Morphological computing is not exclusively concerned with empirical outcomes but incorporates aesthetics and poetics in its outputs, such as quality of movement, colour changes, or by invoking 'the uncanny' (Freud, 2003).

2.18 Vibrant Matter

The term 'vibrant matter' is used to refer to different kinds of chemical assemblages that underpin the technological capabilities of the natural world. These enable Millennial Nature to spontaneously produce effects, or codesign events in collaboration with humans. The term was coined by Bennett as a philosophical and political project to 'to encourage more intelligent and sustainable engagements with vibrant matter and lively things' (Bennett, 2010, p.viii). As such, vibrant matter itself is a theoretical position that has not previously been applied to design practice. Additionally, I have used the term in reference to a model system that simultaneously incorporates materiality, infrastructure and technology as an experimental production platform that may be shaped by morphological computing approaches. Vibrant matter was selected to represent the cultural proposals implicit in my research, which aims to explore how 'deep' material imperatives, or 'sub-natures' (Gissen, 2011) may be harnessed in ways that demonstrate how inanimate materials 'exceed their status as objects and ... manifest traces of independence or aliveness, constituting the outside our own experience' (Bennett, 2010, p.xvi). Vibrant matter therefore refers the agency within and the substance of all material expressions – whether human or non-human, animate or inanimate.

Dynamic chemistries – ones that are far from equilibrium, or not at the lowest possible energy state in their context – are used to represent the design potential of vibrant matter. Far from equilibrium chemistries have been of scientific interest since the mid-19th century as a way of interrogating vitalism – the idea that matter is animated through the infusion of an ephemeral force. My research applies the uniquely lively properties of these agents to develop general design principles that may be applied more widely within architectural practice and include:

- Bütschli droplets: These agents spontaneously form when concentrated alkali solution is added to an olive oil field and exhibit strikingly lifelike qualities.
- Traube cells: These 'artificial plant cells' are produced when a blue crystal of copper II sulphate is dropped into a weak solution of potassium salt to create a seaweed-like, semi-permeable, inorganic membrane.
- Liesegang rings: These periodic patterns are produced when two interacting salt solutions produce weakly soluble precipitates.
- Leduc cells: These are produced when a crystal of calcium chloride is added to a dilute solution of sodium hydrogen carbonate forming a thin calcium carbonate shell.

By working at a level of design that precedes the emergence of systems that are culturally recognized as Nature, I seek to minimize the unconscious, cultural aestheticisms that complicate our observations, applications and expectations of the material world (Morton, 2007). My research attempts to affirm that inorganic substances may exert effects that are meaningful at the human scale in a design context, with characteristics (Bennett, 2011) that are not merely metaphorical relationships but could be explained through physical and chemical interactions – slowness, porosity and inorganic sympathy.

2.18.1 Slowness

The comparative endurance and patience of the non-human world exhibits a relative slowness in terms of its rate of change and may be explained by molecular dynamics.

2.18.2 Porosity

The entangled and cooperating nature of bodies operates according to the laws of physics and chemistry in ways that make them susceptible to invasion or infusion with other bodies.

2.18.3 Inorganic Sympathy

Vibrant matter is driven towards a primaeval inorganic state of existence, which resonates with Erwin Schrödinger's notion of entropy and matter's decay towards equilibrium states (Schrödinger, 1944).

2.19 Vibrant Architecture

'Vibrant architecture' is a stochastic form of architecture that is compatible with a Nature-based method of architectural production. It seeks integration with living systems and emerges where vibrant matter converges with dynamic spatial programs, using ELT and morphological computing techniques. Vibrant architecture distinguishes itself from improved industrial practices such as Neri Oxman's view of 'material ecology' that uses parametrics to integrate material science and digital manufacturing with the environment (Oxman, 2012). It is also distinct from William McDonough and Michael Braungart's 'Cradle to Cradle' naturalistic urbanism (McDonough and Braungart, 2002), where buildings are considered 'alive' owing to the abundance of natural systems in the urban environment. Indeed, vibrant architecture is not 'inspired' by the natural world, like the practice of biomimicry (Beynus, 1997; Armstrong, 2013f), but directly embodies its material processes as an integral aspect of Millennial Nature (see Fig. 2.17).

2.20 Summary

This chapter establishes a portfolio of terms and contexts against which my research into a 21st century ecological design practice may be read and understood.

Figure 2.17: Stone structure at Lyme Regis forged by morphological computation. Photograph, Rachel Armstrong, July 2012.

3 Literature Survey

Architecture is a wonderful means of representing, imposing or arousing states of human nature.
(Eurozine, 2011)

3.1 Overview

My work establishes how the innate potency of the material world may be applied to architectural design practice. I aim to identify experimental models to test this hypothesis and set out the principles that underpin the proposed modes of action that may be practically applied in an architectural design context, through the production of collaborative drawings, models and prototypes. My literature search, therefore, explores the overlapping fields of chemistry, which may be regarded as the 'language' of matter, and architecture, which is concerned with the tactics for developing spatial programs. In juxtaposing these disciplines to find synergies between them, my research aims to identify gaps in the knowledge canons and therefore establish the research opportunities. Omar Khan observes the many historic technological and cultural instances where chemistry is entangled with architecture to create something new in the production of new materials, such as concrete, steel, plastics, rubber, crystals and fluids with mutable structures (Khan, 2011). Architect William McDonough and chemist Michael Braungart combine architecture and chemistry in their 'Cradle to Cradle' manifesto, which proposes that more efficient management and better design of the production process may achieve ecologically intelligent design (McDonough and Braungart, 2002). While I share the material concerns of McDonough and Braungart, my research is not focused on the transformation of industrial processes or 'upcycling' through the cultural (re)usage of waste materials, but in the direct empowerment of matter. Rather, I use an alternative framework to the industrial worldview by applying design principles that are expressed through process philosophy and whose operations are embodied in complexity science.
My literature survey therefore establishes a roadmap of connections between the disciplines of architectural design and chemistry, with the purpose of understanding what kinds of matter may be sufficiently lively to produce effects that can be designed and measured at the human scale. Material systems are identified that embody the principles of vibrant matter and applied in an architectural setting. My reading also draws from a wide range of disciplines such as philosophy, ecology, complexity, origins of life sciences, sociology and physics, to more fully characterize any apparent knowledge 'gaps'. Taking a processes-oriented rather than object-centred perspective, also allows experiments to be described without invoking machine metaphors. A rich range of terms and ideas are invoked that embody the term 'vibrant matter' and collectively shape my approach to developing a unique portfolio of design operations (see Fig. 3.1).

Figure 3.1: This Venice canal reflection captures aspects of the complex interplay between environment and architecture that is continually orchestrated by the physical and chemical forces of the natural world. Photograph, Rachel Armstrong, August 2012.

3.2 Information Sources

A wide variety of information sources are used, ranging from academic journals, reference books, news websites to special interest blogspots. They identify knowledge canons, as well as keeping up to date with the latest cultural and scientific developments. My reading is concerned both with new findings in the literature and the curating and re-positioning of existing ideas with different disciplines, such as the relationship between quantum physics and vitalism in reference to the dynamic nature of materials (Armstrong, 2012c). To quickly get a feel for a subject area I attended international conferences that drew together expert views on the latest research. For example, 'Artificial Life XI' inspired the model system for my studies of vibrant matter following Takashi Ikegami's keynote address on dynamic droplets (Ikegami, 2008), and 'Prototyping Architecture' by Bob Sheil at the Building Centre, London (Murphy, 2013) discussed the relevance of emerging technologies to architectural design practice. I participated in laboratory discussions, attended academic presentations and panel discussions that engaged with overlapping fields of interest, such as Liam Young and Geoff Manaugh's 'Thrilling Wonder Stories' series at the Architectural Association (Thrilling Wonder Stories, not dated). This established the broadest possible context for my evolving architectural design and chemistry research questions, which, in turn, directed my further reading and research.

3.3 Critical Context

3.3.1 Overview

My literature survey draws from four broad fields of study:

- Organicism: establishing the historical context for vibrant matter in the field of architectural design practice.
- Bio Design: identifying how dynamic materials are incorporated into buildings, projects and products.
- Vibrant materials: characterizing the kinds of substrates that embody the principles of vibrant matter.
- Morphological computing: developing a portfolio of manipulation techniques so that vibrant matter can be practically applied and tested in scientific experiments and architectural prototypes.

3.3.2 Organicism

Organicism refers to the design practice that seeks more Nature-like approaches (Capra, 1996) to the production of spatial programs and proposes an alternative paradigm to the machine-based construction that underpins modern architecture. Throughout the ages, architectural design has looked to Nature as an ideal through typological forms such as Marc-Antoine Laugier's sublimated references to essences of the natural world, which were driven by aesthetic values like archetypes (primitive hut), idealized proportioning systems (Golden Section), anthropomorphic imitation (Vitruvian Man) and literal forms of ornamentation, geological stratification (base, middle, top), and materials (undressed stone). Yet morphological abstractions and discrete typologies are mired in cultural aspirations (Morton, 2007) and do not directly embody the dynamic material properties of the living world, such as growth, movement or evolution.

I therefore surveyed 20th and 21st century design movements to establish how lifelike qualities may have been directly incorporated into architectural design practice. The anthologies of architectural theory written by Kate Nesbitt (Nesbitt, 1996) and K. Michael Hays (Hays, 2000) offered accessible surveys of western architectural debates during the late 20th century. George Baird outlines the different concerns within the modern tradition of organicism, such as how natural systems build across scales and different material regimes (Baird, 2003). Sarah Bonnemaison and Philip Beesley further critiqued organicism's main interests, proposing that the work of Richard Buckminster Fuller, Pierre Teilhard de Chardin and Frei Otto best characterized key perspectives within this movement (Beesley and Bonnemaison, 2008). Beesley and Bonnemaison suggest that Buckminster Fuller champions a mathematically inspired vision of Nature (Fuller, 1969), while de Chardin embodies the environmental concerns of New Age philosophy (De Chardin and Huxley, 2008) and Frei Otto represents the process-led, environmentally-sensitive maker movement (Otto, 1996). I was also interested in morphogenetic debates that examined how the relationship between architectural program and structure were being imagined in the production of generative architectures (Frazer, 1995). These discussions embodied contemporary scientific ideas about living systems between neo-Darwinists such as Richard Dawkins who viewed genes as being the primary organizing force of matter (Dawkins, 1976), and biological structuralists such as Brian Goodwin and Conrad Waddington (Waddington, 1957). These key architectural debates also reflect important and parallel scientific discussions regarding the paradox of biology, which appears to contravene the second law of thermodynamics[16] (Schrödinger, 1944) in ways that were not observed in the physical and chemical sciences[17] and even locally resists the decay towards equilibrium. My reading, therefore, investigated how architectural design practices have captured the paradoxical qualities of biological systems, which are lively because they are not at equilibrium states. Indeed, traditional building practices incorporate biology directly into architectural solutions, such as the living bridges of Cherrapunji, where powerful ropey root tendrils of the *Ficus elastica* tree are entwined to span rocky gullies.[18] However, native biological systems pose significant design challenges in urban environments. For example, organisms require a constant flow of nutrients and removal of waste products and – compared with mechanical construction methods – grow slowly, thereby exhibiting 'slowness', one of Bennett's qualities of vibrant matter (Bennett, 2011). Architects have therefore tried to capture the unique essence of living systems through many different approaches to expressions of form, such as Art Nouveau's stylization with its biological motifs carved into traditional materials. Notably, Antonio Gaudí developed a dynamic set of material processes that could capture natural qualities by suspending clay in hanging baskets and letting physics and chemistry, the drivers of biology, work on the setting material. This approach produced primordial organic forms, which he then

---

16 The second law of thermodynamics assumes a closed system, which organisms are definitely not: consider the need for sustenance and how living things respond to changes in their environment.

17 Several authors observed the paradox of the poetry of biology and the mathematics underpinning physical and chemical systems, notably the French philosopher Henri Bergson in L'Evolution Créatrice (1922). Bergson proposed an anti-mechanistic worldview that celebrated Life as an improbably diversity-creating whole, animated by a powerful élan vitale. He also suggested that these forces were responsible for accelerating biological coevolution, which transformed the inert matter of the Earth's surface into living systems in a way that appeared to violate the second law of thermodynamics, which speaks about the conservation of energy in a system (Bergson, 1922, p.44).

18 Like many others in the banyan family, F. elastica (often called the rubber fig) has secondary roots that grow above the soil surface. By guiding these roots across chasms, villagers can slowly grow a strong, permanent bridge that can support 50 people and reach 30 metres in length. Such a bridge may take 15 years to grow and is fine-tuned to its niche environments.

incorporated into the façade of the *Sagrada Família* cathedral in Barcelona, which is still under construction (see Fig. 3.2).

Figure 3.2: Gaudí's La Sagrada Familia cathedral, which is growing 'organically' and still under construction, incorporates organic motifs. Photograph, Rachel Armstrong, November 2011.

By applying this approach Gaudí reversed conventions of order in the process of building an architecture, which usually follows a top-down approach where form is imposed upon matter. Instead, Gaudí let the individual shapes of the set clay inform the final appearance of the building, which was Gaudí's hallmark organic style, which, for example, incorporated the use of mosaics (Burry, 1993) (see Fig. 3.3).

The rise of modern computing and the discovery of the genetic code over the second half of the 20th century enabled architects to experiment with mathematical principles used to describe natural forms (Thompson, 1917), which became an increasingly testable proposition with the advent of graphical user interfaces. Increasingly, biological systems could be considered as soft, wet, efficient, purposeful machines (Nicholson, 2009), which could therefore be assimilated within industrial frameworks. Indeed, the current sustainable architecture movement, which may be thought of as a 'better' kind of industrialization originating in the work of Frank Lloyd Wright (Weintraub and Hess, 2012), Alvar Aalto (Ray, 2005), Ludwig Mies van der Rohe (Zimmerman and Gossel, 2007) and R. Buckminster Fuller (Fuller, 1969), who believed architectural design could increase the material efficiency of the

Figure 3.3: Gaudí's mosaics in Park Guell adopt a 'cellular' approach in their designs. Photograph, Rachel Armstrong. November 2011.

natural world while emulating its forms 'as if' (Vaihinger, 1968) the buildings were organisms. Indeed, Emilio Ambasz proposed that 'green over grey' architectures not only subsumed the artificiality of architecture but also constituted 'artificial natures' (Dean, 2011, p.230). Yet this synthetic Nature is one of form rather than process, being obedient to abstracting geometrical principles from single moments in time that are (re)presented in architectural contexts and guided by Enlightenment ideals. For example, Casa de Retiro Espiritual, Seville (1978), a white cube building set against stark geometric cuts and folds in a carefully domed lawn that rolls against a bucolic landscape, is no more 'artificial' than a garden with a central sculpture. Yet Nature and industry were increasingly conceived as being compatible and even interchangeable, such as in Paolo Soleri's architectural ecologies (arcology) where cities were formally entwined with their landscape (Newitz, 2013a). Using this approach, more ecological outcomes could be achieved by producing more ordered forms of Nature and could be equally applied to the design and manufacture of objects and processes. For example, Robert Frosch and Nicholas Gallopoulos developed the idea of closed-loop exchanges within urban environments, where the waste processes of one system became the raw material for the next (Frosch and Gallopoulos, 1989). Janine Benyus further explored the idea of natural efficiency by promoting the practice of biomimicry (Beynus, 1997), which looks to Nature as a template for problem solving. Biomimicry typically observes original natural structures performing ingenious tasks, such as plant burrs attaching to sheep wool using tiny hooks, and translates these into design solutions like Velcro. It typically adopts a rather one-sided Romanticized view of Nature, which is regarded as providing us with sublime, pre-packaged solutions to help us achieve our collective ecological quest, while generally glossing over or ignoring its counterpart – grotesquery. While biomimicry proposes to learn from Mother Nature (Beynus, 1997) by copying the shapes and functions of biological processes, the practice is mostly concerned with observing abstractions of the end products of hugely sophisticated material systems that are not static, but in continual flux. 'Mimicking' biological outputs ignores the sophisticated chemical hardware and software of natural systems, which are deeply entangled through the dynamic exchanges of metabolism, whose effects are spatialized by the environment. Biomimicry also concentrates on an incredibly small solution set, which is the outcome of many previous 'failures' and pays very selective attention to making its preferences in transforming identified 'solutions' into industrial materials such as plastics, metals and glass, which may then be incorporated into buildings. Biomimicry, therefore, is effectively an aesthetic practice, or a design version of the *Just So Stories* (Moran, 2008) and does not represent a systematic search of a solution space, which Kauffman calls the 'adjacent possible' (Kauffman, 2008, p.64). Rather, it is an ad hoc set of aesthetic or functional preferences made from an incompletely observed and biased set of possible choices. Yet, while biomimicry may inspire more attractive, mechanically efficient buildings, their relationship with the environment exists within the same paradigm as industrial buildings. Indeed, although McDonough and Braungart recommend keeping biological and technical metabolisms separate (McDonough and Braungart, 2002), Rachel Carson's chilling *Silent Spring* had already highlighted the inevitable leakage of invisible pollutants from industrial plants into natural systems to produce insidious effects that were 'not quite fatal' (Carson, 1962, p.29).

While irreconcilable differences between the technosphere and biosphere were sanctioned by the Club of Rome (Meadows et al, 1972), a third way existed which used the new science of complexity as an approach for understanding the extensive, multi-scalar connections between even potentially unlike systems such as industry and the environment. Ecologist Arthur Tansley, who coined the term 'ecosystem' in 1935 (Willis, 1997; Odum, 1971), set out a new paradigm for thinking about material relationships through the existence of functioning wholes, comprised of highly integrated parts that could not merely be disentangled and dissected into dichotomous, deterministic states. While cities have historically been likened to 'organisms' since Aristotle's time, over the course of the late 20th century, cities have been increasingly viewed as complex structures, and the term 'ecosystem' has been used to indicate the multiple urban relationships that form our metropolitan environments and give rise to architectural events. Yet, the appreciation of the systemic nature of cities has done little to dispel the fundamental dichotomy between human development and Nature. Rem Koolhaas notes that not only has 'the sum total of current architectural knowledge [not] grown beyond this opposition' (Koolhaas, 1995)[19] but also observes the downside of excessive formal solutions within social spaces, where glamorous forms alienate their populations by eroding cultural and historical values.[20] Stewart Brand, in particular, identified a new emerging order of technology in modern computing which was compatible with ecological systems that could potentially evolve alongside each other. The mutually overlapping frameworks of cities, ecosystems and the fluid medium of 'cyberspace' shared the same complex, fundamental organizing principles. Digital computing rapidly became the platform for cybernetic, 'whole-systems' technologies, which could potentially align the goals of technology and the natural world (Brand, 2009). Cyberspace enabled the complexity of our reality not only to be conceived but also visualized according to the principles of complex systems, by revealing the relationships between otherwise intangible elements, such as the map of the Internet. Indeed, envisioning digital technologies as being compatible with complex processes such as natural systems through the field of cybernetics was one of the most important ways in which the Whole Earth discourse helped prepare the way for the digital revolution of the 1980s. However, the language of cybernetics is based on the metaphor of machines, which, coupled with the notion of Spaceship Earth, could begin to describe the Earth's systems in mechanical and geometric terms (Latour, 2013). The notion of a complex, terrestrial Nature-machine was enabled in cyberspace, whose visual software is informed by Euclidian geometry. Indeed, the digital domain enabled a new Cambrian explosion of morphological species whose possibilities were explored in the work of architects such as John Frazer (1995), Marcos Novak (1992), Neil Spiller (Pearce and Spiller, 1995), Greg Lynn (Lynn, 2000) and Lars Spuybroek (Spuybroek, 2011). This new language of possibility and interconnectivity enabled designers to imagine relationships between environments, objects and processes that were previously unrealizable, and has underpinned the practice of parametrics (Schumacher, 2009). The fluidity of cyberspace has also enabled academics such as Bruno Latour and David Harvey to describe the host of technological and natural elements that make up our urban environments as being the outcome of complex multi-scalar relationships (Latour, 1993; Harvey, 1996). Visualization of these concepts has informed the way that cities are (re)imagined through process-led relationships in which urban 'ecological' systems may be formed by the spatial configurations of all kinds (Hillier and Hanson, 1984), that are realized through horizontal couplings between agents through exchanges of culture, nature, power and capital. While

---

19 Koolhaas is referring to the cultural opposition of artifice and Nature.

20 Koolhaas notes that Singapore has kept up with its rapidly expanding populations by applying a tabula rasa developmental logic, which has subtracted any perceivable contextual background, adding only glamorous foreground.

Philippe Rahm's 'Hormonorium' establishes 'continuity between the living and the non-living' in the production of social spaces through new material relationships that operate through 'assemblage[s] of physiological devices acting on the endocrine and neurovegetative systems', its construction remains within the hierarchical traditions of architect as sole designer of systems (Philippe Rahm Architects, 2002). The new, systems-based ordering paradigm not only challenges the status of objects and geometries as the primary ordering units of architecture, but also invites agentized materials to inhabit non-classical methods of architectural production, as exemplified by Alison and Peter Smithson's 'new Brutalism' (Gissen, 2011) and post-structuralist architects like Bernard Tschumi (Tschumi, 2012), Gordon Pask (Pask, 1995), Cedric Price (Jacobs, 2011) and Lebbeus Woods (2012b). Further decentralization in the production of architecture is exemplified in naïve building projects such as postman Ferdinand Cheval's Palais Idéal (Dannies, not dated:a) and the steel Watts towers of Sam Rodia (Watts Towers, 2006–2013); on a smaller scale, caddis-fly larvae have been used by artist Hubert Duprat to produce ornate jewellery (Open Culture, 2013). More recently, growing artisan maker movements are further enabling distributed forms of architectural production (Morin, 2013) as well as artistic projects such as Roger Hiorn's 'Seizure' installation, a violet-blue crystal-coated building interior, grown in situ from solution (Searle, 2008). Furthermore, deanthropocentrized methods of architectural production, as proposed by Friedensreich Hundertwasser's manifestos (Hundertwasser, not dated; Hundertwasser, 1976),[21] are increasing creativity at all levels of material organization, which are not the exclusive domain of architectural designers but may be engaged by many actants as a fundamental quality of the material realm. Indeed, Hundertwasser proposed that a 'decomposing solution should be poured over ... glass walls and smooth concrete surfaces, so the moulding process can set in' (Hundertwasser, 1976). By subverting the expectations of architecture, new materials, processes and forms of social organization may be free to produce new methods of architectural production. These are essential for the development of truly subversive forms of material creativity that open up a convergent space for the entanglement of artificial and natural processes. These alternative modes of architectural production erode the classical dichotomy between architecture and landscape and transform urban landscapes into post-natural fabrics, which are new forms of ecology in which technology is seamlessly and coherently entwined with natural systems (see Fig. 3.4).

Indeed, post-natural architects are not master planners, but expert facilitators and collaborators. They work with the innate creativity of many heterogeneous actants to midwife spatial programs into material expressions that reinvent conditions for living. To be fully realized, such possibilities require an alternative technological platform that is not constrained by the aestheticisms of Nature, or the geometry of machines.

---

21 Read by Hundertwasser at Seckau Abbey, Styria, 4 July 1958.

Figure 3.4: This fountain in Plaça de Catalunya, Barcelona suggest provokes the idea of post-natural landscapes actively evolving within urban communities. Photographs and collage, Rachel Armstrong, November 2011.

David Gissen proposes that such a fabric is 'sub-natural', being deeper, messier and more primordial than our cultural expectations of Nature. He suggests that such subversive materiality may constitute the 'architectural reconstruction of nature' where 'we can be surrounded by things that are absolutely alive without transforming them into simplistic expressions of life' (Gissen, 2011). Yet, form-finding alone in the reconstruction of material paradigms is not sufficient to build mutually reinforcing relationships between actants. Indeed, Christian Groothuizen proposes that: 'Artificial ecologies must be able to sense and respond to a variety of natural and artificial stimuli at a multitude of scales, and provide a platform for real ecologies to re-establish' (Pohl, 2011). In appreciating that the difference between natural and technological landscapes is underpinned by cultural aesthetic preferences (Morton, 2007), the coupling of systems and the kinds of social organizations produced by these entanglements become more important in bringing changes in the production of architecture than any specific contributions by the participating technologies. For example, Liam Young proposes that near-future landscapes will be entanglements of robotics, biotechnology and ubiquitous computing, which will give rise to new specimens of biotech creatures (Young, 2011). This forward-looking, propositional view of 'architecture of the future' is also shared by R&Sie(n), who anticipate 'unimaginable post-human landscapes and "things" that are hybrid mixtures of organic and inorganic material' (Frichot, 2011). Such multitudinous, heterogeneous couplings between actants and assemblages do not coexist without difficulties or conflicts, but through managing the different relationships between the many scales of existence, in which architectural design has the potential to play an increasing role. Latour notes that such groupings deal with 'life' and occupy space in unique ways that define their specific territories. The borders of these terrains are not destined to be harmonious and infinite but are locally defined and may indeed provoke conflict between neighbouring systems. Latour proposes that our 21st century challenges relate to limiting the prospective wars of our world over the ordering of space through the actions and needs of different territories and different bodies (Latour, 2013).

To establish what kinds of materials, infrastructures and technologies may literally, rather than propositionally, give rise to post-natural fabrics, I directed my reading towards investigating practices that were working with lifelike materials. I was particularly looking for agents that spoke to the idea of 'sub-natural' materials (Gissen, 2011) and could propose how such raw forms of matter could be applied technologically within an architectural design context.

3.3.3 Bio Design

> *... the romantic tradition ... depicts nature in terms of a sublime pantheism ... the scientific tradition ... depicts nature in terms of a mundane positivism.* (Bok, 2001, p.93)

Nature may be regarded as a Romantic technology (Bowie, 1995) that has experienced a very slow evolution owing to conservative views that characterize our cultural expectations of Nature (Morton, 2007). Yet, recent developments in the field of biotechnology have changed the practice of biology and the possibilities of working with natural systems from being a backward-looking, descriptive practice – one that simply describes the way that things are – into a forward-looking approach that imagines what they could be, with the goal of engineering living systems to produce modified biological or lifelike systems. The rise of modern biotechnology began in 1985 with the invention of the polymerase chain reaction (PCR) technique by Kary B. Mullis (Smithsonian Institution Archives, 2004), which became the catalytic technology needed to fuel the rapidly developing insights into genetics. PCR allowed scientists to make millions of copies of even the tiniest samples of DNA and greatly accelerated the rate at which previously painstaking genetic experiments could take place. This enabled the entire genome of a simple organism, *Caenorhabditis elegans*, to be sequenced (Hodgkin et al, 2011). As more organisms joined the ranks of genetically decoded creatures, it was evident that the sequencing of the three billion base pairs of human genetic code was within sight, and international research groups collectively worked towards decoding the entire human genome. This monumental venture was called the Human Genome Project, and generated intense competition between research groups to find more efficient ways of reaching their goal. New biotechnologies seemed to be invented almost daily, and it was already apparent in the early 1990s that the full DNA sequence of the human genome would soon be known (Jablonka and Lamb, 2006, p.1).

The formal sequencing of the human genome in 2003, and recent biotechnological engineering feat by J. Craig Venter's team who produced the world's first organism with a synthetic genome (Gibson et al, 2010), dubbed 'Synthia' by the popular press, have heralded the current age of synthetic biology, which represents a new manufacturing process that does not rely on petroleum as a substrate but directly harnesses natural processes by commanding their cellular machinery (Venter, 2007). Although the advent of molecular biology and the discovery of DNA (Watson and Crick, 1953) provided techniques that enabled access to the microscale, designers have not historically had access to this space owing to institutional, conceptual and technological challenges and therefore these tools and approaches have not been central to the organicist debates.

Yet, recent changes in policy with the advent of Artscience projects championed by organizations such as the Wellcome Foundation (Wellcome Trust, not dated) and Harvard University (The Laboratory at Harvard, 2013), coupled with a high-profile campaign for public engagement in science (*Guardian*, 2012), have provided a new generation of designers with access to tools, materials and associated methods that are expanding designers' reach (Armstrong, 2008a; Armstrong, 2008b; Armstrong, 2010d). Biotechnology enables designers to work directly with the natural world, and frees them from the obligation to mimic its forms and processes (see Fig. 3.5).

Bio Design (Myers and Antonelli, 2013) is an emergent, experimental design space that incorporates ideas implicit in modern biotechnology and views biological systems as the subject, canvas and medium for design across a range of contexts (Gambino, 2013). This movement proffers an ecological agenda that seeks alternatives to industrial methods of production, with beneficial environment impacts. Yet, this design platform is still emerging and the specific technologies and methods that characterize the practice are not formalized. Consequently, Bio Design projects are often not fully realized and invoke possible as well as actual designs. This presents a research opportunity in evaluating the current approaches and techniques to identify technological and methodological gaps in the field.

Bio Design operates across a very broad spectrum of practices that includes the direct manipulation of native biology, such as 'Cattedrale Vegetale' by Giuliano Mauri, which fashions firs, hazel and chestnuts into an avenue of arches (Myers and Antonelli, 2013, pp.32–35). Greg Bosquet's 'Harmonia 57' incorporates plants directly into more traditional architectural materials such as porous concrete (Myers and Antonelli, 2013, pp.22–25), and Henk Jonkers 'Bioconcrete' seeds natural extremophile bacteria, which can tolerate extreme alkalinity, within traditional concrete mixes to bestow

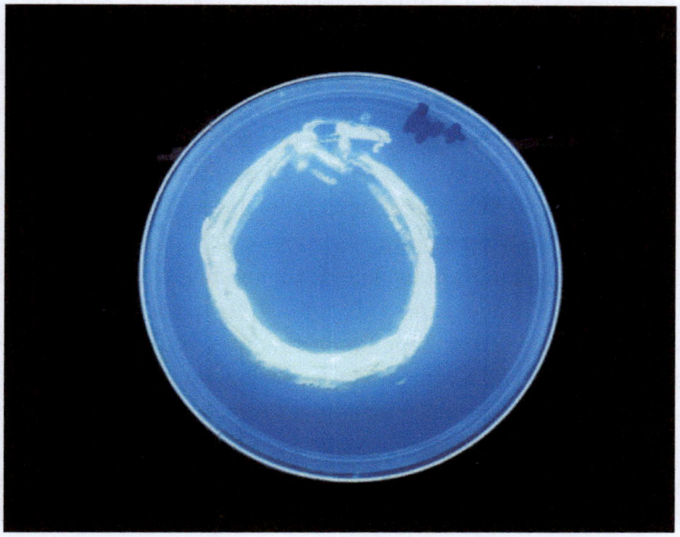

Figure 3.5: This drawing of an ouroboros was made using a sterile loop to apply a genetically modified strain of *E. coli* to a nutrient medium. The bacteria had been transformed by inserting a gene that expressed a green fluorescent protein, which produces bioluminescence under dark light. The experiment was carried out at the Cantacuzino National Institute for Research and Development in Microbiology and Immunology (NIRDMI), Bucharest, Romania, with senior research scientist Alexandru Vladimirescu, supported by a Darwin Now Award from the British Council. Photograph of bacterial drawing, Rachel Armstrong, July 2009.

them with self-healing properties.[22] The hardy organisms mixed into the cement are activated when tiny cracks in the concrete let in water and produce a calcified sealant that prevents further progression of the micro-fractures (Mandel, 2013; Myers and Antonelli, 2013, pp.80–81). In fact, bacterial metabolisms are frequently used in Bio Design as 'workhorse' alternatives to industrial processes, which would usually require large amounts of heat to perform tasks such as brick manufacturing. Ginger Krieg Dosier's 'Bio Bricks', however, uses no heat and encourages bacteria to transform sand grains into sandstone (Myers and Antonelli, 2013, pp.78–79). Other microorganisms, such as mycelia,[23] are used for making biological building materials like Eben Bayer's 'Ecocradle' product packaging (Myers and Antonelli, 2013, pp.106–107) and Suzanne Lee's microbial cellulose cloth, which is produced by a blend of yeast, green tea and sugar (Myers and Antonelli, 2013, pp.108–111). Bio Design also incorporates more speculative work such as 'Meat House' by Terreform ONE that

22 Concrete is an extremely alkaline material; however, Jonkers' extremophiles can endure these conditions despite this noxious environment that is lethal to most other life forms.

23 Mycelium is the term used to describe the branching networks of fungi.

envisages growing leathery skins around buildings using tissue culture and large-scale three-dimensional printing techniques. In practice, however, tissue culture is prohibitively expensive for architectural projects, costing around $1,000 for three square centimetres of skin (Armstrong and Spiller, 2011). Moreover, cultured tissues are not yet scalable, since traditional buildings do not possess suitable infrastructures to support and nurture living tissues and need to be kept in a sterile environment to protect it from bacterial infection, which is impractical. Indeed, Bio Design considers a range of uses for genetic manipulation, such as Alberto T. Estevez's 'Genetic Barcelona', where cities are imagined to be lit by genetically modified trees that glow with green fluorescent protein (GFP) (Myers and Antonelli, 2013, pp.68–69).

Such propositions of incorporating synthetic organisms into social spaces raise important questions about public health and safety, as well as starting important cultural conversations about how living spaces with 'transformed' genetic materials may be inhabited. Other architectural projects speculate even further into the future, such as Conny Freyer and Sebastian Noel's project that proposes strange, transformed plants, which are reminiscent of an applied 'Parallel Botany' (Lionni, 1977), will manage the health of urban environments (Myers and Antonelli, 2013, p.178–181). Indeed, Bio Design actively engages with the technological and cultural challenges implicit in the advent of modern biotechnology, where lively materials may not only impact on our lives through the kinds of substances they produce, but may also become entangled with our cultural, moral and ethical systems. For example, SymbioticA's 'Victimless Leather' (Schwartz, 2008) offers another way of producing meat than by slaughtering animals and raises questions about how meat-yielding tissue cultures should be raised (Armstrong, 2012a). Although Bio Design agitates for a new paradigm of practice, there is a significant risk that it will fail to do so. This is partly due to the immaturity and expense of biotechnologies outside of a medical or industrial setting, but also because of the difficulty in reaching a critical escape velocity by distinguishing itself from a dominant machine-led culture, which may simply assimilate Bio Design into its system as 'little soft machinery'.[24]

This particular literature survey identified that, within the field of Bio Design, there are opportunities to establish new ways of working with lifelike materials in both real and speculative ways. Yet, I wanted to differentiate these approaches from neo-environmental, industrial systems that regard Nature as a robust consumable. So, I directed my reading towards identifying a discrete set of materials and approaches that encapsulated the principles of vibrant matter (Bennett, 2011).

---

24 The term 'little soft machinery' was architecturally coined by Neil Spiller (2007, pp.202–224) to refer to William Burrough's novel about how control mechanisms invade the human body (Burroughs, 1961).

3.3.4 Vibrant Matter

Bennett's theory of vibrant matter establishes a philosophical and cultural context that proposes that we might achieve more ecological outcomes if we observe how events may change if we give the force of 'things' more due (Bennett, 2010, p.viii). While Bennett draws on the vitalist tradition to attribute power to the material world (Bennett, 2010, pp.83–84), my research asserts the capacity for matter to act firmly within the material realm as an expression of the paradoxical laws of quantum physics. These powers may be viewed as 'sub-natural' forces (Gissen, 2011), which are encapsulated in the origins of life sciences that are concerned with the twilight zone between animate and inanimate systems (Luisi, 2010). The innate forces of vibrant matter may be further explored through associated technological research platforms in scientific fields such as combinatorial chemistry (Kauffman, 2011) and synthetic biology. I sought to identify a suitable system that could embody the cultural agendas implicit in vibrant matter that exhibited spontaneously lively behaviour which could be observed at the human scale. My aim was to establish an experimental platform which could be safely and ethically studied in the laboratory as a way of investigating the principles of vibrant matter as a material substrate for architectural design.

I began my search for these materials in the origins of life sciences, where Fredrich Wöhler first demonstrated the continuity between living and non-living systems by synthesizing biological molecules from non-biological ones (Wöhler, 1828). This set the scene for a revolution in chemistry, namely the quest to build living systems from non-living ingredients. This went in opposition to the work of the alchemists, who sought simplification through the distillation of essences that made up the world; for example, *aqua vitae* – the essence of life. Since the Enlightenment, chemists and gentleman scientists such as Johann Rudolf Glauber (Glauber, 1651), Frederic Ferdinand Runge (Runge, 1850), Moritz Traube (Traube, 1867), Raphael E. Liesegang (Liesegang, 1869), Otto Bütschli (Bütschli, 1892), Boris Pavlovich Belousov (Belousov, 1959) and Vladimir Yevgenyevich Zhabotinsky (Zhabotinsky, 1964) began investigating the complex material processes that underpinned the transition from non-living to living matter and challenged vitalistic perspectives[25] by establishing that chemical systems possessed biotic qualities (see Fig. 3.6).

I first encountered the most striking and experimentally appropriate system of chemical self-assembly during the Artificial Life XI conference in 2008 (Artificial Life XI, 2008). Takashi Ikegami's keynote talk outlined how self-assembling droplets could be produced by a carrier oil system with a metabolism (Hanczyc et al, 2007) (see Fig. 3.7). Although these droplets were not technically alive, they exhibited a range of behaviours such as movement, sensitivity and the production of microstructures.

25 Vitalism proposed that living things are fundamentally different to non-living ones because they possess an intangible essence such as soul or consciousness.

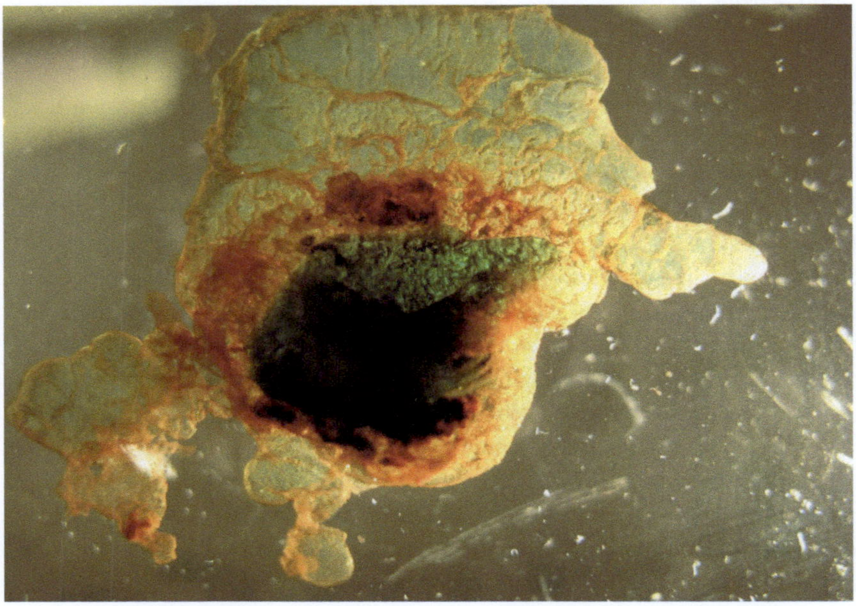

Figure 3.6: This 'artificial plant-like' Traube cell was grown by adding a crystal of copper II sulphate to a weak 0.1 M solution of potassium hexacyanoferrate. Photograph, courtesy Philip Beesley Architect, July 2009.

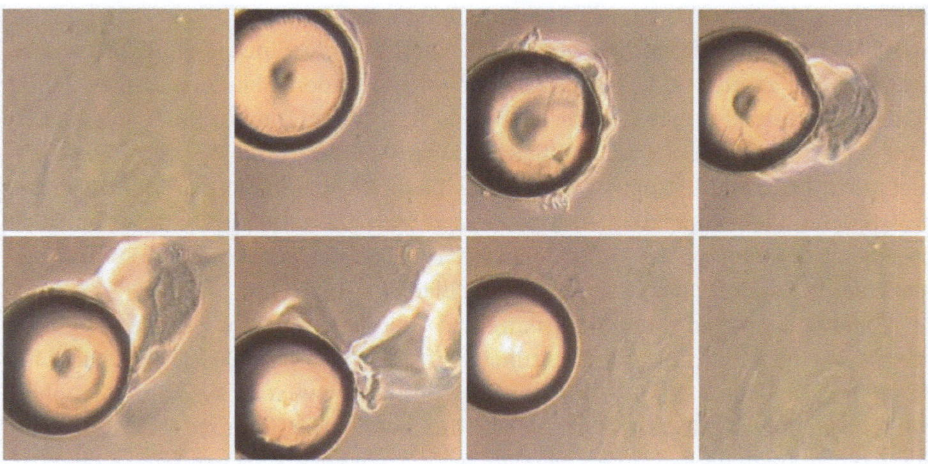

Figure 3.7: This series of images of a benzene droplet with a simple anhydride metabolism is exhibiting complex lifelike behaviour including movement and the shedding of a chemical 'skin'. Movie stills taken at intervals of 30 s, following the addition of an oil droplet to an aqueous solution, courtesy Martin Hanczyc, August 2007.

Yet, dynamic droplets had no formal applications other than being considered as a possible platform for 'soft robots'.[26] This suggested dynamic droplets possessed technological potential and could conceivably work in a qualitatively different way to machines. Indeed, a focused literature survey revealed that these dynamic droplets (Hanczyc et al, 2007; Toyota et al, 2009) could spatially distribute matter, respond to changes in their environment and produce a chemical skin that was vigorously shed like a synthetic amniotic sac and, therefore, may be of architectural relevance. Martin Hanczyc outlined the scientific history of the droplets that had been developed as research tools in the origins of life sciences (Hanczyc, 2008). Other research groups were also investigating dynamic chemical systems using amphiphiles (oils with detergent-like qualities) such as reverse micelles (water in oil droplets stabilized by a surfactant) (Pileni, 2006) and oil droplets in aqueous media (Hanczyc et al, 2007; Toyota et al, 2009); but, although these dynamic chemical systems produced lifelike effects, they had not been used as a technological platform. I therefore sought to establish a variety of practical approaches for working with dynamic, lifelike chemical systems as a possible form of technology and model systems for demonstrating the principles and potential practice of using vibrant matter in architectural design.

3.3.5 Morphological Computing

Since dynamic droplets had not been architecturally examined, a literature survey identified ways of working with these agents in a technological capacity so they could be incorporated into a portfolio of design tactics that could execute spatial programs. Of particular interest were a range of alternative forms of computing that used complex, chemical systems as the programming language and hardware called 'natural computing', following Alan Turing's interest in the computational power of Nature (Turing, 1936). 'Natural' computers have been made using DNA, light, modular bricks and paper, and therefore provided an appropriate context for exploring how non-mechanical, physical systems might operate as technological platforms (Denning, 2007; Zyga, 2013). Currently the field includes a broad range of mutually supportive, overlapping practices which include approaches such as digital modelling of biological systems (Paun, 2005) and the manipulation of chemical patterning systems (Aron, 2011). Within the field of natural computing, 'morphological computing' (Pfeifer and Iida, 2005) was identified as a suitable platform to deal with the computational powers of material processes and could potentially respond to spatial programs to achieve directed effects. Owing to the early stages of development

---

26 Soft robotics is being a material movement within robotics that does not use 'hard' bodies that are composed of metallic structures on 'conventional bearings' that do not exist in Nature but a portfolio of softer, more flexible materials (Whitesides Research Group, 2011).

of this research field, the design and engineering tactics of morphological computing are diverse and have not been formalized into a particular rule set. The knowledge gap identified was, therefore, how dynamic droplets could be used as a material, and infrastructure and technology that could function as an alternative production platform to machines.

3.4 Summary

My literature survey draws from overlapping, multidisciplinary research fields and practices. It suggests that by using dynamic chemistries at far from equilibrium states potential applications of lively materials for architectural design may be systematically interrogated. Dynamic droplets appear to provide a viable model system for establishing the possibility of a practice that incorporates the principles of vibrant matter with precedents within organicism and Bio Design. These may be meaningfully applied in an architectural design context and manipulated using morphological computing techniques. The following chapters explore the research opportunities identified in my literature survey and develop them towards testable design propositions.

4 Method

An encounter that does not explain but produces. (Stengers, not dated)

4.1 Overview

This chapter outlines my research methodology, which is based on the knowledge gaps identified through an extensive literature search that established the architectural and scientific contexts for my practice. My research questions were formulated under the supervision of Neil Spiller, who guided my architectural design investigations, Martin Hancyzc, who oversaw my chemical studies and Philip Beesley, who I collaborated with in the Hylozoic Ground installation, which informed much of my practical architectural design work. Their continued guidance and support have enabled me to navigate the theoretical and practical aspects of my inquiry. These research questions became the building blocks for further speculative proposals about the possibilities of this multidisciplinary architectural design method and helped me reflect on its broader potential for architectural design practice in the 21st century.

In formulating my research questions, I developed a combined methodology that is informed by a number of mutually supporting research methods:

— Multidisciplinary approach: This enabled me to identify synergies between different knowledge fields through shared interests, so that it was possible to broaden knowledge sets and explore previously unknown and unchartered territories.
— Action research: Action research methods enabled me to develop new skills, ideas and approaches by immersing myself in new disciplines or context for example, during field trips, conducting laboratory experiments, or building installations.
— Speculative practice: The limits of possibility may be extended through speculative practices that enable new relationships between otherwise unconnected disciplines or skill sets, to be explored conceptually and practically.
— Morphological computing: This provided a platform for testing new ideas and embodying them without recourse to machines.

4.2 Multidisciplinary Approach

A first principles basis for the acquisition of knowledge was adopted using foundational propositions or assumptions that could not be deduced from other knowledge sets. This approach is consistent with my research questions, which require reimagining the nature of matter through process philosophy and applying these ideas in testable ways to the field of architectural design and chemistry. Since I was attempting to work across Two Cultures (Snow, 1959), it was essential to create partnerships and build new knowledge from the synthesis between these different disciplines. Working across knowledge fields has helped me distil common concepts that could be dealt with through mutual re-synthesis, such as designing the Hylozoic Ground chemistries, which was equally influenced by a working knowledge of chemistry and architectural design. Despite the multidisciplinary nature of my research, it is fundamentally an architectural study concerned with the construction of spatial programs and the development of design tactics. Specifically, the questions I was asking regarding the nature of materiality in architectural design is situated within the realms of experimental architectural inquiry pioneered by cyberneticist Gordon Pask, who built a chemical ear that responded to street sounds (Bird and Di Paolo, 2008, pp.185–212), and Stafford Beer's use of daphnia and pond ecologies (Beer, 1994), who was investigating alternative substrates for the production of architecture.

Teams of multidisciplinary practitioners were curated to inform my practice in many ways – ranging from expanding my subject matter reading to developing new experimental techniques such as Unconventional Computing and Architecture (Building Centre, 2010), Artificial Life and Architecture (Artificial Life XII, 2009) and WETFab (Adams, 2012). From the outset, I have aimed to keep my objectives clear, communicate frequently and be very specific about authorship and how work is credited, such as developing drawings for the Vibrant Venice project with Christian Kerrigan and GMJ. Collaborative practice is not simply a meeting of minds but a much more extensive system of how relationships are built and actively managed (Driver, Peralta and Moultrie, 2011). This has been key to completing my work, which has required me to work in different countries and organize research time in studios outside my own university such as, arranging laboratory time in a busy scientific laboratory at the University of Southern Denmark, working with the European Centre for Living Technology in Venice, or constructing an installation for the 2010 Venice Architecture Biennale.

Adopting a multidisciplinary approach has enabled me to identify synergies and enable fusions to take place across subject boundaries. These can be explored as a research conversation, or practical engagement, between subject fields and collaborators. My approach is consistent with Isabelle Stengers' constructivist, multidisciplinary method, which operates through building 'an ecology of practices' (Stengers, 2000) where different knowledge sets, or participants, create their own new understanding of a subject by working alongside each other. Stengers looks to the outcomes of synthetic practices to produce ideas that can reinform us about reality rather than proposing endless critiques of the various methods used by the contributing disciplines. This constructivist method enables researchers to work in a non-hierarchical way (Stengers, 2000, p.87) so they can find commonalities between disciplines and change assumptions by coming into contact with new views. These, in turn, feed back into the method to further refine practices (Resnick, 1989). Roy Ascott uses the term 'syncretic' to describe a particular kind of synthesis that does not provoke homogenized practice but enables collaborating disciplines to retain characteristics of their discipline of origin (Ascott, 2005). The convergent approaches that I have applied in my research may not only give rise to synthesis, in which disparate things meld into a homogenous whole and lose their individual distinction, but may also produce syncretic or entirely novel events. My research does not value one kind of outcome over another, but rather aims to utilize the various events appropriately, in a design context.

The value of convergence through multidisciplinary collaborations in producing novelty has been recognized in an NSF report (Roco and Bainbridge, 2003, p.9), which proposed that these approaches have the potential to bring significant economic and human benefits, particularly through the emergence of new, combined technologies.[27] This so-called NBIC (Nano Bio Info Cogno) convergence has led to funded 'sandpits' both by the NSF and EU, where multidisciplinary practitioners, predominantly scientists, collaboratively addressed 'grand' challenges such as artificial photosynthesis, and has resulted in projects such as the cyberplasm robot.[28] Convergence between the Two Cultures is also gathering support as a method of innovation from central funding sources, and currently the UK government is supporting STEAM (science, technology, engineering, arts and mathematics) in which the contributory role of arts is supported and recognized (Else, 2012).

4.3 Action Research

My studies are consistent with action research where an investigator is immersed in their subject matter – specifically, architectural design and chemistry, as well as their associated practices (Whitelaw et al, 2003; Creswell, 2009). This approach is frequently used in clinical practice and anthropology, where researchers explore theoretical issues by working within the research environment (Diamond, 2012; Latour and Woolgar, 1979; Hird, 2009; Goodall, 1969). Action research enables researchers to learn from their collaborators and environment to acquire new

---

27 The non-deterministic, forward-looking, speculative nature of these constructivist scientific approaches does not amount to 'science fiction' (Bassett, Steinmueller and Voss, 2013). Steve Fuller proposes that the literary genre has relevance to understanding the impacts of new technological developments on society. Fuller notes that the 'exclusion of science fiction from sociology pertains less to its content than its institutionalization. In other words, à la Karl Popper (Popper, 1959), if we were to treat science-fictional propositions as revisable hypotheses rather than stand-alone fantasy worlds, then they could quite quickly form a kind of sociology' (Fuller, 2011, p.45), so its character is primarily sociological rather than scientific or technological. Yet science fiction does not provide testable hypotheses but deals with a technological 'fait accompli', with little attention paid to the detailed procedural aspects of its existence.

28 The 'cyberplasm' robot is an example of the kind of projects that have arisen from the NBIC 'sandpits', which is a melange of biological and mechanical systems (Cyberplasm Team, 2010).

knowledge and make repeated, informal evaluations and judgements. In achieving my research aims my abilities were broadened specifically in tasks, such as careful planning, practical skills, sharpened observation and listening, evaluation, and critical reflection. In an architectural design context, by working with Philip Beesley and with Martin Hanczyc, action research helped me develop the Hylozoic Ground chemistries. These experiences enabled the cross-referencing of different ideas and practices to synthesize new ideas and approaches, such as developing 'Carbon Eater Flasks', which removed dissolved carbon dioxide and fixed it into mineral form. I also participated directly in the scientific research community by joining in conversations on advanced materials and technologies such as synthetic biology and morphological computing.

The new skills I acquired have been applicable to other related activities such as building an installation at the Synth-ethic group show at the Natural History Museum in Vienna in April 2012 (Synth-ethic, 2011) and for the 'En Vie/ALIVE' group show at the Espace Foundation EDF in Paris, in April–August 2013 (Textile Futures Research Centre, not dated; Meyer, 2000). By engaging myself in field research in the city of Venice, I was able to imagine how the city might find a means of surviving the relentless assaults on its fabric by responding differently to them and found direct evidence to support my hypothesis through field work. Photographic surveys of the waterways and lagoon-side experiments were particularly useful in informing how programmable droplets could become a city-scale morphological computer that would meaningfully produce a form of vibrant architecture within Venice's waterways. By visiting the site during different seasons, I was able to appreciate the different synthetic cycles of the marine wildlife (which were more vigorous in the spring and summer seasons) as well as observe how the accretions grew and even changed colour in niche-specific ways.

Action research methods have enabled me to practically engage with my subject matter through creating new work and reflecting on the outcomes to produce new knowledge, which was transferable to other areas of research. For example, I have provided students in Canada, England, Italy and New York with recipes to make Hylozoic Ground chemistries and have also been able to reach new target audiences, like using a modified 'hygroscopic' form of the chemistry in public demonstrations such as for the UK ArtScience Prize (Ignite, 2013) and also for the Glenfiddich Pioneers event (Future Laboratory, 2010). The downside of action research is the risk of introducing bias into the research and requires steps to be taken to minimize this. However, I did not use action research approaches for scientific experiments and my influence within the systems under study was anticipated and even desired. For example, in documenting the lifelike behaviour of Bütschli droplets, I created the conditions in which it was possible to influence Beesley's decision to use Bütschli droplets as an integral system within the Hylozoic Ground installation.

4.4 Speculative Practice

Speculative approaches are consistent with the theme of probability that underpins my experimental work. As such, to explore the 'adjacent possible' available to vibrant matter in architectural design practice I make frequent use of propositions that are conceptually plausible, yet experimentally unproven. Taking a propositional approach to working with emergent systems has helped me to resist over-directing the outcomes, which carries the risk of skewing, constraining or biasing the research. Instead, speculative proposals help keep solution spaces open for exploration and therefore preserve access to a broad range of possible applications, rather than seeking to advance any particular formal solutions at such an early stage of its characterization. Throughout my research I used a range of speculative approaches including, Design Fiction (Sterling, 2011; Bosch, 2012; Koch, 2013; Sterling, 2013), Science Fiction (Fuller, 2011), 'fictionalism'[29] and 'post-normal' scientific research (Hulme, 2007; Funtowicz and Ravetz, 1992). Additionally, I produced speculative fiction narratives to imaginatively explore issues that were raised during the research period, so that the implications of the emerging materials, methods and technologies could be considered in much broader contexts beyond the laboratory and field (see Fig. 4.1).

Figure 4.1: The use of reflections in my photographic documentation helped me make a transition between objective material study and propositional designs. Photograph, Rachel Armstrong, August 2012.

29 Fictionalism is the philosophical view that a serious intellectual inquiry need not aim at truth (Kalderon, 2005).

Yet these speculative propositions did not stand alone as imaginary constructs but were reflected back on my research findings and design explorations to inform new possibilities; for example, possibilities raised in Chapter 9 informed the production of 'protopearls'. Indeed, speculation is widely used as an architectural research method and includes: the European Commission-funded VISIONS project (Funtowicz and Ravetz, 1992), architectural prototyping (Davies and Vercruysse, 2012) and also Bio Design (Myers and Antonelli, 2013). Indeed, the National Endowment for Science, Technology and the Arts (NESTA) proposes that there is a link between science fiction and innovation (Turney, 2013; Bassett, Steinmueller and Voss, 2013), where speculative approaches build mutual relationships between scientific and design practices – not to predict the future, but to increase the probability that desired outcomes will come true, since storytelling prepares societies for change. Even with no ready means of testing the proposals, speculative explorations can be used as models to represent and compare ideas or to explore incomplete knowledge sets. This allows increasingly more reasoned trajectories to be developed, which may eventually be experimentally testable and gain the status of scientific hypotheses.

4.5 Morphological Computing

> *I long for the day when we can see objects forming, like pools of mud, flowers on a wall or clouds in the sky, as pure products in a context of pure productivity, without any intermediaries. These will be no desires, no opinions, no critics, no designers, just pure flourishing.* (Spuybroek, 2011, p.333)

Architect Lars Spuybroek's vision of a new technological platform that 'is not simply the means to the made, it is the construction of a vast horizontal plane of making' (Spuybroek, 2011, p.332) sounds like a form of Nature. While Spuybroek imagines this design utopia being located in digital computing, my research proposes that it is possible to work with matter directly using a different kind of technological platform to produce lifelike systems that are consistent with the characteristics of Millennial Nature. This is forged by vibrant matter whose fundamental agents form assemblages and can be shaped by internal and external influences. Coordination of these events is called morphological computing, which provides a very different technological platform to machines. Some key differences are summarized in Table 4.1.

Morphological computing is a form of natural computing that originates from a branch of robotics where the physical composition of the system directly contributes to its performance. Where digital computing employs the actions of subatomic particles, called electrons, to embody its binary outputs, the decision-making abilities of morphological computing take place upwards from the atomic to molecular to macroscale events, which are capable of parallel processing. In the context of

Table 4.1: A comparison between mechanical system and natural computer

	Mechanical system	Natural computer
Component	Object	Agent
Order	Series	Parallel
Power structure	Hierarchical system	Non-hierarchical
Functional system	Machine	Assemblage
Energy	Extrinsic	Intrinsic and extrinsic – spontaneous operations may be prolonged with resource supply
Control	Hard	Soft
Transformation	Binary – on/off	Variable states. Generally conservative but may behave unpredictably and collapse or transform at tipping points
Influence	Internal	Internal and external (environmentally sensitive)

vibrant matter, morphological computing enables chemistry to exceed the traditional expectations of materials by tapping into its environmental responsiveness and ability to autonomously produce effects. The aim of morphological computation is to work with, rather than dampen out, the unpredictability of the material realm.[30]

The fundamental design units of morphological computing – atoms, molecules, complex chemical assemblages – couple information flow (droplets, proteins, DNA) and energy transfer with matter during the computational process. Morphological computing is therefore a leaky material system, which is capable of adaptation and change. It does not use top-down instructive programming such as genetic modification, but orchestrates the intrinsic power of matter by promoting the horizontal coupling between heterogeneous agents and working with the passage of time as a creative agency (Prigogine, 1997). In other words, morphological computing establishes the conditions in which the material world – and by implication, architecture – can begin to perceive the world, not merely be acted upon by other agencies. Morphological computing also uses the properties of matter as effectors at many scales, and can read and respond to physical changes by virtue of its parallel processing abilities. It reveals and provides a way of working with the strangeness and dynamic potential of a material world that we thought we already knew by

30 Ilya Prigogine observes that his insights into the irreversible effect of time on a system provided a means to extend the theory of classical dynamics, not invalidate it, noting, 'that we now need to extend classical mechanics . . is quite unanticipated. Even more unexpected is the realisation that this revision of classical mechanics can guide us in extending quantum theory' (Prigogine, 1997, p.109).

opening up new design possibilities.[31] A table comparing and contrasting between a process-led view of matter that underpins assemblage technology and the atomistic worldview that informs machines is presented in Table 4.2.

Harnessing the key qualities of vibrant matter in new ways is essential in establishing a platform for a new kind of material synthesis. Key to my research inquiry is that the agency inferred in vibrant matter is not ephemeral, or vitalistic, but is real and takes the form of non-classical behaviour of matter at far from equilibrium states. This is essential for the operationalization of the 'assemblage' platform as a technology, which produces its effects by 'horizontally' coupling complex phenomena together to produce its effects and even generate novelty. Similar processes are observed in natural systems, where, for example, the protein myelin exploits phase changes in its structural system to produce self-assembly (Hewitt, 2013). Rather than viewing myelin as a self-assembling machine, which does not operationalize non-linear phenomena but dissipates them, the protein may be usefully considered as an assemblage whose work is a function of phase transitions that can be provoked by the collective agency of actants. Using different concepts, such as assemblage, to consider Nature-like solutions may open up new strategies for working with materials in architectural design contexts. Importantly, non-classical phenomena may require different infrastructures and initial conditions to machines to harness complexity in their systems. For example, morphological computing operations require appropriate infrastructure such as water or soils, which keep the flow of matter and their chemical processes open, spatialized, multidimensional and temporalized. These tactics enable the system to couple and build on the non-classical dynamics of these systems so they evade the decay towards equilibrium (Schrödinger, 1944) and may even become evolvable.[32]

31 Timothy Morton calls this the 'beautiful soul syndrome', which is based on Wilhelm Friedrich Hegel's formulation of an aspiration to attain moral sensitivity and purity of vision that severs humans from the world of nature and disempowers people from taking action during ecological crisis (Morton, 2007, pp.182–183).

32 Ilya Prigogine notes that entropy is the price for structure, the idea that life might evade the production of entropy by distributing its structures temporally and spatially to resist the production of architectures, but once they occur in themselves they provide the means to continue to resist equilibrium states by increasing the time and space between sets of interacting molecules. For example, biological cells have a long 'endoplasmic reticulum,' which is an internal system of tubes that can maximize flow through the cells and delay the sudden consumption of nutrients. This speculative proposition may be investigated through the technologies of morphological computing. 'Smart' droplets systems demonstrate this possibility by being able to distribute matter in time and space and appear to resist equilibrium, which can be observed as a set of 'living' characteristics (Armstrong and Hanczyc, 2014). Ultimately of course, the system reaches thermodynamic equilibrium but the structures produced by these dynamic acts of resistance may tell us something about the spatial nature of living processes and what kinds of architecture may be desirable to provoke 'living' qualities in technological or synthetic systems.

Table 4.2: Vibrant matter compared with atomism

Characteristic	Vibrant matter	Atomism
Model of reality	Probabilistic	Deterministic
Entropy	Exists at relative non-equilibrium	Operates according to the principles of relative equilibrium.
Physics	Quantum, 'spooky action at a distance' (*Science Daily*, 2013), paradoxical behaviours, not linearly scalable but follows continuity principle (Bohm, 1980)	Underpins classical Enlightenment perspectives, or Galilean 'spell' (Kauffman, 2008, pp.129–149)
Properties	Dynamic. Operates within fields of definable probability that possess limits. Exhibits lifelike characteristics, such as unpredictability, robustness, flexibility, adaptability, growth, movement, sensitivity and evolvability	Fixed, geometric, hierarchical, object-oriented
Boundaries	Permeable	Discrete, fixed
Geometry	Non-Euclidean, chaos, dissipative, organic, manifolds, 'abhors' straight lines (Hundertwasser, not dated)	Euclidean, straight lines
Qualities	Potency, fertility, transformation at tipping points	Brute, inert, requires instruction
Principles	Operates through networks, relationships and flows	Exists as objects, hierarchies with 'essential' qualities
System	Open. Evades decay towards equilibrium. Requires nutrients, energy and removal of waste products	Closed, operates at equilibrium states
Relationship with environment	Contextualized. Continual negotiation with environmental conditions to inform, sustain and augment performance	Belligerent to context
Behaviour	Stochastic	Predictable
Interactions	Cooperative. Constantly negotiated. Formation of assemblages extends fields of operation, which are contextualized according to environmental changes	Insular
Agency	Autonomous force	Requires instruction
Technological system	Assemblage	Machine
Nature	Couples with and transforms natural systems (which also exist at relative non-equilibrium) into subversive, relentlessly material bodies and combines with them as synthetic ecology and post-natural fabrics	Works in binary opposition to Nature

Morphological computing is not exclusively concerned with 'empirical' outcomes and may also include aesthetic (Bateson, 1979) or poetic effects such as Spuybroek's 'entangled knots of mutual feeling and action' (Spuybroek, 2011, p.322). These outputs are created through the collaboration and codesignership between a population of design agents and human designer, or 'programmer', and may produce effects by:

— Modifying an environment (such as changing its acidity to induce movement (Hanczyc et al, 2007))
— Acting as a carrier system to move matter through time and space (Toyota et al, 2009)
— Integrating with other systems to produce highly contingent, synthetic outputs (Armstrong and Beesley, 2011)

Although morphological computing proposes a terrain in which 'life' is a possible event, it does not aim to be a form of artificial life per se. Yet, the emergence of fully 'alive' systems from a morphological computing platform is a possibility, since the solution space in which autocatalytic sets of interacting agents (Kauffman, 2008, p.55) could conceivably become fully autopoietic. Indeed, there are many types of evolution (Gould, 1994) and morphological computing could be considered as a way of conducting experiments in replaying 'the tape of life' (Morris and Gould, 1998), which may ultimately result in artificial biological systems, or new kinds of Nature.[33] Indeed, morphological computing could be considered as a hylozoic system of quintessentially ecological acts, which possess the capacity to produce lifelike events and even artificial 'life' itself.

4.6 Summary

My research develops a practical approach to exploring the principles of vibrant matter and establishes the conditions for its realization within architectural design practice. By working across disciplines, immersing myself within a field of discovery, applying the judicious use of speculation and facilitating convergent approaches, it may be possible to discover how vibrant matter, ELT and morphological computing can be practically applied to influence design outcomes.

33 Timothy Morton proposes that we need to challenge contemporary notions of 'green' practice and how our ideas about 'Nature' influence both cultural and scientific outcomes. Instead, he suggests working with the material conditions that exist in the present, rather than trying to recapitulate ideas from other locations and ambient spaces that are not dominated by humans, such as 'the wilderness'. Morton remarks that 'Nature is always eluding being conceptualized – not because it transcends the material realm – but because it is relentlessly material' (Morton, 2007, p.70).

5 Vibrant Matter in Practice

One of the things that's fascinating about research into the world of the ultra-tiny is the way it changes the way you perceive every object around you. Instead of a table being solid, it's a bunch of molecules whose electrons are, as Schwartzberg put it, 'sloshing around.' Even an atom no longer feels like a solid thing. 'When you're looking at atoms, you're just measuring densities of electrons,' Schwartzberg said. Everything is fluid; everything is moving in particles and waves. (Newitz, 2013b)

5.1 Overview

The cosmos is composed of many different species of stardust (Sagan, 2007) and despite our advanced, secular knowledge, we imagine these primordial substances give rise to a universe, fashioned in our own image, in which Nature and the technological expressions of the human mind, are cleaved. This chapter proposes that the material agency within vibrant matter is a real, physical phenomenon and ultimately a testable proposal, which is based on a new set of ideas about reality that paint a portrait of our universe. This is far more autonomous, lively and sensitive than the one that has historically framed western discourses that have relied on human reason for instruction. The lively character of vibrant matter is examined within this context and its agency attributed to real fundamental forces, grounded in quantum physics, rather than being the product of vitalistic theories (Bergson, 1922, p.44; Driesch, 1914, p.vi; Spuybroek, 2011; Bennett, 2010, pp.83–84), or psychological responses to material groupings (Bennett, 2010, p.4). My research seeks to empower matter by recognizing that its innate vitality resides within molecular bonds, which forge assemblages that can interact with and respond to their surroundings. Locating the liveliness of vibrant matter within the material realm provides the basis for an experimentally testable set of observations that may enable its better characterization and identify possible applications in architectural design.

5.2 The Dynamic Nature of Matter

All matter squirms. This is the fundamental reality that underpins our cosmic fabric[34] (see Fig. 5.1).

34 This NASA simulation of galaxy formation during the first two billion years of the universe (YouTube, 2013) illustrates the dynamic character of Millennial Nature.

Figure 5.1: This spiral galaxy is a rotating disk of stars, gas and dust that gives rise to about 60% of our local universe. Photograph, courtesy Space Telescope Science Institute (STSci).

During the Enlightenment, when Galileo and Newton set out the mathematical principles of the universe as 'natural laws' (Kauffman, 2008, p.xi), ideas that shaped our world as an expression of brute materiality that required rational instruction were forged. Rene Descartes compounded these notions by severing our experience of the world into encounters between an ephemeral substance (soul/mind) and mechanical body (Descartes, 1983). Classical science therefore developed the habit of describing matter in passive terms, which was infused by an active agency whose physical substance could be completely contained within and defined as an expression of geometry. Yet there is nothing simple about matter. Material forces and flows are not best imagined as deterministic systems, since they are reactive and capable of transformation. For example, when two gases, hydrogen and oxygen, are combined they produce liquid water. Material vigour does not originate from a geometric understanding of materials, nor their essential qualities, but through molecular processes that forge their interactions from which novelty emerges.[35] While the atomic

35 In Intensive Science and Virtual Philosophy, DeLanda argues that Deleuze's concept of multiplicities is designed to replace the old philosophical concepts of essences, and that things, substances, or objects are to be explained in terms of how they are produced, rather than in terms of their essence. As

model proposes that matter might be reduced to fundamental units, called atoms, modern science has theoretically and experimentally demonstrated that these are not simply irreducible objects, or particles (see Fig. 5.2).

Figure 5.2: These white dwarf stars are composed of electron-degenerate matter and are thought to be the end stage of 97% of all stars in the Milky Way. Photograph, courtesy Space Telescope Science Institute (STSci).

At the subatomic level matter is more complex and exists as probabilistic clouds of muons, leptons and hadrons that behave as waves and particles (Bitbol, 1996; Lederman and Teresi, 1993, p.168). David Bohm suggested that atoms act as amplifiers of information contained in quantum waves (Bohm, 1973), which are connected within a universal network of subatomic particles called the 'Implicate Order' (Bohm, 1980), and within this cosmic fabric, atoms do not always appear to obey Newton's laws. Quantum physics therefore uses an alternative conceptual framework to understand these new characteristics and raises challenging epistemological issues about

DeLanda puts it, '[i]n a Deleuzian ontology [...] a species (or any other natural kind) is not defined by its essential traits but rather by the morphogenetic process that gave rise to it' (DeLanda, 2002, p.10).

materiality, such as in the Heisenberg Uncertainty Principle, which proposes that matter is deeply entangled with our measurement and observation of it (Heisenberg, 1927, pp.62–84). Although particle physics might be dismissed as being relevant only to very small scales of operation, the Correspondence Principle asserts that conflicting micro and macro realities cannot coexist (Nielsen, 1976). Yet, John Haldane observed the importance of the size of an object, institution or animal in establishing their structure, as molecular relationships need to strengthen more than proportionately as the size of systems increased if the structures were to remain robust and resilient (Haldane, 1926). So, if matter possesses fundamental liveliness, then this should be discernable at the macroscale and not just at the level of particles and atoms. Indeed, quantum phenomena shape our experience of the world, which we perceive as fundamental properties of matter, such as colour and the effects of gravity. However, the modern world wrestles with polarities of being, where natural events can be imagined and explained as the product of dualistic forms of existence and geometric functions, whose ordering can be manipulated through binary representations using digital computing methods. Indeed, dualism is contained within the very language and codes we think with and even embody our ideas about 'life' (Armstrong, 2009b).

Recently, Markus Covert compiled data from more than 900 scientific papers to account for every molecular interaction that takes place in the life cycle of *Mycoplasma genitalium* – the world's smallest free-living bacterium – to create a digital model of how it works (*Stanford News*, 2012). The digitalization of living systems into a binary sea of zeroes and ones is encapsulated by J.C. Venter's team building an artificial genetic code using a form of biological computing called 'systems genomics'. This incredible feat of engineering produced 'Synthia', an organism which was extracted from its binary soup and animated in a 'ghost' yeast body to become the world's most ambitiously manufactured synthetic organism (Gibson, 2010). Both Synthia and Covert's organizational topology embody an Enlightenment view of matter, which is inert, insensitive to its environment and requires external instruction to perform tasks. It springs from a mechanical worldview where brute matter can be built from identified parts and infused with ephemeral instructions encoded in digital or analogue information streams. They are subsequently transformed into a lively configuration that is assembled atom by atom. Attempts to vitalize this base image of matter, such as Hans Driesch's view of 'entelechy'[36] or Henri Bergson's 'vital' principle (Bergson, 1922, p.44), only reinforce the idea of a material world in which agency is an afterthought. As Latour notes, 'to be both material and social[37] is not a way for objects

36 Aristotle's entelechy is neither material nor spatial, and manifests its action by a diversity of operations of the organism. Hans Driesch describes it as 'intensive manifoldness, which orders each part to the whole and gives to the organism its reality as a living being. It is the principle of life, the ordering "form" of the living body. Finally, it acts as the final cause' (Driesch, 1929, pp.1–113).

37 Here, the reader may substitute terms that imply other intangible agencies, such as mental, vital, language, etc., for the term 'social' in this quote, which is Latour's particular subject of interest.

to exist: it is simply a way for them to be artificially cut off and to have their specific agency rendered utterly mysterious' (Latour, 2005, p.83).

Researchers such as Vladimir Vernadsky (Vernadsky, 1945; Vernadsky, 1998; Vernadsky, 2007), Ilya Prigogine (Prigogine and Stengers, 1984), and Stuart Kauffman (Edge, 2008) reject the brute nature of matter by observing the world through another scientific lens described by Ludwig von Bertalanffy as 'general systems theory' (Von Bertalanffy, 1950), more commonly encountered as 'complexity'. This breaks away from the geometric discourses of the Galilean Spell (Kauffman, 2008, pp.129–149) by considering reality as a network of relationships and flows, which are never fixed and always under construction. Latour identifies actants (Latour, 2005, p.54; Bennett, 2010, p.9), not atoms, as the fundamental organizing agents within the complex worldview of ANT. This challenges the anthropocentric assumptions at the heart of western philosophy by drawing attention to the subversive agency within non-human bodies (Latour, 1996). Actants are considered from a materialist perspective as probabilistic clouds of action that are both objects and processes, which work through groups of interassociating bodies, or assemblages. Since the individual influence of actants is weak, their phenomenological, social, cultural and even political power is a consequence of their group activity. The effects of actants are not constrained by the geometric limits of their materials but they exert their effect by building real relationships that define their field of activity through a spectrum of types and strengths of molecular interactions. Actants therefore operate through the production of assemblages at different scales and exert varying degrees of influence on their surroundings, such as light at an air/water surface (see Fig. 5.3). Their influence not only draws from Newtonian laws of cause and effect but also embrace counter-intuitive, strange behaviours such as quantum entanglement.[38] Actants may therefore coherently inhabit both complex and classical scientific frameworks and may be observed as objects and as systems.

Yet our cultural conditioning in observing the world around us as one thing or another, leaves us habitually trying to resolve apparent contradictions. In fact, as in the case of photons, these paradoxes are an intrinsic to the fabric of reality. While architectural design has consciously appreciated the centrality of objects according to the classical scientific model, it has not developed formal strategies for designing with the potentially stranger role of materials as process. Identifying the materials, infrastructures, technologies and expectations of matter as process could help catalyse shifts in the performance and expectations of buildings. Yet the diversity and abundance of its processes, which are expressed through 'metabolism', characterize the natural world. Even the grand objects placed in the urban landscape that constitute contemporary architecture are swathed in organic seas of microsystems that inhabit

38 Albert Einstein called quantum entanglement – two particles in different locations, even on other sides of the universe, influencing each other – 'spooky action at a distance' (Science Daily, 2013).

Figure 5.3: The dynamic nature of matter is clearly visible at the interface between water and light. This theme of interface, such as air/water or oil/water, as a site for transformation is a recurrent theme in my research. Photograph, Rachel Armstrong, August 2012.

their inert edifices like plaque. These relentless assemblages work tirelessly to corrode, transform and decompose even our most resilient buildings and remind us that we inhabit a dynamic, microbial world. Earth's micro-communities have given rise to our ancestry over billions of years, having produced the soils and the atmosphere that we breathe (Su et al, 2011; Royal Society of Chemistry, 2011). However, architectural conventions, with their calibration systems set at the human scale, not only overlook the importance of our micro-ecologies but also ignore the largest portion of Earth's natural living history in which humans are an extremely late arrival. We are entangled within a non-human reality, whose living and non-living actants, when considered en masse, prove to be a lively force to be reckoned with.

Vibrant matter becomes creative and convincing as a real phenomenon when it is actively forging assemblages through the persistent material interactions that enable heterogeneous groups of actants to horizontally couple together and produce potentially surprising outcomes (see Fig. 5.4).

These restless networks of molecular interactions are not simply a mass of wriggling, indeterminate forces that render a designer's task as futile as nailing jelly to a wall – they possess a structural quality, which is forged by organizing hubs that create linkages between and within the open relations of matter. For example, haematite crystals appear to undergo lifelike cycles of growth and decay in response to chemical gradients and blue light, which are quenched when the lights go out (Palacci et al, 2013). Materials that are at far from equilibrium states may be uncannily lifelike, with striking characteristics such as movement, innate intelligence, environmental sensitivity and change with the passage of time. 'Living' systems possess some of the

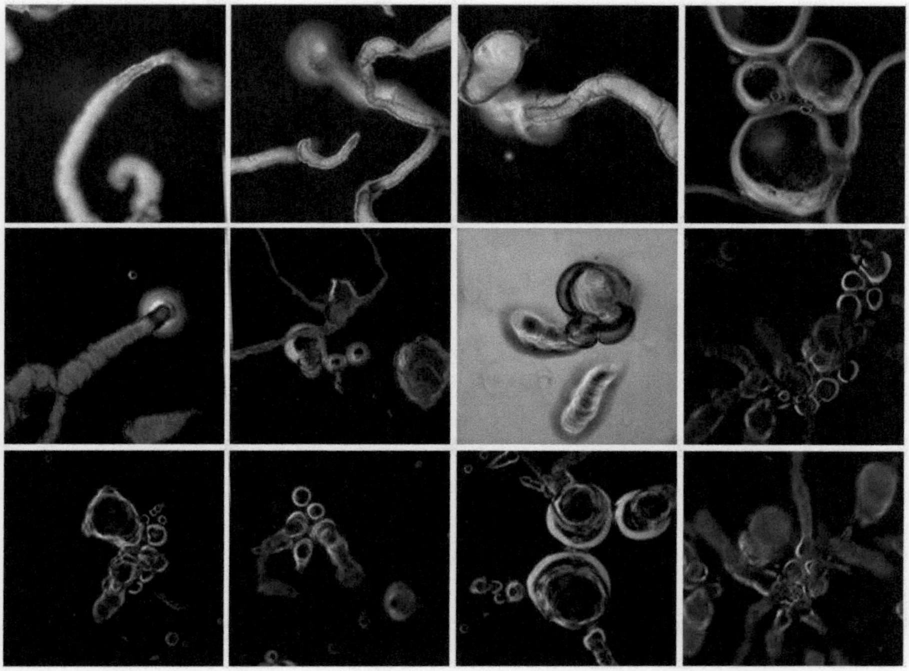

Figure 5.4: Simple chemical systems, such as Bütschli droplets, may exert effects on their surroundings. Micrographs and collage, magnification 4×, Rachel Armstrong, February 2010.

qualities of fully alive agents, such as growth, movement or sensitivity, but may have not been given the full status of 'life'. Yet they share the same chemical language as the biological world so that, as Richard Lewontin observes, organisms and their environments ultimately 'co-evolve' (Moran, 2008). Indeed, Vernadsky argued that no living organism exists in a free state on Earth and proposed a conception of Nature that combined geochemistry and biogeochemistry, which embraced both non-living and living systems that continuously connected them to the biosphere through processes such as feeding and breathing (Vernadsky, 1945).

Although minerals are more limited in their capacity to generate change by forming new compounds or forging assemblages when compared with biotic materials, they still offer a portfolio of choices from which Charles Darwin's notion of Natural Selection may operate (Darwin, 1999) and may even possess an innate evolvability (University of Central Florida, 2013). The mineral world has a structural memory that can be observed through the behaviour of metal alloys (Phys.org, 2010), or the production of crystals (Nobelprize.org, 2013) and compounds, which also have the capacity to shape events and possess technological potential. Stéphane Leduc (Leduc, 1911), who coined the term 'synthetic biology', used a variety of mineral systems to demonstrate their lifelike character, and A.G. Cairns-Smith proposed an even more intimate connection between the mineral domain and living matter by proposing that the first life forms originated from clay minerals (Cairns-Smith, 1971).

Contemporary research has confirmed that clays, such as montmorillonite, may have been key to biogenesis (Hanczyc, Fujikawa and Szostak, 2003), while William Bryant Logan regards 'the clay code' as being 'more complex than either the genetic code or human language' (Logan, 2007, p.127). Minerals have contributed significantly to setting the conditions for life to flourish and continue to do so. The role of iron pyrites in removing sulphur from the ocean and producing atmospheric oxygen has recently been established as being much more important than was previously thought (Weizmann Institute of Science News, 2012). The shared ontology of matter that Bennett takes such care to establish is written into the history of life on Earth, where living and non-living systems have persistently worked to enrich chemical networks through countless environmental acts that are enjoined in the process of evolution. Perhaps 'evolution' is not a single event but infinite, incessant, co-evolutionary, collaborative acts of material exchange with the biosphere, whose collective effects are recognized by human observers at a single moment in time (Gould, 1994).

With the awareness that our environment is unstable, with no divine obligation to support us, we are experiencing a cultural change away from the Enlightenment ideals in the way we imagine the ordering of the world. We live in a world of definable probability that is physically entangled in networks of continuous exchange in which life and matter incessantly evolve – and designers are beginning to construct new ways of dealing with this ambiguous state of affairs. For example, Xandra van der Eijk works with decay, using a device that enabled her to develop a self-painting apparatus that is active during 4–6 day cycles (Meta.morf, 2012a) (see Fig. 5.5).

The prints that van der Eijk lifts off the water surface speak of an archaeology of process that captures forms of self-organization, which mark our own decay towards entropic equilibrium. Indeed, boundaries that were once consolidated by duality – machine/human, man/woman, and organic/inorganic – are now incontrovertibly blurred, and links between different worlds – such as quantum and macroscale realities – can even be mapped as these borders are breached (Nimmrichter and Hornberger, 2013). Increasingly, these ideas are becoming culturally adopted, such as through Henri Lefebvre's *Rhythmanalysis* (Lefebvre, 2004), which embodies aspects of the Implicate Order.

Yet the precedents for recognizing material agency draw from encounters with Nature where natural processes reveal deep truths about the potential for construction that lies within the materials and elements that form the world in which we are immersed (Ruskin, 1989). While mountains such as the Matterhorn take hundreds of thousands of years to be ground down by the elements, other formations are more evanescent. For example, 'ice-flowers', which are bacterially rich, strange structures, grow under the most particular conditions during the polar winter and then disappear again as quickly as they formed, leaving no trace of their existence

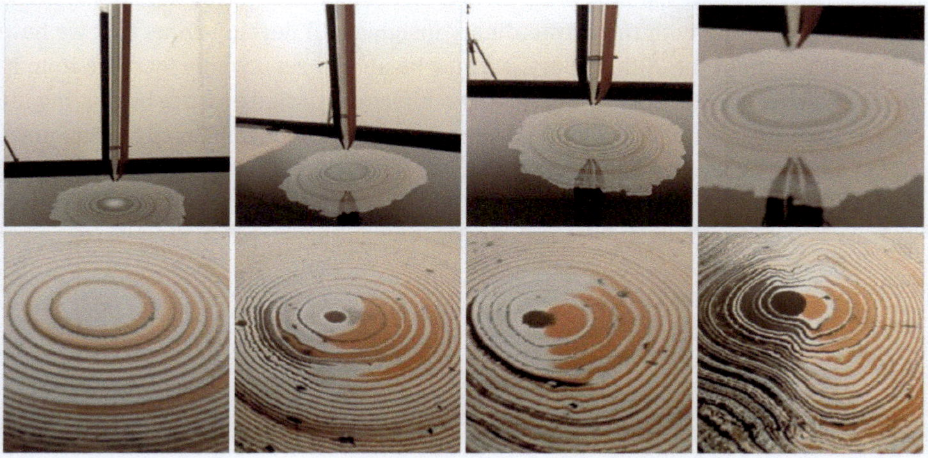

Figure 5.5: Van der Eijk's automatic drawing system harnesses the flow of paint and water to produce a self-evolving pigment series that is recorded through time-based prints. Photograph collage, Rachel Armstrong, September 2012.

– like a polar microbial Brigadoon.[39] They are composed from frozen atmospheric moisture and brine drawn through capillary action from the surface of the sea ice, which concentrates salt and bacteria from the sea (Dias, 2012).

Self-assembling processes can also be precipitated under controlled conditions, in gallery settings. Architect Tetsuo Kondo has used cloud technology to experiment with new types of spatial effects for the Tokyo Museum of Contemporary Art (Quay, 2013) where visitors can climb the stairs in a transparent room above the interior cloud; while Dutch artist Berndnaut Smilde (Design Boom, 2012) created an eerie and whimsical illusion of a cloud floating within a gallery space, which was produced using smoke, moisture and backlighting. This ephemeral formation often lasted only for a moment before it dissipated entirely. Indeed, material assemblages offer a new kind of technological platform, with an intriguing portfolio of opportunities that collaborate with designers in codesigned acts of creativity.

5.3 Summary

The magic of our reality is not that absolutely anything is possible, but that there is a great deal of untapped potential that already exists within the material world, as a real, not imaginary, phenomenon. By framing our understanding of matter so

39 Brigadoon is a mythological village, a magical structure, which only appears every hundred years (YouTube, 2008).

that it is in keeping with the extraordinary insights and developments that we have gained throughout the 20th century, we may be able to get a whole lot more from it. But we will not achieve new paradigms in architectural design practice by treating the material world as being full of inert things to be controlled or consumed by machines, but by liberating its innate potential and co-evolving our living spaces in partnership with vibrant matter – which promises to be a conservative force with revolutionary potential.

6 Dynamic Droplets

... a computer of such infinite and subtle complexity that organic life itself shall form part of its operational matrix. (Adams, 1995, p.129)

6.1 Overview

This chapter establishes the Bütschli droplet system as an experimental model to interrogate the technological potential of vibrant matter. The original recipe developed by zoologist Otto Bütschli in 1892 (Bütschli, 1892) was recreated to fully characterize the system. A series of around 300 experiments were conducted in a laboratory setting under the supervision of Associate Professor Martin Hanczyc at the Center for Fundamental Living Technology (FLinT), at the University of Southern Denmark. Each experiment was photographically recorded using a Nikon Eclipse TE2000-S inverted microscope with a Photometrics Cascade II 512 camera and in-house software. Having become familiar with the limits and range of the dynamic chemical system, ways to influence its outputs were explored and examined from technical and graphical perspectives. This could be achieved by manipulating internal and external conditions of the system, and suggested that Bütschli droplets could be applied in both technological and drawing contexts. Finally, the ontological and epistemological implications of a non-mechanical[40] technology were considered for its potential application within architectural design-led experiments.

6.2 Identifying a Suitable Model System for Vibrant Matter

A testable model for vibrant matter that is relevant to architectural design practice needs to exhibit observable behaviours at the human scale. My literature survey identified 'dissipative' systems (Prigogine, 1976; Prigogine and Stengers, 1984; Prigogine, 1997) as suitable candidates, since they exist at many scales and exhibit lifelike properties, which include cosmic phenomena (Prigogine, 1997; Smolin, 1997; Langton, 1980), weather patterns (Prigogine and Stengers, 1984; Prigogine, 1997) and even microscale events (Prigogine, 1997; Max Planck Institute for Dynamics and Self-Organization, 2003–2013). Dissipative systems have recognizable forms of organization such as vortices, and although they possess structure, they are not objects but are shaped by a constant flow of energy and matter (Prigogine, 1972). Dissipative structures therefore possess both object-centred and process-led qualities.

---

40 I also use the term 'non-linear technology' owing to the Deleuzian concepts that shape process-led events (DeLanda, 2000).

Even before Prigogine coined the term 'dissipative structures', such far from equilibrium self-organizing chemical systems have been studied since the Enlightenment in phenomena such as Glauber's chemical gardens (Glauber, 1651), Runge's dynamic chemical patterns (Runge, 1850), Moritz Traube's 'artificial' plant cells (Traube, 1867), Liesegang's self-organizing rings (Liesegang, 1869), Otto Bütschli's protozoan-like chemical system (Bütschli, 1892), Stephane Leduc's 'fungal' osmotic structures (Leduc, 1911, pp.123–146) and Belousov and Zhabotinsky's vibrant periodic chemistry (Belousov, 1959; Zhabotinsky, 1964). More recently, a range of different species of dynamic droplets (Hanczyc, 2007; Toyota, 2009) and iCHELLS (Cooper et al, 2011)[41] have also been observed to embody lively processes associated with living systems. Each was demonstrated experimentally and explored to establish their suitability as a model system for vibrant matter. The outcome of these experimental demonstrations was that dynamic droplet systems were a preferred experimental model for vibrant matter.

6.3 Dynamic Droplets as Vibrant Matter

A range of dynamic droplet systems exhibited striking, immediate and sometimes sustained effects that were observable at the human scale, which appeared suitable for use in an architectural design context. It was therefore important to identify a species that could be safely applied within social settings.

Dynamic droplets are self-assembling agents that are based on the chemistry of oil and water. They arise from a spontaneous field of self-organizing energy and can exist as oil droplets in a water medium, or water-based droplets in an oil medium. They exist as a range of different kinds of 'species' being composed from different recipes. Where oil/water interfaces occur, there is a spontaneous self-assembly of molecules owing to the chemical basis for energy exchange at the droplet interface. The consequences of mass interactions are observed in the system as emergent phenomena that typically exhibit lifelike behaviour such as movement. Even when the initial conditions are the same, the various droplet species show a range of possible types in any given environment because of the emergence in the system, and these can be characterized. Dynamic droplets can be influenced by internal and external factors and, therefore, are suitable systems for engaging with design principles. Dynamic droplets are restless, inherently creative agents that ceaselessly patrol and reposition their chemical networks. As dissipative structures, they throw out energy and materials to resist the decay towards equilibrium towards which they will eventually succumb in their mayfly-like existence, which lasts between several seconds to many weeks, depending on their chemical composition and context. It is

41 iCHELLS are 'inorganic chemical cells'.

possible to read the activity of a dynamic droplet through the environmental traces that are left as microstructures and crystals that may become the site for further droplet activity, resulting in complex constructions that can be seen with the naked eye. A range of preparations, including decanol/decanoate oil in water droplets and the Bütschli water in oil droplet system, were explored in a laboratory setting where it was possible to make a cursory assessment of the systems with respect to their technological potential and their suitability for architectural design contexts. The Bütschli system was examined in further detail as it produced the most vigorous agents from inexpensive ingredients.

6.4 Characterizing the Bütschli Dynamic Droplet System

Otto Bütschli first described a dynamic water in oil droplet system using potash and olive oil as reactants, in which he observed the genesis of an 'artificial' amoeba with pseudopodia (cytoplasmic extensions) that behaved in a lifelike manner (Bütschli, 1892). His aim was to make a simplified experimental model to explain the plasticity of body morphology and movement, based purely on physical and chemical processes such as fluid dynamics and changes in surface tension (Belousov, 1959). Bütschli's original experiment was documented with hand drawings. Although various research groups are investigating other dynamic chemical systems that use amphiphiles, such as reverse micelles (water in oil droplets stabilized by a surfactant) (Pileni, 2006) and the behaviour of oil droplets in aqueous media (Hanczyc et al, 2007; Toyota et al, 2009), no photographic documentation of the Bütschli system appears to exist in the contemporary literature.

For vibrant matter to be applied to problem solving, it needs to be operationalized. The Bütschli system therefore requires full characterization before its technological potential can be evaluated. The behaviour and morphology of this system was observed under light microscope in approximately 300 replicate experiments. It qualifies as an example of ELT through its formation of discrete dynamic droplets during a variable window of time (from 30 s to 30 min after the addition of alkaline water to the oil phase) that are characterized by their lifelike behaviour patterns. Self-organizing patterns are observed during this dynamic, embodied phase that provide a means of introducing temporal and spatial order into the system and offer the potential for further chemical programmability.

6.4.1 Bütschli System Preparation

The experimental design followed was a modern interpretation of Bütschli's original ingredients (potash and fresh olive oil). A 0.2 ml drop of 3 M sodium hydroxide was added to olive oil in a 3 cm diameter petri dish, which was filled to a depth of 0.5 cm with extra virgin olive oil. These ingredients combine through a saponification reaction, in which the trigylerides of the olive oil are cleaved to produce free fatty acids and glycerol. The main ingredient of olive oil is oleic acid, which constitutes around 61.09% to 72.78% depending on the source (Matthäus and Özcan, 2011). The same brand of oil, Monini extra virgin from Spoleto, Italy, was used exclusively in this experiment, although it is not known whether different bottles came from the same production batch. All ingredients were used at room temperature. Systems that included a titration of sodium hydroxide were also performed.

Controls included adding a 0.2 ml drop of water to a 3 cm diameter petri dish filled to 0.5 cm deep with olive oil, and also by adding 0.2 ml 3 M sodium hydroxide to a 3 cm diameter glass-bottomed petri dish filled to 0.5 cm deep with canola oil (rapeseed), from Cargill Oil Packers, which is around 85% oleic acid (Zarinabadi, Kharrat and Yazdi, 2010). The behaviour of the system was characterized in detail using a Nikon Eclipse TE2000-S inverted microscope with a Photometrics Cascade II 512 camera and in-house software.

6.4.2 Bütschli System General Observations

The breaking up of the alkaline droplet in the oil could be clearly seen with the naked eye, as shown in Fig. 6.1, Fig. 6.2 and Movie 6.1. The active chemical field produced smaller droplets whose diameters splayed variably between a millimetre and a centimetre, generating turbid deposits of soap in the dish. In the case of the water in canola oil control, no breaking up of the droplet was observed, and in the case of adding 3 M sodium hydroxide to canola oil,[42] the alkali droplet dispersed into smaller droplets but did not show the asymmetric pattern progression, dynamism or production of material observed in the Bütschli system.

Additional experiments were also carried out under the same conditions to establish the concentration range of sodium hydroxide that would produce the characteristic Bütschli pattern formation, which was established to lie within the 3 M to 5 M range. At concentrations of less than 3 M, the droplets possessed little dynamism or visible crystal formation and, although the droplet gradually broke up over a course of several minutes (around 3–10 min) to form droplets, the characteristic sequence of patterns typically observed at higher molarity was not observed. At concentrations of greater than 5 M, the system quickly became inert and instantly formed a crystal layer

42 Sodium hydroxide, at 0.5 M increments from 1–5 M, was used to test the reactivity of the canola oil control but the droplets produced did not produce lifelike behaviours across the whole spectrum of this range. The optimum range for the break-up of surface tension in canola oil was 3 M, which produced most (non-dynamic) droplets. This concentration of sodium hydroxide was therefore used as a standardized 'control' against which to compare the reactivity of the Bütschli system.

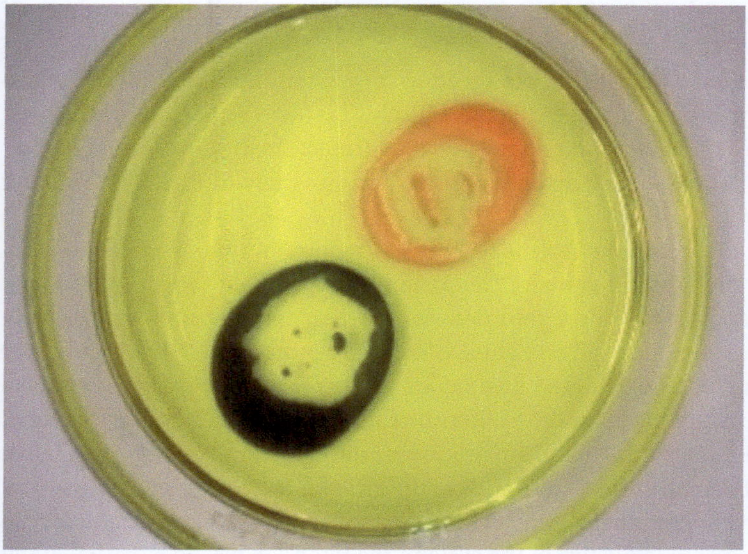

Figure 6.1: High-energy field: Bütschli droplets form when droplets of 3 M sodium hydroxide containing 1% v/v food colouring (red and blue) are added to a 3.5 cm glass dish of olive oil. Movie still, courtesy Martin Hanczyc, February 2009.

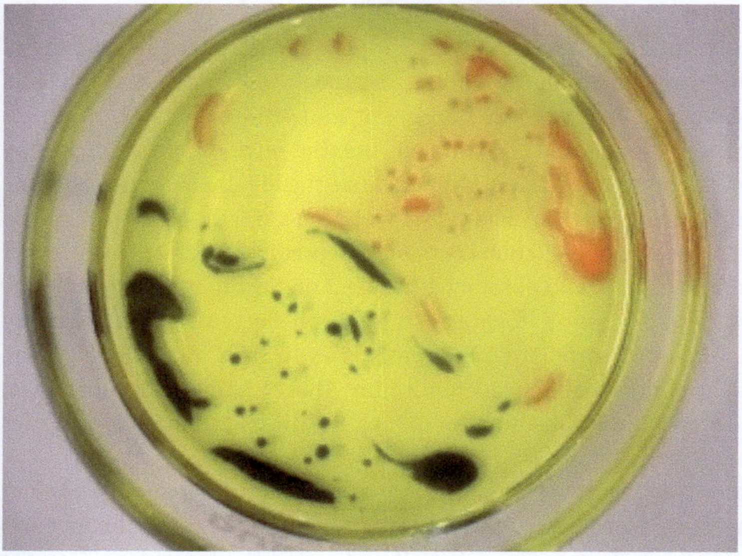

Fig 6.2: Bütschli droplets are spontaneously produced when fields of olive oil and alkali overlap. Movie still, courtesy Martin Hanczyc, February 2009.

at the oil/water interface, quenching the reaction and preventing the appearance of dynamic patterns.

When canola oil was used as the medium for sodium hydroxide in the active range for pattern production seen in the Bütschli system (3–5 M), the activating droplet broke up immediately into smaller, regular droplets in the oil field, but neither was any sequential organizing activity observed nor was any formation of product visible.

In the study group of experiments (0.2 ml 3 M sodium hydroxide added to extra virgin olive oil), the Bütschli system demonstrated a repeatable sequence of events with identifiable characteristics, recorded in still photography and movies. The Bütschli droplets were observed and studied in a similar manner to that is currently used to study and report on single-celled organisms such as protozoa or bacteria. No staining was necessary to observe the Bütschli droplets, due to their refractive index, and they ranged from the microscale to around a centimetre in diameter. The lifelike qualities of the Bütschli system were sufficiently striking to appropriate the use of a method of observation normally applied in a natural history context as useful for the study of the living characteristics of the system; the intention was to experimentally consider what kinds of organizing principles appeared to be at work in the transition from inert to living matter.

Bütschli droplets possess a primary metabolism, saponification, which spontaneously exists at the interface where strong alkali water and olive oil meet. This reaction releases both energy and products in the form of surfactants that modify the oil/water interface. This reaction is responsible both for the lowering of surface tension allowing the droplet to deform, and the flow of liquid, which results in droplet morphological fluctuations, movement and splitting. As the droplets move through their environment, they can consume the olive oil, processing it by the saponification reaction. In addition, they also use the alkali reactant within the droplet as fuel. Bütschli droplet movements last between several seconds to around 20 minutes. The activity of any particular droplet is not predictable and the success of creating the system is variable and possibly dependent on the quality of ingredients, with additives or degradation products in the olive oil decreasing the reactivity of the system. As the active droplet system progresses in time, the activity of the system slows as it approaches chemical equilibrium. Due to both the accumulation of inhibitory products and the consumption of fuel, the droplet eventually becomes inactive.

Typically, water droplets in oil self-assemble and do not dissipate due to their hydrophobic properties. However, in the Bütschli system, once the saponification reaction begins at the interface between the oil and water, the tension holding the droplet intact relaxes considerably and the droplet begins to distort and spread with increasing surface area. The droplets contain enough energy to split up into smaller droplets that are then able to move about in the olive oil environment. Notably, a control with water at neutral pH produces a spherical droplet in the olive oil that does not react, spread, split or behave like the alkaline droplet, and an alkaline droplet in canola oil splits into smaller droplets but without pattern formation.

In the reactive system, chemical potential is combined with physical instabilities and fluid dynamics, resulting in the movement of droplets associated with the production of a soapy crystalline deposit that spontaneously forms at the oil/water interface. Distinct phases characterize the progression of the ingredients from a highly energetic dissipative system to one that has reached equilibrium. During this progression, mass interactions are observed in the system as emergent phenomena where droplets and populations of droplets typically exhibit lifelike behaviour such as movement and the production of microstructures. Even when baseline conditions are uniform (temperature, pressure), these agents show a range of distinct characteristics that lend themselves to classification through distinct morphological and behavioural types that emerge from the self-organizing field (see Movie 6.2).

6.4.3 Stages of Bütschli Droplet Development

Osmotic growths like living things may be said to have an evolutionary existence, the analogy holding good down to the smallest detail. In their early youth, at the beginning of life, the phenomena of exchange, of growth, and of organization are very intense. As they grow older, these exchanges gradually slow down, and growth is arrested. With age the exchanges still continue, but more slowly, and these then gradually fail and are finally completely arrested. The osmotic growth is dead, and little by little it decays, losing its structure and its form. (Leduc, 1911, p.151)

In 1911, Stephane Leduc studied the behaviour of chemical solutions mixed together. He noted they produced strikingly lifelike results that he described as 'evolutionary'. Leduc likened the behaviour of these chemistries to living systems, associating the behaviour of the chemistry with terminology that is normally associated with the 'life cycle' of an organism. This section builds on Leduc's analogy and proposes a progression of events in the Bütschli system that alludes to possible consideration of and reflection on natural phenomena.

The stages of the lifespan of Bütschli droplets are summarized in Tables 6.1–6.4. They are described with reference to figures and movies that are organized into different phases of pattern progression and in their evolution through three distinct stages:

— Birth (0–5 min)
— Life (30 s–30 min)
— Death (0–30 min)

Table 6.1: Birth – from 0 s to 5 min following addition of sodium hydroxide droplet to olive oil phase

Figure ref.	Time after addition of alkali to oil phase	Movie	Pattern morphology	Comments
6.1	20 s	6.1	3.5 cm petri dish. Early movement dispersion of droplet and breaking up of the chemical wavefront due to changes in surface tension	Macroscopic view of Bütschli system
6.2	50 s	6.2	3.5 cm petri dish. Progressive movement and dispersion of droplet and breaking up of the chemical wavefront due to changes in surface tension	Same preparation as in Fig. 6.1 after the passage of 30 seconds
6.3	2 min 40 s	6.3	6 mm width of micrograph. Polarized field of 'fire' and 'ice'. The leading 'fire' edge is facing downwards and the trailing 'ice' edge is facing upwards in the micrograph	
6.4	8 s	6.4	6 mm width of micrograph. Turbulent, shell-like droplets that appear as a series of sequentially emerging manifolds	Some 'shells' collapse while others self-organize into droplets with lifelike properties such as movement

6.4.3.1 Birth: Field of Fire and Ice

When the alkaline droplet first breaks up in the oil field it self-organizes into a polarized, dynamic field with a characteristic appearance. The active, leading front end of the field moves outwards, away from the point at which the water droplet enters the oil field, and produces ripples as it moves through the oil media, producing a flame-like appearance. The leading edge is where oil molecules are consumed in the metabolism of the droplet. The trailing back end accumulates the product soap crystals that are swept backwards by the movement of the system, and in the case of sodium oleate, appear like ice crystals. In this initial dynamic and energetic stage, smaller droplets can break off from the moving front and then continue to display the same reactive motion. In the initial phase of self-organization, these fields look like moving islands of 'fire and ice', where it is possible to determine which direction the field is moving in by its morphology as shown in Fig. 6.3 and Movie 6.3.

Table 6.2: Life – primary morphologies from 0–30 s following addition of 0.2 ml droplet of sodium hydroxide to oil phase

Figure ref.	Time after addition of droplet to oil phase	Movie	Pattern morphology	Comments
6.5	2 min 30 s	6.5	300 micron width of micrograph. Motile droplet derived from the chaotic chemical field	Crystalline material is visible accumulating at the oil/water interface at the posterior pole
6.6	3 min	6.6	6 mm width of micrograph. Droplet with osmotic crystalline deposit	Crystalline material is visible as an osmotic microstructure attached to the droplet at its posterior pole
6.7 & 6.8	8 min	6.7	300 micron width of each micrograph. Osmotic structure seen with and without fluoroscopy in which the Bütschli droplet has just detached from an osmotic structure	Figs. 6.7 & 6.8 are the same structure
6.9	10 min	6.8	6 mm width of micrograph. Bütschli droplets produce deposits of sodium oleate at the trailing end of the motile droplet	Oleate crystals accumulate and extend to form fluid-filled 'osmotic' microstructures
6.10	2 min	6.9	6mm width of micrograph. Bütschli droplets before fusion	Fusion events are spontaneous and may be the generative agency for the production of compound, complex, osmotic microstructures

6.4.3.2 Birth: Shells

As the polarized field of self-organizing activity progresses, it starts to break up due to lowered surface tension and fluid dynamics as a consequence of saponification and the presence of soap crystals. The first recognizable 'structures' that appear are turbulent, shell-like morphologies and probably represent 'dissipative' structures that are literally throwing away energy to remain stable, as shown in Fig. 6.4 and Movie 6.4. These kinds of non-equilibrium phenomena were noted by chemist Ilya Prigogine (Glansdorff and Prigogine, 1971), who observed their occurrence in nature being characterized in structures such as snowflakes and vortices (cyclones and

Table 6.3: Life – primary behaviours from 0–30 s following addition of 0.2 ml droplet of sodium hydroxide to oil phase

Figure ref.	Time after addition of droplet to oil phase	Movie	Pattern morphology	Comments
6.11	8 min	6.10	300 micron width of micrograph. Two Bütschli droplets engage active interfaces generating various dynamic points of contact. They continue to make contact until the product (sodium oleate crystals) obstructs the interface between them	Interfaces between droplets persistently osculate
6.12	12 min	6.11	6 mm width of micrograph. Bütschli droplets 'mirroring' one another	
6.13	12 min	6.12	6 mm width of micrograph. A smaller Bütschli droplet is interfacing with a much larger one	The droplets remain in close proximity with each other until the build-up of soap crystals occludes the oil/water interface
6.14	8 min	6.13 & 6.14	6 mm width of micrograph. Bütschli droplets in a simple chain formation	Periodic oscillations are observed in agents during a chain-forming event
6.15	10 min	6.15	6 mm width of micrograph. Bütschli droplets in a complex chain formation	'Protocell roses'
6.16 & 6.17	15 min	6.16 & 6.17	6 mm width of micrograph. Two droplet assemblages merge and suddenly change behaviour and morphology	Phase change behaviour observed during the formation of an assemblage when a 'tipping' point is reached. Such events were observed on separate occasions

whirlpools); and they are also found in living systems. Video footage suggests that the droplet shells are manifolds, rather than chaotic spheres of activity, which burst out of themselves like Russian matryoshka dolls, suggesting that these droplets are in a high-energy state. Some shells suddenly collapse and form crystalline deposits, while others eventually stop splitting and bursting out of themselves and enter a new

Table 6.4: Death – from 0–30 s following addition of 0.2 ml droplet of sodium hydroxide to oil phase

Figure ref.	Time after addition of droplet to oil phase	Movie	Pattern morphology	Comments
6.18	20 min	6.18 & 6.19	300 micron width of micrograph. Fine crystals of sodium oleate accumulate at the oil/water interface	Crystal deposits accrue at the 'posterior' pole of the droplet

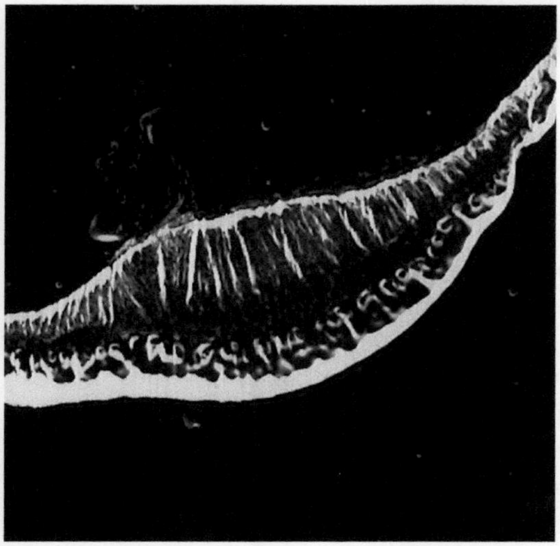

Figure 6.3: The leading edge of the polarized Bütschli droplet field is reminiscent of 'fire'. Its trailing edge, laced by forming soap crystals, is suggestive of 'ice'. Micrograph, magnification 4×, Rachel Armstrong, February 2009.

phase of organization as lifelike droplets. It is not possible to predict which shell-like formations, or even what proportion of them, will become self-organizing droplets, as their distribution is outside of the field of view of the microscope.

6.4.3.3 Life: Organizing Droplets

Post chaotic formation phase, the resulting droplets are able to move around, sense their environment, modify their surroundings, produce complex structures and even interact with each other. The interactions and systems are complex and it is not possible to predict the outcomes of the various droplet types. Yet, there are definitive

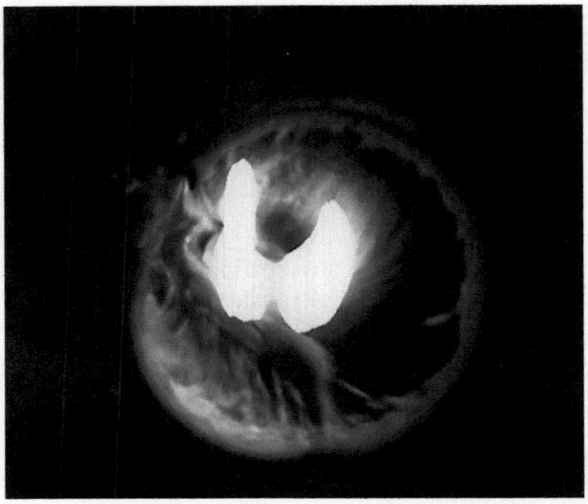

Figure 6.4: Turbulent, shell-like structures are observed at the early, high-energy stages of formation of the Bütschli system. These are indicative of dissipative structure formation, which is characteristic of living systems (Prigogine, 1997). Micrograph, magnification 4×, Rachel Armstrong, February 2009.

patterns of behaviour and interactions that offer a pedagogical view of the system. These characteristics will be discussed in the context of:

— Primary morphologies: Structural characteristics encapsulating the state of the system: droplets, droplets with product, droplet with extended 'osmotic' crystalline structures, polyps, compound structures.
— Primary behaviours: Dynamic interactions that lead to more complex phenomena: interfacing, mirroring, population dynamics.

6.4.3.3.1 Primary Morphologies

The primary morphologies of the Bütschli system are summarized in Table 6.2.

6.4.3.3.1.1 Droplet

The first form that an organized dynamic droplet adopts is a polarized, free-moving droplet, like the one shown in Fig. 6.5, which possesses a fundamental direction partially conferred by its original position in the primary field of 'fire and ice'. Propelled by its primary metabolism, the droplet moves in a given trajectory away from where the original droplet met the oil field, influenced by inhibitors or attractants in the medium, as shown in Movie 6.5. It appears that dynamic droplets modify their surroundings as they pass through a medium (Horibe, Hanczyc and Ikegami, 2011;

Hanczyc, 2011a) and create chemical changes in the field that dynamic droplets sense, which have not yet been characterized.

6.4.3.3.1.2 Droplet with Osmotic Product

Depending on the speed of the chemical reaction and the environmental conditions, a small deposit of crystals appears at the trailing end of the active droplet as the metabolism progresses, as shown in Fig. 6.6 and Movie 6.6. The physical properties of the crystals cause downstream effects on the body of the droplet that influence its locomotion, and ripples can be observed as it drags the gradually increasing load behind the active front. This gives rise to jellyfish or worm-like morphologies and different kinds of movement behaviours such as peristalsis-like locomotion.

6.4.3.3.1.3 Droplet with Extended Osmotic Product

Bütschli droplets undergo progressive physical changes as they continue to consume their primary metabolism and interact with environmental cues, resulting in the production of osmotic microstructures. These are similar in character to the forms Leduc produced on mixing various solutions (Leduc, 1911, p.151), which grow at the trailing end as the droplet moves around the environment. Droplets can also break

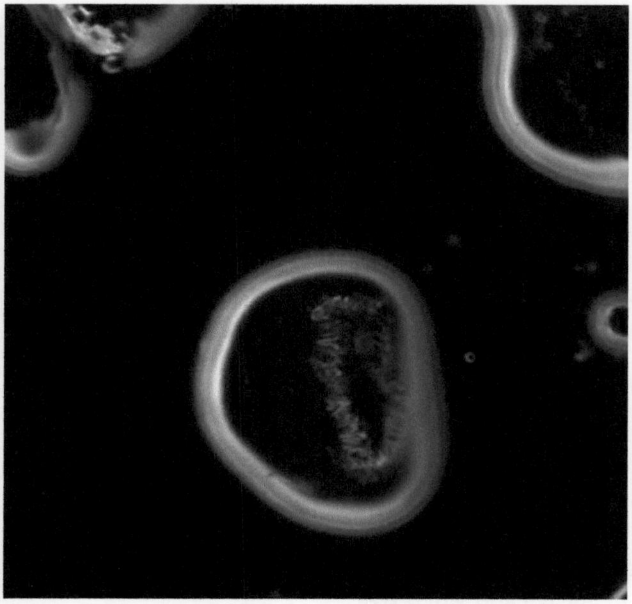

Figure 6.5: Polarized, free-moving droplet. Micrograph, magnification 40×, Rachel Armstrong, February, 2009.

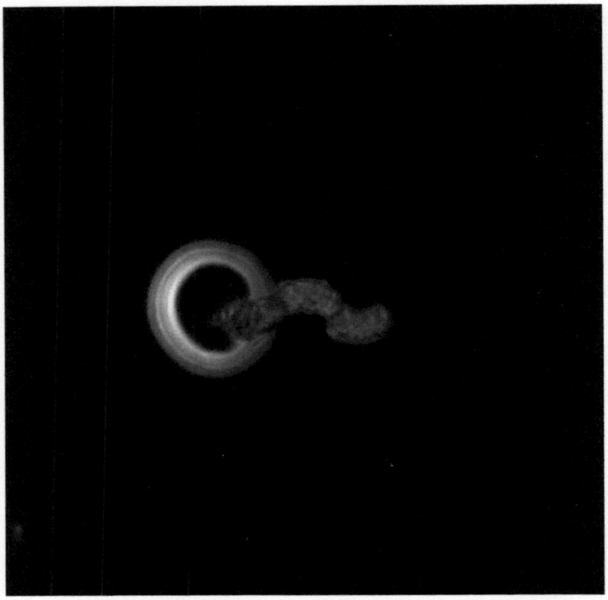

Figure 6.6: Osmotic structures may be produced at the posterior pole of free-moving droplets. Micrograph, magnification 4×, Rachel Armstrong, February 2009.

free from osmotic structures leaving behind them residues that consist of soap crystal 'skins' that are wrapped around an inner core of aqueous (alkaline) media. These structures are visible via fluorescence microscopy, by adding a hydrophilic dye to the droplet at a concentration of 0.25% fluorescein by weight. In Fig. 6.7 and Movie 6.7, a Bütschli droplet deposits a large osmotic residue. When observed under fluorescence microscopy, as in Fig. 6.8 and the latter part of Movie 6.7 (from 1 min 30 s), the fluorescence shows the aqueous phase, which is present in both the residue and the droplet. Bütschli droplets consume themselves as they metabolize and produce soap crystals during this process that travel to the back end of the droplet and accumulate at such a speed and density that they form a tubular, tail-like extension of material.

6.4.3.3.1.4 Microtubes

The character of simple osmotic products may be striking. Under very highly alkaline conditions that approach 4–5 M solutions of sodium hydroxide, the Bütschli droplets respond in a characteristic way in the oil field by producing long, thin, tapering tubes of crystalline product that are shaped by the direction of motion and size of the droplet producing them, as shown in Fig. 6.9 and Movie 6.8.

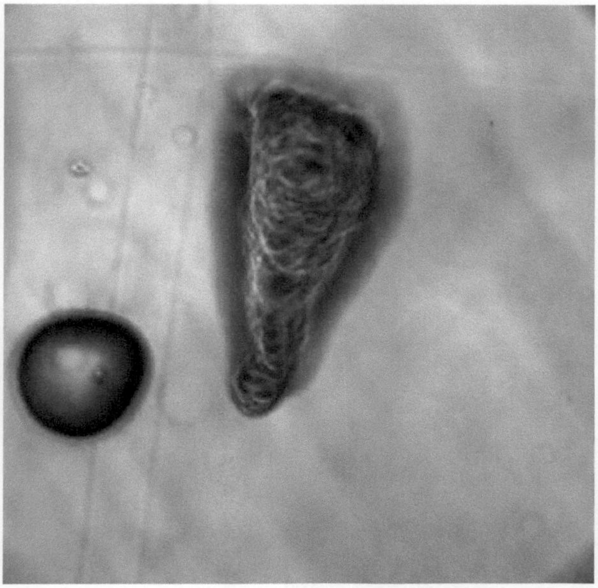

Figure 6.7: Osmotic casts may be produced by dynamic droplets from which they may break free. Micrograph, magnification 40×, Rachel Armstrong, February 2009.

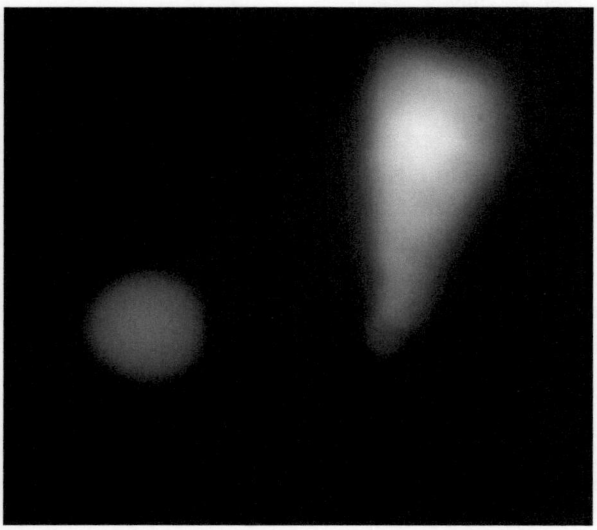

Figure 6.8: An osmotic cast is observed under fluoroscopy from the droplet in Fig. 6.7, which has been pre-stained using a fluorescent dye (fluorescein 0.01 M at pH 9). The structures are observed under a red light filter to pick up the green light emitted by the stain. The images show that the residual osmotic structures appear to be soap crystal skins that encase an aqueous inner core. Micrograph, magnification 40×, Rachel Armstrong, February 2009.

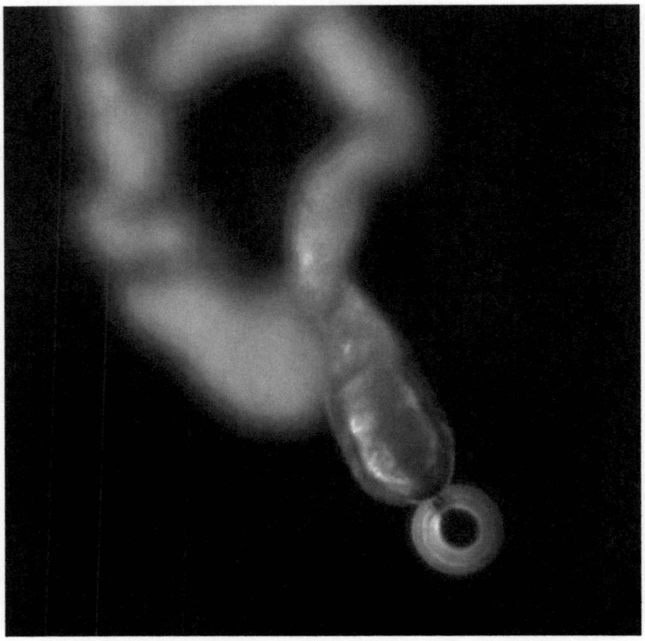

Figure 6.9: A Bütschli droplet is producing a polyp-like osmotic microstructure. Micrograph, magnification 4×, Rachel Armstrong, February 2009.

6.4.3.3.1.5 Compound Structures

Droplets can produce compound osmotic structures when their bodies fuse and skins combine as a new growth point, as in Fig. 6.10, which was taken from Movie 6.9 at the moment when two Bütschli droplets fused to produce a new growth point for an osmotic structure. The short osmotic structure of one droplet meets a longer branched one to produce a compound microstructure, which is just out of focus. A spiral structure is also clearly visible, which has most likely been produced by another droplet passing through the oil field twisting and advancing simultaneously.

6.4.3.3.2 Primary Behaviours

The primary behaviours of the Bütschli system are summarized in Table 6.3.

6.4.3.3.2.1 Interfacing

Bütschli droplets are chemically attracted to each other and when they meet, they do not usually fuse. Instead, they align their interfaces, producing a very dynamic, oscillating, yet loose relationship between the oil/water boundaries of adjacent

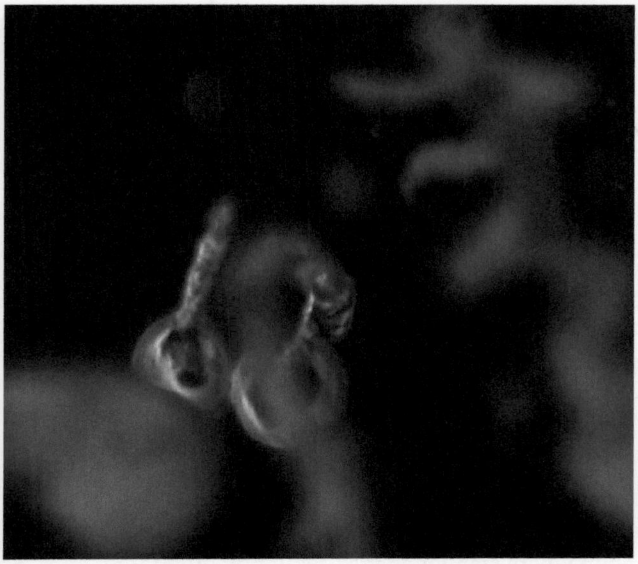

Figure 6.10: Two droplets building microstructures come into close proximity moments before they fuse. Micrograph, magnification 4×, Rachel Armstrong, February 2009.

droplets. These dynamic interface connections seem to exert influence on droplet behaviour and generate different outcomes depending on the number of participating agents. It was not possible to determine the exact number of agents required to produce a systemically different kind of interaction between small and larger groups from these experiments. It is not known if there are specific thresholds, or tipping points, for the emergence of different patterns of interaction. More research is needed to further characterize the observed effects. A more precise delivery system for the production of discrete numbers of droplets, such as 3D printing, is hoped to be useful.
Different kinds of 'interfacing' behaviours are observed:

- Between individuals (2).
- In small groups (3–5).
- With larger populations (6 or more).

Individual droplets moving independently can collect together, forming a shared contact area. The contact zone is unstable and the droplets continually change their interaction points, as they are persistently osculating, as shown in Fig. 6.11 and Movie 6.10, where a small Bütschli droplet is situated between two larger ones, where an active interface exchange is constructed between them. There is another point of contact between the two larger droplets below the small one. In general, Bütschli droplets appear to make multiple points of contact at an interface zone. It is not clear if any material is exchanged during this process, but the intensity of the contact decreases as product builds up and the metabolism, which provides the energy for interaction, runs down.

6.4.3.3.2.2 Mirroring

Bütschli droplets that establish an early connection have been observed to mirror each other's appearance and behaviour. In Fig. 6.12, two agents have established an active interface connection and have produced similar broad-based osmotic structures that anchor them. Smaller droplets appear to be attracted to this site of intense activity and a second site of interfacing has been established between the two large droplets by a smaller one, as shown in Movie 6.11.

6.4.3.3.2.3 Satellites

Bütschli droplets appear to be attracted towards sites of intense metabolism. Large droplets appear to be able to strongly attract smaller ones, resulting in a commonly observed satellite phenomenon where smaller agents frequently orbit larger ones, as shown in Fig. 6.13 and Movie 6.12. It is likely that a product of the primary metabolism is acting as a chemical attractant, though this has not been scientifically verified.

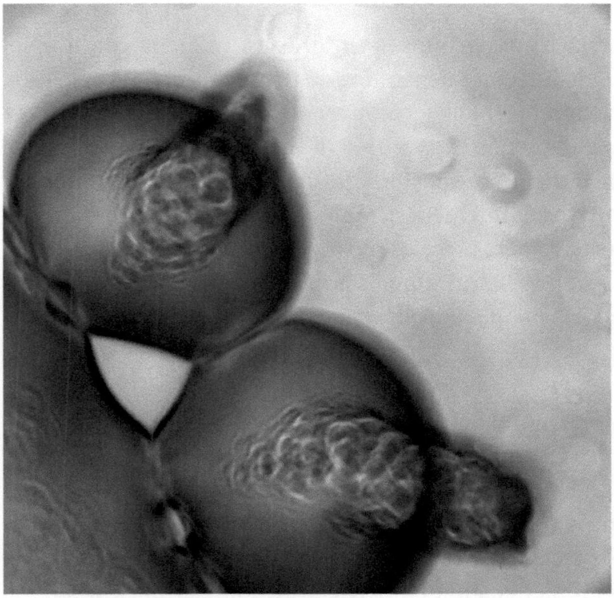

Figure 6.11: High magnification of the active interface between two Bütschli droplets in close proximity. Micrograph, magnification 40×, Rachel Armstrong, February 2009.

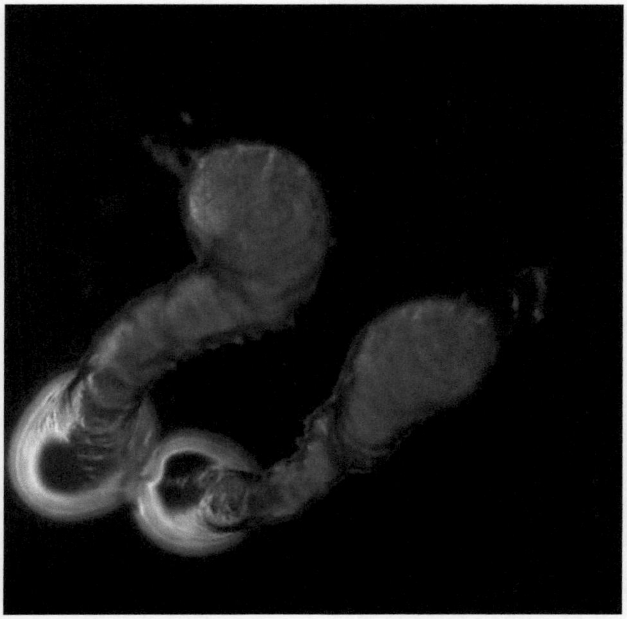

Figure 6.12: Bütschli droplets morphologically 'mirroring' each other. Micrograph, magnification 4×, Rachel Armstrong, February 2009.

6.4.3.3.2.4 Chains

Chains of interfacing Bütschli droplets are frequently the first formations that can be seen in the early self-organization process. This occurs where individual droplets have stopped travelling but which are engaged in intense, phased activity at their interfaces with neighbouring droplets. These chains appear to stimulate the metabolism of participating droplets and rapidly encase the active interface with crystals as shown in Fig. 6.14, also in Movie 6.13 and Movie 6.14.

6.4.3.3.2.5 Populations

As Bütschli droplets are drawn towards each other, they form larger populations. They then undergo a range of interactions that result in both a change in the behaviour of the individual agents as well as their appearance. Behavioural changes are likely to occur as the result of metabolic products that attract and/or repel individual droplets as well as the accumulation of product that progressively reduces the amount of available area that the droplets have available as an active interface, as shown in Fig. 6.15 and Movie 6.15. It is likely that a product created by the metabolism causes droplets to be attracted to each other and may be responsible for characteristic emergent behavioural differences observed between small populations (around 2–6 interacting droplets) and larger groups (more than 6 droplets). These numbers are

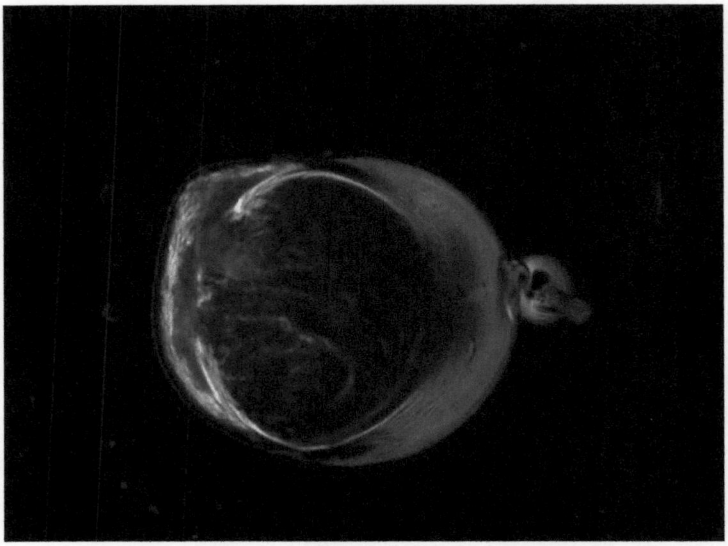

Figure 6.13: Satellite phenomenon, where a smaller Bütschli droplet appears to 'orbit' a large one. Micrograph, magnification 4×, Rachel Armstrong, February 2009.

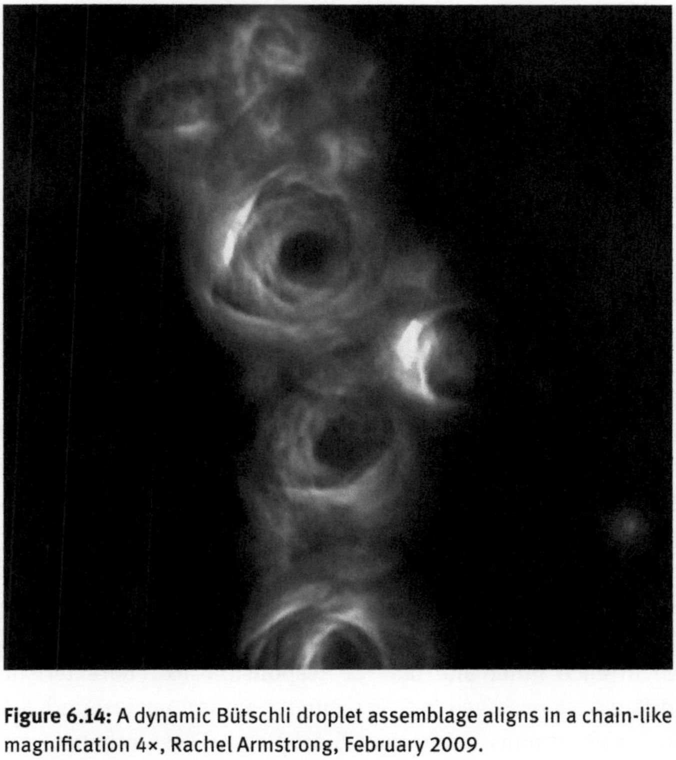

Figure 6.14: A dynamic Bütschli droplet assemblage aligns in a chain-like formation. Micrograph, magnification 4×, Rachel Armstrong, February 2009.

a guideline, based on observation and familiarity of working with the constantly changing system. They are estimated from the frequency of observation of transient, multiple formations of interacting droplets that have been observed during the active phase of the Bütschli system. Finer control of delivery is unlikely to create specificity within this constantly changing system until the Bütschli system itself has been better characterized.

Bütschli droplets appear to possess both attractants/stimulants and inhibitors/repellents of droplet activity. Synchronous group behaviour has been occasionally observed, which results from the recruitment of a number of droplets in proximity. In larger groups, a different, emergent quality has been observed several times, characterized by sudden group behaviours such as scattering, as shown in Fig. 6.16 and Fig. 6.17, as well as Movie S16 and Movie S17, which were independently captured events. These group interactions could be likened to 'quorum' sensing (Nealson, Platt and Hastings, 1970) that occur in certain species of bacteria when, at a threshold number of communicating bacteria, a signal is passed between members and causes a change in the products expressed by the colony. However, unlike quorum sensing where the signal attracts other agents to the site, Bütschli droplets appear to be producing a repellent.

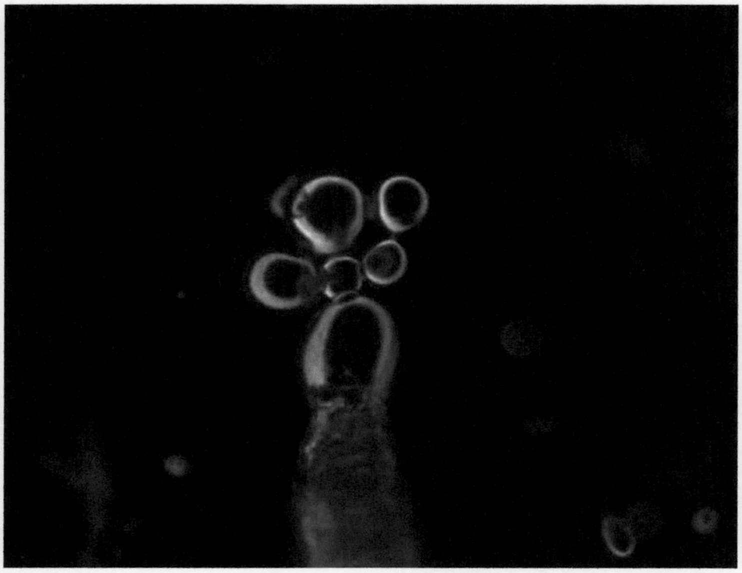

Figure 6.15: Assemblage of dynamic droplets around an osmotic structure. Micrograph, magnification 4×, Rachel Armstrong, February 2009.

Figure 6.16: Spontaneous phase change in morphology and behaviour in an assemblage of dynamic droplets that reach an unknown chemical 'tipping point' in the system. Micrograph, magnification 4×, Rachel Armstrong, February 2009.

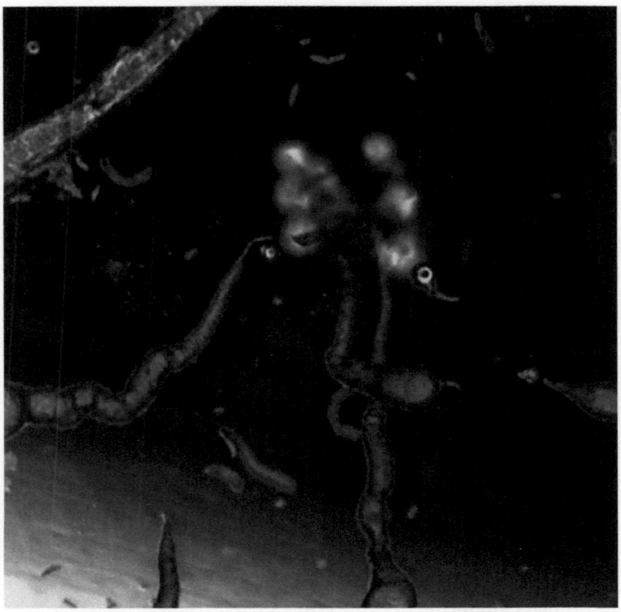

Figure 6.17: A droplet assemblage reaches a chemical 'tipping point' and undergoes a phase change in its morphology and behaviour. Micrograph, magnification 4×, Rachel Armstrong, February 2009.

6.4.3.4 Death: Quiescence

As the metabolism of the Bütschli droplets consumes its body and surroundings, it leaves skins of crystalline materials behind, breaking free of the structures when they produce too much drag. A couple of examples are summarized in Table 6.4. Over time, the metabolism is less vigorous, the droplet moves more slowly and more crystals accumulate over a larger region of the oil/water interface, partially occluding it and reducing the amount of product. The droplet enters a stage of chemical oscillations, where it appears to pulse until it finally stops moving, when all the area available to act as a site of chemical exchange is occluded entirely by crystals. This constitutes a chemical form of 'death' as shown in Fig. 6.18, also in Movie 6.18 and Movie 6.19.

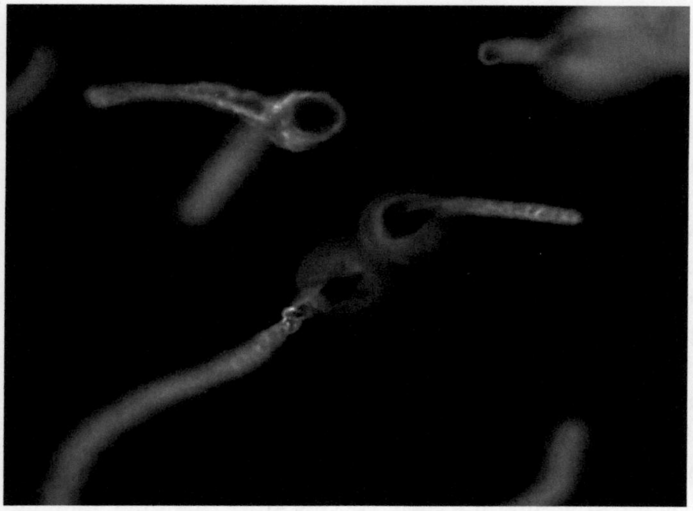

Figure 6.18: Dynamic droplets reach quiescence as their active interfaces are occluded by product. Micrograph, magnification 4×, Rachel Armstrong, February 2009.

6.5 Exploring the Technological Potential of Bütschli Droplets

This section develops the idea of vibrant matter as an alternative production platform to machines with distinctive operational principles. While the language that conveys the technological potential of this platform is well established in the fields of process philosophy (Whitehead, 1979; Agar, 1936; Seibt, 2012), new materialism (Deleuze and Guattari, 1979, pp.3–28) and vibrant matter (Bennett, 2010), the technological capacity has not been explicitly referenced beyond descriptive encounters such as Whitehead's primordial nature of God (Whitehead, 1979) or Deleuze and Guattari's creative use of metaphors like rhizomes (Deleuze and Guattari, 1983, p.2).

My research proposed to investigate whether the Bütschli system:
- Could be manipulated using morphological computing techniques
- Provided any unexpected insights into the nature of vibrant matter

6.5.1 Manipulating the Bütschli System Using Morphological Computing Techniques

Having established a set of performance characteristics for the Bütschli system, I designed a series of exploratory experiments that incorporated morphological computing techniques to manipulate the droplets, which are summarized in Table 6.5 and Table 6.6:
- Internal conditions
- External conditions

6.5.1.1 Changing Internal Conditions

Bütschli droplets can be designed to create a range of different products by adding different chemistries to the system, which spontaneously fuse with their bodies. The droplet of aqueous inorganic salt is added to a field of Bütschli droplets and reacts on contact with their alkaline bodies. In this way, the Bütschli droplets can be engineered to make 'secondary' forms and metapatterns (Volk, 1995) that are deposited at the oil/water interface, using different kinds of ingredients. For example, insoluble, magnetic 'magnetite' crystals can be produced within osmotic structures by creating a layer of Bütschli droplets at an interface of olive oil and diethyl phenyl phthalate (DEPP) by adding 0.2 ml drops of iron II/iron III salts prepared according to an aqueous ferrofluid recipe with a molar ratio of $Fe3+$:$Fe2+$ of 2:1 (Berger et al, 1999) and produce magnetite on fusion. The droplets are at around the same specific gravity as DEPP and may either form crystals within the droplet bodies, or produce organic-looking growths as they pass through them under the influence of gravity. The movement of the droplets through the oil medium and their subsequent interactions produce sculptural forms, as shown in Fig. 6.19 and Table 6.5.

It is of note that the Bütschli droplets produced at an olive oil/DEPP interface do not exhibit lifelike behaviours, as it appears the contact between the droplet and a surface such as glass is a critical ingredient and warrants further investigation. Indeed, when droplets at an olive oil/DEPP interface are observed under 4× magnification, they exhibit disorganized, vigorous movement, which can be seen in Fig. 6.20 and Movie 6.20.

Figure 6.19: Macroscopic view of modified Bütschli droplets following the addition of a 0.2 ml drop of aqueous ferrofluid. Photograph, Rachel Armstrong, February 2009.

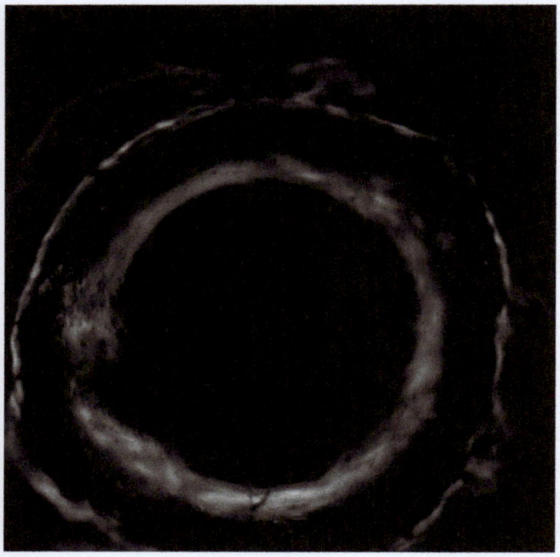

Figure 6.20: Chaotic chemical activity is reminiscent of a 'solar flare' and is observed in the absence of surface contact between the Bütschli system and a solid surface such as glass. Micrograph, magnification 4×, Rachel Armstrong, February 2009.

Table 6.5: Manipulation of Bütschli system: Internal programming

Figure ref.	Time after addition of droplet to oil phase	Description of pattern morphology	Comments
6.19	45 s	Layer of modified Bütschli droplets suspended over 'osmotic stalagmites' composed of iron II and iron III precipitates. 3 cm width of photograph	Macroscopic view of Bütschli system following the addition of a 0.2 ml drop of aqueous iron salt solutions into a modified Bütschli system

Table 6.6: Manipulation of Bütschli system: External programming

Figure ref.	Time after addition of droplet to oil phase	Movie	Description of pattern morphology	Comments
6.20	Any	6.20	6 mm width of photograph. Disorganized chemical activity in the Bütschli system appears like a 'solar flare'	Chaotic chemical activity is observed in absence of surface contact. This appears to play an important role in the emergence of lifelike characteristics
6.21	Any	6.21	6 mm width of photograph. The addition of acetone to the oil field increases droplet aggregation	Droplets vigorously move towards the source of acetone
6.22	Any	6.22	6 mm width of photograph. Ethanol causes rapid movement and agitation of droplets that are short-lived	Droplets migrate rapidly towards the source of ethanol
6.23	Any	6.23	300 micron width of photograph. Droplets appear to form assemblages more readily in the presence of butan-1-ol	Activity appears prolonged in the presence of butan-1-ol

6.5.1.2 Changing External Conditions

Changing the external conditions of the medium alters the behaviour of the Bütschli system and the results from a series of experiments are summarized in Table 6.6. However, the chemical basis of chemotaxis in dynamic droplets has not been established as the whole system is driven by the complex dynamics of composite molecular self-assembly, autonomous movement and interactions between droplets (Toyota, 2009). A series of experiments were conducted to examine the influence on the Bütschli system of organic solvents added to the olive oil field (viscosity 103 mPa at 20°C) (Bürkle GmbH, 2011) such as ethanol, 1-butanol and acetone. These were observed using a Nikon Eclipse TE2000-S inverted microscope with a Photometrics Cascade II 512 camera and in-house software. The chemical basis for the observed complex movement and assemblage formation in the Bütschli system is outside the focus of my research, but warrants further scientific research and analysis. Initial observations are, however, provided here for the following substances:

— Acetone
— Ethanol
— Butan-1-ol
— 2-propanol
— 1-octanol

6.5.1.2.1 Acetone

A 4 cm diameter glass dish of olive oil was prepared and 0.2 ml 3 M sodium hydroxide was added to produce Bütschli droplets. As the field of sodium hydroxide began to spread out and break up into millimetre droplets, 0.2 ml acetone (viscosity 0.3040 mPa at 20°C) (Physical Properties of Liquids, not dated) was added to the field of olive oil by trickling it down the side of the glass dish. The Bütschli droplets responded vigorously to the diffusion wave and rapidly moved towards the high concentration gradient. The spontaneous dynamic activity of the droplets rapidly ceased and their tendency to form assemblages was remarkably increased, which was confirmed by observing the system at 4× and 10× magnification and is shown in Fig. 6.21 and Movie 6.21. Around 25 such preparations were conducted and the spontaneous activity in the system ranged from 30 s to 3 min following the addition of acetone.

As a small molecule, acetone quickly diffuses through the oil field and establishes a polarity in surface tension of the droplets, which may at least contribute to their chemotaxis, as well as locally decreasing the viscosity of the oil field. This may also play a role in the increased tendency for the droplets to aggregate as assemblages, which would imply that the 'interfacing' is at least in part provoked by surface tension dynamics. Also, the acetone can diffuse through the oil field and react with residual sodium hydroxide in the Bütschli droplets. Acetone undergoes the highly vigorous 'aldol condensation' (aldehyde and alcohol) reaction (Nielsen and Houlihan, 1968) in the presence of concentrated sodium hydroxide, to produce ethanal. This vigour may

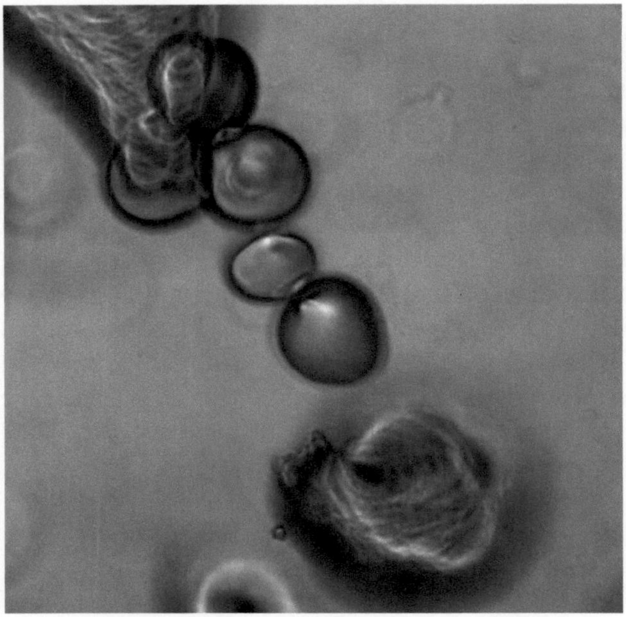

Figure 6.21: Acetone increases droplet aggregation. Micrograph, magnification 40×, Rachel Armstrong, February 2009.

also at least partly account for the system dynamics, which are initially 'explosive', then rapidly 'quenched'.

6.5.1.2.2 Ethanol

A 4 cm diameter glass dish of olive oil was prepared with 3 M sodium hydroxide to produce Bütschli droplets. As the field of sodium hydroxide began to spread out and break up into millimetre droplets, 0.2 ml 100% ethanol (viscosity 1.078 mPa at 20°C) (Physical Properties of Liquids, 2013) was added to the field of olive oil by trickling it down the side of the glass dish. The droplets responded vigorously to the diffusion wave and rapidly moved towards the high concentration gradient. Following vigorous movement towards the source of ethanol, the droplets rapidly formed large assemblages. This grouping, which initially appeared to increase spontaneous dynamic activity of the Bütschli droplets, was rapidly quenched and was observed at 4× and 10× magnification as seen in Fig. 6.22 and Movie 6.22. Around 40 such preparations were conducted and the spontaneous activity in the system ranged from 30 s to 5 min following the addition of butan-1-ol.

As in the case of acetone, ethanol is a small molecule that diffuses rapidly through the olive oil and comes into contact with sodium hydroxide, where it reacts to produce water and sodium ethanoate, which is an ester. However, the ethoxide

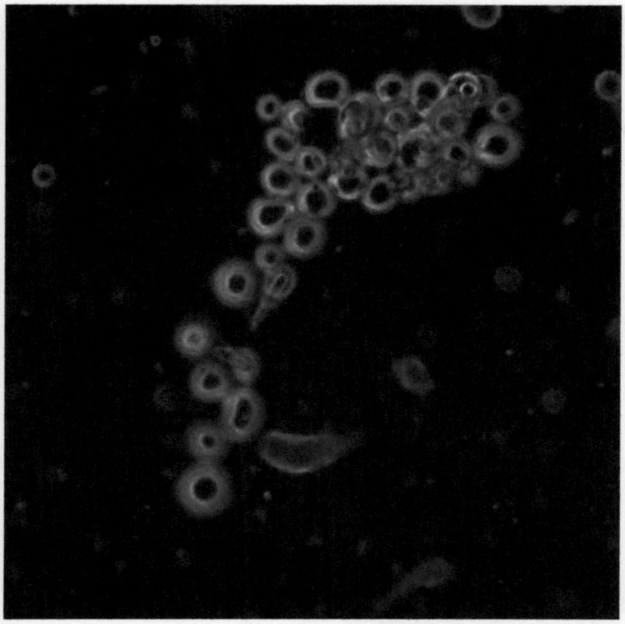

Figure 6.22: Ethanol causes rapid movement and agitation of droplets, which is short-lived. Micrograph, magnification 4×, Rachel Armstrong, February 2009.

ion of the sodium ethanoate also reacts with water, re-forming the alcohol. These are in equilibrium under normal laboratory conditions, so effectively, the net reaction is no reaction because the ethanol re-forms. In the case of ethanol it is possible that the sudden movement is caused by dramatic, polarized changes in surface tension that promote movement of the droplet dynamics but also, perhaps more significantly, by reducing the viscosity of the olive oil. These surface tension changes may be responsible for the increased tendency to form large assemblages, although these observations are speculative and need further formal scientific analysis.

6.5.1.2.3 Butan-1-ol

0.2 ml 3 M sodium hydroxide was added to a 4 cm diameter glass dish of olive oil to produce Bütschli droplets. As the sodium hydroxide field began to spread out and break up into millimetre droplets, 0.2 ml 10% butan-1-ol (viscosity 2.593 mPa at 20°C) (Physical Properties of Liquids, 2013) was added to the field of olive oil by dribbling it down the side of the glass dish. The droplets responded rapidly to the diffusion wave and travelled towards the high concentration gradient, where their spontaneous dynamic behaviour produced small but multiple droplet assemblages. Unlike the cases of acetone and ethanol, the clusters persisted for many minutes before their activity gradually ceased, as seen in Fig. 6.23 and Movie 6.23. This was just visible with the naked eye but was confirmed by observing the system at 4× and 10× magnification. Around 10 such preparations were conducted and the spontaneous activity in the system ranged from 30 s to 20 min following the addition of butan-1-ol.

Butan-1-ol is a fairly small molecule that diffuses through the olive oil and comes into contact with sodium hydroxide, where it reacts to produce water and the ester sodium butyrate. In the case of butan-1-ol, it is possible that the movement is caused by the chemical changes at the droplet surface which speed up the consumption of the sodium hydroxide in the Bütschli droplet, as well as polarized changes in surface tension that promote movement of the droplet dynamics and reduce the viscosity of the olive oil, but less so than the smaller molecules such as acetone and ethanol, so the resultant dynamic changes are less vigorous. These surface tension changes may be responsible for the increased numbers of assemblages observed and their persistence, but these observations are speculative and need further formal scientific analysis.

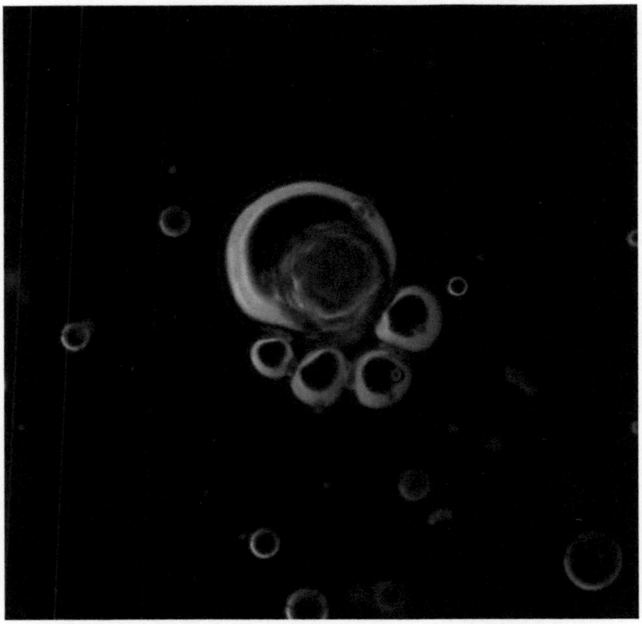

Figure 6.23: Bütschli droplets appear to form assemblages more readily in the presence of butan-1-ol and remain active for longer. Micrograph, magnification 4×, Rachel Armstrong, February 2009.

6.5.1.2.4 2-propanol

0.2 ml 3 M sodium hydroxide was added to a 4 cm diameter glass dish of olive oil to produce Bütschli droplets. As the dissipating field of sodium hydroxide spread out and broke up into millimetre droplets, 0.2 ml 2-propanol (viscosity 2.073 mPa at 20°C) (Physical Properties of Liquids, 2013) was added to the field of olive oil by dropping it down the side of the glass dish. The droplets barely responded to the diffusion wave and little, if no, increased aggregation was observed. The spontaneous activity of the droplets was observed under the microscope at 4× and 10× magnification and appeared to have reduced general activity, although still dynamic. Around five such preparations were conducted and the spontaneous activity in the system ranged from 30 s to 25 min following the addition of 2-propanol.

2-propanol is a relatively small organic molecule that appears to diffuse slowly through the olive oil. It is assumed that, like the other organic solvents, 2-propanol comes into contact with sodium hydroxide, where it reacts to produce water and the ester sodium propanoate. In the case of 2-propanoate, it appears to attenuate the normal activity observed at the droplet surface, which may slow down consumption of the sodium hydroxide in the Bütschli droplet and explains the apparent reduced production of product in the system. Changes in surface tension may also affect the movement of the droplet dynamics by altering the viscosity of the olive oil. These observations are speculative and need further formal scientific analysis.

6.5.1.2.5 1-octanol

Bütschli droplets were produced using 0.2 ml sodium hydroxide in a 4 cm field of olive oil. As the field of sodium hydroxide began to spread out and break up into millimetre droplets, 0.2 ml 1-octanol (viscosity 7.21 mPa at 25°C) (Viswanatha et al, 2007, p.144) was added to the field of olive oil by trickling it down the side of the glass dish. The droplets responded very slowly to the diffusion wave. The spontaneous activity of the droplets was observed under the microscope at 4× and 10× magnification and appeared to have very slightly reduced general activity, although the droplets formed many small clusters and were still dynamic. Around 10 such preparations were conducted and the spontaneous activity in the system ranged from 30 s to 30 min following the addition of 1-octanol.

1-octanol is a fatty alcohol that diffuses slowly through the olive oil, and possesses an amphiphilic character due to its non-polar, lipophilic carbon chain with a polar, hydrophilic hydroxyl group that confers surface activity upon it. Fatty alcohols undergo a wide variety of reactions in the presence of concentrated alkali (Condea, 2000), which is why they are widely used in the soap industry.

6.6 Unexpected Insights Into the Nature of Vibrant Matter

During the research period, the Bütschli system was expected to exhibit interesting qualities relevant to the potential technological performance of the system, which were characterized as follows:
— Locomotion
— Self-replication

6.6.1 Locomotion

The Bütschli system is sensitive to its context and changes its morphology and behaviour in space, time and according to the nature of its metabolism. The forces driving Bütschli pattern formation are therefore different to the formation of bubbles, whose patterns emerge as a consequence of the amphiphilic bilayer interface being supported by internal air pressure and is not fuelled by a specific chemistry. Over their active lifespan, Bütschli droplets may undergo a wide range of changes, where the dynamics of form and movement are entangled. Each actant experiences a different set of forces and conditions that shape the behaviour and morphology of its assemblages. However, the Bütschli system exhibits a minimum complexity, and it has been possible to observe repeatable patterns appearing when both complex systems interact with each other. In Movie 6.6, an individual Bütschli droplet changes its appearance as it grows a crystalline skin at the posterior pole. This causes drag and causes the agent to alter its form of locomotion, since it appears to crawl over the bottom of the petri dish dragging the weight of the crystalline osmotic structure behind it. Indeed, contact between Bütschli droplets and a surface appears to be critical for movement. If a thin layer of dense, clear oil such as DEPP is added to a petri dish, preventing the alkali droplet from touching the base of the container, then the self-organization that drives this behaviour is chaotic and directed movement does not occur, as shown in Fig. 6.20 and Movie 6.20. The degree of plasticity and behavioural change in this system is remarkable, as it does not require any central programming from an organizational molecule such as DNA to initiate this state change. This behaviour suggests that rapid morphological changes without DNA are not only possible but may occur rapidly in systems that possess only a few interacting chemistries, when compared with timescales associated with more complex biological ones.

6.6.2 Self-replication

The Bütschli system does not replicate, and although droplets are observed to divide and fuse, they do pass any specific chemical information to other droplets such as nucleotide polymers that can replicate. This adds more intrigue to the indeterminate identity of Bütschli droplets between living and non-living states, as they have a very low degree of autonomy. Their technological potential is therefore very susceptible to human and non-human influences, as well as requiring significant infrastructural support.

6.7 Summary of Findings Related to the Technological Potential of the Bütschli System

The continuing search for increasingly lifelike materials in the practice of the built environment raises new opportunities in the development of the ELT portfolio. Materials that can deal with continual real-time changes in their surroundings by harnessing living properties, without needing to be pre-programmed with an all-embracing palette of future possibilities, raises the possibility of exploring the production of qualitatively different kinds of spatial program and design tactics in the production of space. The analysis of the Bütschli droplets suggests that this rudimentary chemical system offers a potentially rich, experimental platform, not only for artificial life investigations but also for possible real-world applications of vibrant matter in architectural practice. Enabled by the parallel processing capabilities of chemical systems, Bütschli droplets may simultaneously respond to multiple, overlapping chemical programs that produce behavioural effects such as chemotaxis, attraction or repulsion and morphological outcomes such as the production of casts, tails or sculptural formations. Such opportunities also present new architectural and technological challenges, which require an understanding of how it is possible to spatialize chemical programs and design with emergent phenomena.

As a technology, the self-organizing Bütschli system exhibits a recognizable series of chemical patterns that result from the process of saponification and are visible to the naked eye. Closer examination under the microscope provides further information about the morphology of the chemical waves that shape the evolution of the droplets. The technological potential of the system exists during the lively phase of the reaction (which exists from between 30 s and 30 min after formation), when the droplets are sensitive to chemical and physical fluctuations in their surroundings. For example, during this phase, Bütschli droplets can produce spatially distributed mineral deposits with sculptural qualities when they come into contact with discrete chemistries such as aqueous ferrofluids. It is anticipated that applying precision-guided devices, such as 3D modelling software coupled to 3D printing devices, will provide opportunities to design and engineer with bottom-up chemical solutions to provide a development platform for dynamic, chemistry-based ELT (Adams, 2012) with potential architectural applications. Yet, the actions of Bütschli droplets can be orchestrated by manipulating flows of chemical information and instructed to consume or produce selectively in a given environment, as shown with other droplet systems (Hanczyc, 2007).

Bütschli droplets embody the principles of assemblage formation that underpin the effects of vibrant matter through population-scale behaviour and in resisting fusion with adjacent droplets through dynamic boundary interactions (Latour, 1996; Deleuze and Guattari, 1979; Bennett, 2010). This study suggested that interacting droplets exhibit as yet uncharacterized chemical periodicity through cycles of attraction and repulsion at the oil/water interface. This appears to maintain the 'body' of the assemblage by preventing even densely packed groups of agents from fusing. The periodic 'interfacing' between Bütschli droplets also enables them to remain mobile and sensitive to environmental changes. Additionally, it appears that assemblage formation in these systems can be induced by the addition of organic solvents to the olive oil field, except for 2-propanol, although this requires further testing since the sample size was small. Yet, these initial observations are intriguing and, ultimately, may be valuable in understanding how to orchestrate complex technologies. Indeed, as lifelike, chemically programmable delivery systems for a variety of materials that can also respond to environmental conditions, Bütschli droplets may have future real-world applications that are relevant to the practice of the built environment, such as smart paints, or surface coatings with the potential to fix carbon dioxide into inorganic carbonate in response to environmental cues (Armstrong, 2010d; Armstrong, 2011b). From a technological viewpoint, the Bütschli droplet system provides a model system that is sufficiently robust to begin to establish a set of design and engineering principles that could be used in architectural design practice.

It is envisaged that droplet technology may also become part of a larger production process suggested in recent work at the University of Oxford by Gabriel Villar and colleagues, who used vesicles within a 3D printing system to form microscale structures (Villar, Graham and Bayley, 2013) and also by Klaus Peter Zauner's group at the University of Southampton, who are producing dynamic vesicle systems through microfluidics devices (University of Southampton, not dated; Palmer, 2010).

6.8 Bütschli Droplets as a Potential Drawing Technology

> *... what I am searching for is a way to turn ... a mode of analysis into one of synthesis.*
> (Kipnis and Leeser, 1997, p.8)

The technological potential of Bütschli droplets was explored with the context of drawing practice tactics forged through collaboration between human and non-human codesigners, as an 'ecology of drawing'. While there is no formal definition for drawing, the practice requires coherence, integrity and artfulness (Kimmelman, 1992). Sigmund Abeles viewed drawing as a 'touching at a distance' where human desire is entangled with physical experience. These remote and immediate relationships form networks of interactions that engage the technological potential of the material realm and converge in the production of a material effect that may be read as a drawing. The use of dynamic droplets in the practice of drawing relates to a long history of spontaneous forms of drawing where matter possesses various degrees of non-human agency. For example, acheiropoieta, like the Turin shroud (Charney, 2012), and automatic drawing processes, such as exquisite cadavers and frottage, engage the material realm beyond direct human conscious control. The traces produced from the interactions between the assemblages may be culturally interpreted and ascribed meaning as drawings.

Dynamic droplets constitute a 'wet' drawing method where both medium and traces evolve during the production process. Such practices are in keeping with water-based printing where pigments are distributed over the surface of water using soaps and stains and lifting them on to paper, as well as – in Xandra van der Eijk's self-drawing process – where paint is dropped down a pendulum to produce a wet, continuous, self-organizing drawing (Van der Eijk, 2013). They also forge a direct relationship with the physical world that is directly expressed as an 'ecology' of interactions. Dynamic droplets may also be considered as a form of prototyping, with functional similarities to 3D (Armstrong, 2012g) and 4D (TED.com, 2013b) printing processes (where wet materials are processed into 'dry' media), rapid prototyping (where resinous materials are cast into 'dry' forms) and film-based photography (where chemistry and light capture a moment of complex interior and exterior relations). Since dynamic droplets do not exist in the natural world, their engagement in a drawing process is a deliberate intervention precipitated by a human agent. Therefore, the artistic pursuit within the drawing system is in establishing the conditions for the drawing and then shaping the subsequent interactions. The quality of recordings are influenced by the medium and the chemistry of the agents, where drawings are produced through casts and chemical traces as non-linear graphical recordings, which also have a sculptural quality. They may either be left to follow their own trajectory, or chemically persuaded to adopt new behaviours and trajectories by adding chemical cues. This method of drawing production can be likened to a form of frottage, where otherwise invisible chemical landscapes are graphically revealed through the production of material traces. Dynamic droplet drawings observe a very proximate relationship between agents that are expressed at the molecular level through chemical encounters. In this manner, the droplet draws by revealing invisible cues present in the environment with a level of sensitivity and precision that cannot be apprehended directly by the human senses. By viewing the dynamic droplets as a technological extension of our body – similarly to the way a hammer exerts the force that a body can exert on its surroundings, or how Google Glass goggles enable us to experience virtual spaces in the real world – then dynamic droplets may also be thought of as an extended sensory system that enables us to graphically read molecular palimpsests in our surroundings (Armstrong, 2012b). Although we cannot see these relationships, the experience provides us with the kind of sensory detail that exceeds the capabilities of our unaided vision. Dynamic droplet recordings are 3D soft, wet, 'osmotic' skins that can be seen with the naked eye as residues that produce unconventional, dissipative geometries and evanescent structures during their trajectories. Since the Bütschli system is effectively 'closed', most of the dynamic droplet drawing activity reaches equilibrium between 30 s and 30 min, during which time they reveal chemical information in landscapes from the microscale to the megascale. After several hours, unmodified Bütschli droplets gradually dissipate and decay into a fine, soapy precipitate in the petri dish.

When Bütschli dynamic droplets are used as drawing agents, they produce a variety of outputs whose limits may be established by a human designer by altering the internal and external conditions of the system using morphological computing techniques. Architects working with dynamic droplets produce drawings through a continuous process by shaping the conditions in which the emergence of a drawing is increasingly likely. Each drawing is unique and moulded by the internal and local conditions to produce traces that are emergent, contingent and permanent. The following drawings in Figs. 6.24–6.27 offer some examples of the graphical range of the system.

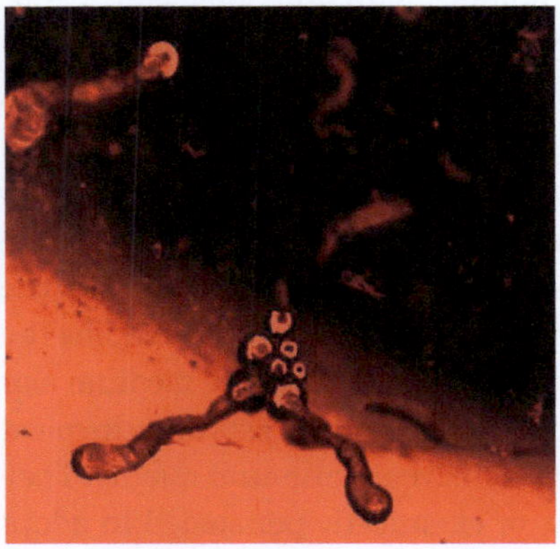

Figure 6.24: Landscape produced by droplet assemblages. Micrograph, magnification 4×, Rachel Armstrong, February 2010.

Figure 6.25: Landscape produced by droplet assemblages. Micrograph, magnification 4×, Rachel Armstrong, February 2010.

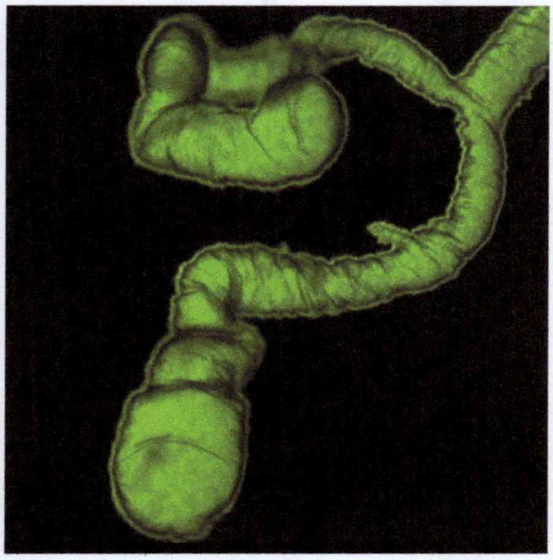

Figure 6.26: Organic structures composed of osmotic skins produced by droplet assemblages. Micrograph, magnification 4×, Rachel Armstrong, February 2010.

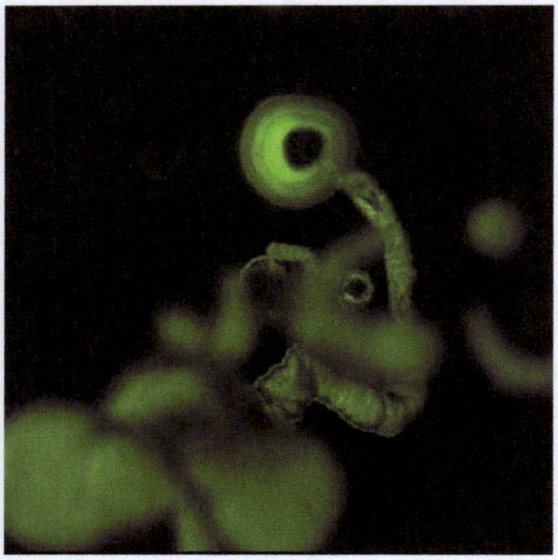

Figure 6.27: Landscape produced by droplet assemblages. Micrograph, magnification 4×, Rachel Armstrong, February 2010.

Dynamic droplet drawings summon improbable forms into existence and provoke new juxtapositions between agents to reveal hidden chemical landscapes. Although the technique is at an early stage of development, the co-development of 3D printing as a delivery platform for these agents may offer architects greater precision when setting up drawing fields. Such potential convergence may give rise to new design and drawing platforms that can further articulate 'ecologies of drawings'. Potentially these combined platforms could provide new materials, methods and technologies that enable architects to effectively design more adeptly with probabilistic systems.

6.9 Ontological and Epistemological Issues Raised by Bütschli Droplets

The Bütschli system potentially offers a new technological platform that exhibits non-classical behaviours that invoke a distinct set of concepts that are different to those of machines and can be practically demonstrated through the formation of chemical assemblages. As an expression of vibrant matter, Bütschli droplets pose a particular challenge to the structuring of knowledge. The complexity, organizational diversity, extreme environmental responsiveness and physical entanglements with other ontologically distinct systems, such as machines, means that trying to describe the potential of the Bütschli system within a portfolio of architectural design tactics is not only a constantly moving physical target, but a conceptual one too! While Bütschli droplets may be framed within the language of process philosophy and scientifically characterized through the principles of complexity, observing the system is inevitably mired in linguistic and aesthetic expectations (Morton, 2007). This makes it difficult to view and describe the constantly changing Bütschli system without trying to establish its performance within pre-existing knowledge sets. Yet this is exactly what needs to be done if the full potential of this emerging technology is to be fully explored and imagined. Indeed, the Bütschli system may yet prove to be 'post-epistemological', or unclassifiable in any coherent, meaningful way using traditional modes of classification such as the Linnæan system (Latour, 2013).

Although man-made, and in that sense 'artificial', the lifelike performance of the Bütschli system provides an opportunity to consider the emergent characteristics as a subset of living qualities in order to construct a more thorough understanding of the system as a whole. However, there is no classification system to characterize dynamic lifelike chemistries. Yet, Carl Linnæus imposed an order on natural systems that included three domains, animal, vegetable and mineral, which therefore embraced both living and non-living materials and facilitated a comparative understanding of these systems by appreciating similarities and differences (Linnæus, 1735). Of interest is Linnæus' taxonomy of stones, which he asserted possessed some of the properties of living things. In particular, Linnaeus asserted that stones grew by way of an accretion process, such as when sand aggregated and became sandstone, or when the apparent clumping of clay particles formed limestone. He also included the formation of quartz in his classification system, which he proposed was due to a 'parasitic' mechanism. However, minerals were dropped from taxonomic classification during the 18th century and are absent from Lamarck's 1809 classification scheme, *Zoological Philosophy* (Lamarck, 1999), which focuses exclusively on the cataloguing of animals. Additionally, Ernst Haeckel's famous 1866 'Tree of Life' (Haeckel, 1866) based on Charles Darwin's taxonomic diagram (Darwin, 1859) equated phylogeny with the story of evolution and excluded the mineral world from phylogenetic ordering systems. It is possible that the omission of minerals from a scientific ordering of the natural world may also have been influenced, at least in part, by the popularization of Louis Pasteur's germ theory (Pasteur, 1866), which refuted a widespread belief in spontaneous generation, where life was thought to be created directly from inert matter (Armstrong and Hanczyc, 2013).

The approach taken in reporting the observations is relevant to current systems of classification used in biology and natural history, which may help to relate non-living phenomena to biological systems through a description of the pattern morphology. There is much to be learned through comparative analysis and my research attempts not only to observe, but also to construct, an understanding of the characteristic of the lifelike properties of the Bütschli system as the basis for further study. An examination of this system also aims to establish some guiding features and principles that also identify its potential for development towards ELT.

Conventionally, dynamic systems are described by recognizing geometric domains within them such as patterns and metapatterns. Yet, there are semantic problems with such an approach, since pattern recognition, through identifying particular kinds of morphology, reveals nothing about the process of production, which is closer to an algorithm that represents a set of rules than any particular geometry, which encapsulates one particular time frame in a sequence of events. For example, complex structures such as the cephalopod and mammalian eye (Serb and Eernisse, 2008) may result from convergent evolution of structures (Doolittle, 1994). Additionally, very similar patterns may be generated within different media, such as DNA-producing mollusc shells (physical systems) and the graphical modelling of shell-like structures on a computer screen (virtual systems) (Tyson, 1994). Moreover, there is semantic and philosophical incongruity in the very practice of using 'geometric' criteria as the conceptual framework for non-linear systems, since they are ontologically distinct.

Ideas that are consistent with an idea of 'process ontology' may be used to observe and interpret the experimental findings of the Bütschli system, as a way of characterizing a potential non-linear epistemology. The aim is not to formalize an approach but to begin to reflect upon the possible systems of reference for the development of non-linear technologies. The hope is that subsequent descriptions, expectations and criteria for success, may not be unconsciously constrained by the expectations of working with machines. Matt Lee uses the term 'oceanic ontology' to refer to the 'contingency of being that relies upon an empiricist property of the sensible as a continuous, connected and open whole' (Lee, 2011, p.14), which is inspired by Friedrich Nietzsche's ontology of forces, being photographically represented in Fig. 6.28.

Oceanic ontology produces maps rather than theories of concepts (Lee, 2011, p.27) and can be understood as an emergent process, which 'produces a model – both implicitly and explicitly, of the process, of which it is a part' (Lee, 2011, p.44). The importance of using a unique ontology is to embody the ideas that it represents. For example, Manuel DeLanda proposes that complexity and non-linear dynamics have shaped human civilization. He rereads human history to examine this idea by embodying the feedback loops and organizing fields that he proposes has shaped

Figure 6.28: The changeable nature of oceanic landscapes is revealed by this photographic recording of the complex interplay between light, wind and water on the surface of the Venetian lagoon. Photograph, Rachel Armstrong, August 2012.

our culture. For example, he views the development of (unplanned) cities as 'arising from the flow of matter-energy' that inhabit a variety of flows and constraints as 'self-organised meshworks of diverse elements' (DeLanda, 2000, p.32) Similarly, the Bütschli system may literally and/or figuratively provide of a way of reading events of transformation that are related to non-linear fields of action, rather than a series of any particular events caused by specific individuals or agents. The pedagogical challenges of such complex systems may benefit from reading unfolding events through an oceanic ontology that does not require the observer to choose between fields of action, local events, or actors as organizing hubs of activity, but can simultaneously consider them. Lee proposes that the process of knowledge acquisition within non-linear systems can be studied through actors (Lee, 2011, pp.27–28). While ANT (Latour, 1996) refers to the agency of elements, which may be human or non-human, as 'actors', Lee takes the idea literally and observes how thespians can make sense of highly unstable environments, which may be a play, stage or text. 'The actor presents us with ... a way of learning ... that isn't subject centred but created through the movement of transformation ... that opens a space of process that is a form of understanding but one radically distinct from the subject centred model' (Lee, 2011, p.130). This idea may be applied to the Bütschli system, where dynamic droplets may be considered as 'actors' within a constantly changing, non-linear field of chemical activity. The interactions between different actors, and also with their complex environment, produces events that leave physical traces which help construct a reading of the 'plot', which may be considered as a form of (micro)architecture having been produced by the events within a space on an 'ever-changing stage' (Tschumi, 2012, p.28).

An oceanic ontology of Bütschli droplets was constructed, in collaboration with Simone Ferracina, by viewing them as actors that simultaneously embody 'space, event and movement' (Tschumi, 2012, p.28) within their complex chemical fields of activity. Drawing from Tschumi's notion that relationships are what give architecture meaning, a diagram was produced (see Fig. 6.29) which represents the contextualization of (meta)events between actors (droplets) with time within a complex field of activity. The stage is not a single reading of events but reflects multiple possibilities where the 'plot', or field of activity, is constructed through exploratory, graphical approaches. The resultant diagram maps relationships in the system rather than invoking the classical 'tree' metaphor of classification systems, which focuses on differences rather than similarities between actors.

The graphic is centred at time zero, from which concentric circles radiate, representing an exponentially increasing series of time intervals. This logarithmically increasing function encapsulates the intense self-organizing activity that happens early on in the chemical reaction and falls off rapidly with the passage of time. An

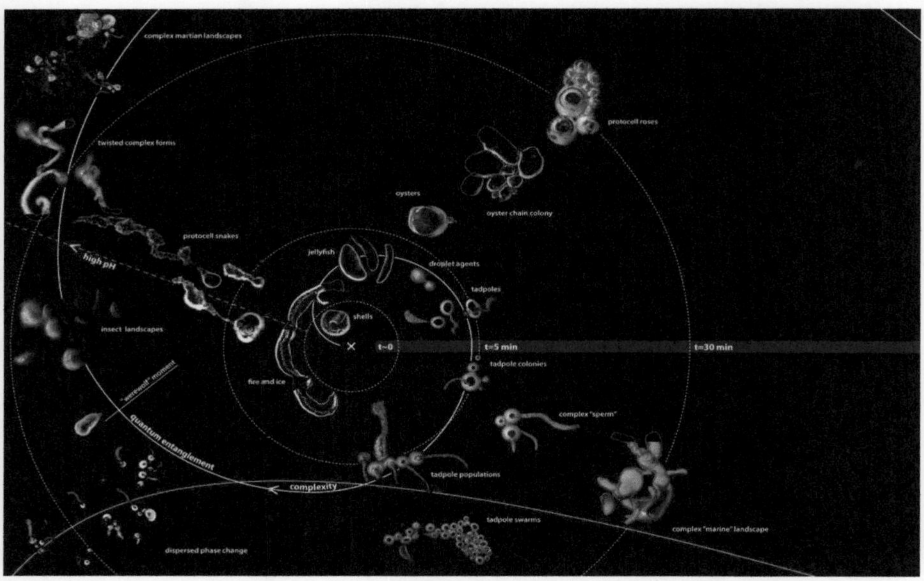

Figure 6.29: This diagram depicts dynamic droplets as 'actors' that operate within the many variable influences encountered in their oil field as an ontolological 'map' of events. While the diagram is drawn as a 2D topology, the possible events within the field are manifold and open up multidimensional spaces through their interactions with continuous, multiple contingencies that shape the evolution of the system. Diagram designed by Rachel Armstrong and drawn by Simone Ferracina, July 2012.

estimated 90% of chemical activity is completed within five minutes of activation of the system, although individual droplets have been observed to be active as long as an hour after their genesis. A spiral that represents complexity also radiates from the origin and depicts the high frequency of events around the start of the reaction, which become less frequent as time unfolds. The various morphologies and behaviours that indicate change in the system are grouped subjectively according to the authors' experimental findings and interpretations. For example, the complex oyster chains are distinct in appearance but only differ in degree from the complex marine landscapes. Specifically, 'oysters' produce a large mass of material and their soft bodies bulge from their material shell-like tethers, which anchor them, as shown in Fig. 6.30.

In contrast, 'marine landscapes' are composed of a variety of largely inert forms that have been produced by droplets that would previously have been described as 'oysters'. However, the undulating droplets are long gone, leaving only a trail of residues behind them, as in Fig. 6.31.

The diagram also indicates the impact of chance events from a source external to the system, such as an incidental trajectory that intersects with the fundamental progressive vectors of the Bütschli system. It represents disturbances in the environment, like changes in ambient temperature, or physical disturbances. This external vector also touches the spiral of complexity and, in this case, may cause agents within the system to reach tipping points.

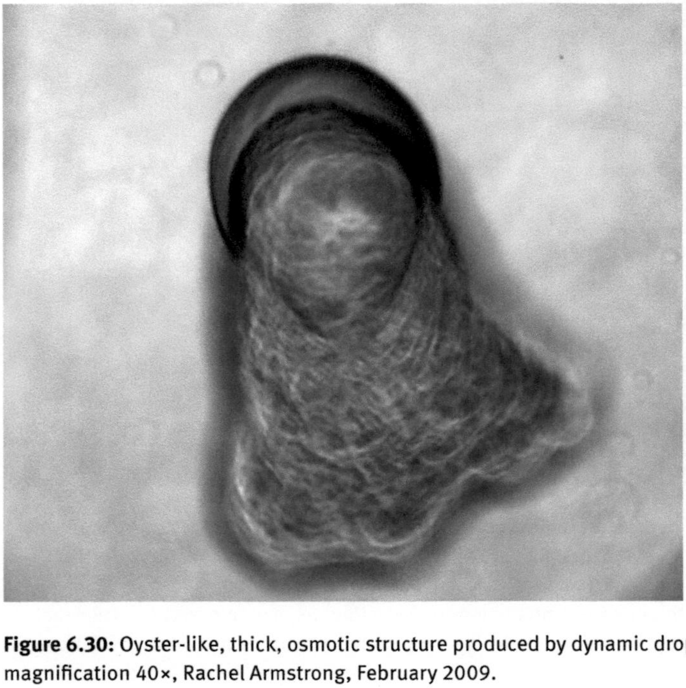

Figure 6.30: Oyster-like, thick, osmotic structure produced by dynamic droplets. Micrograph, magnification 40×, Rachel Armstrong, February 2009.

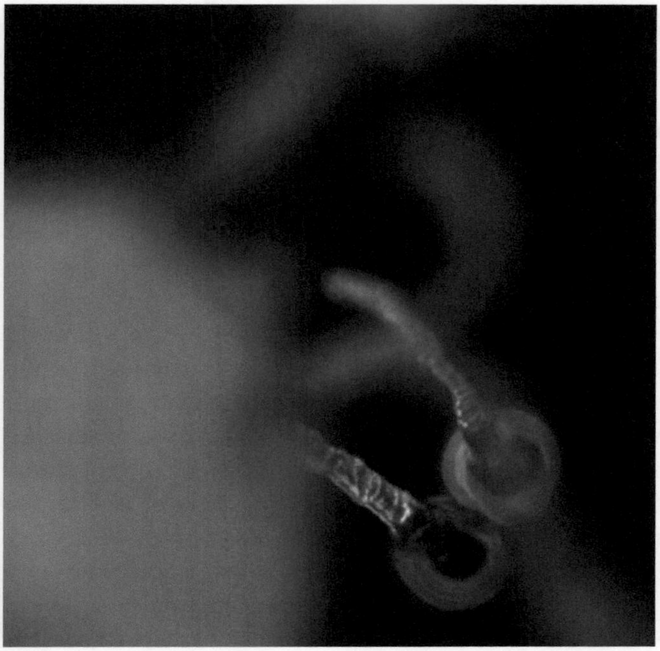

Figure 6.31: Thick, osmotic structures being produced by dynamic droplets that are moving away from their deposits and producing 'marine landscapes'. Micrograph, magnification 4×, Rachel Armstrong, February 2009.

The diagram also employs metaphor to convey dynamic qualities and complex attributes of the Bütschli system. For example, the 'werewolf moment' is a droplet event that is characterized by extreme agitation and simultaneous rapid production of residue, which gives the agent a rather 'hairy' appearance. This striking event is most likely precipitated by the ratio between droplet surface area and the volume of the droplet that is optimized and therefore rapidly consumes the dynamic agent. The rapid precipitation of product over the droplet surface causes drag that precipitates erratic movement in the system owing to the uneven distribution of surface deposits. The chemical excitement phase typically lasts for around a few minutes as it produces a large amount of residue that is swept to the posterior end of the droplet by molecular action and physical forces, where it is suggestive of a 'tail'. This structure contributes to the physical changes in the system as it exerts a great deal of drag on the system. These complex events immediately precede droplet inertia as the dense precipitation extinguishes the droplet metabolism by completely occluding the interface.

6.10 Observations Made with Respect to the Ontology and Epistemological Issues of Bütschli Droplets

Drawing from my experimental observations and insights gained during mapping the oceanic ontology of the Bütschli system, the following observations were made:

— Oceanic ontologies are not tools to solve specific challenges through a process of reductive thinking, but may be useful pedagogies to build assemblages of relationships or concepts (Lee, 2011, p.27) which help navigate complex challenges and terrains.

— A process-based, oceanic ontology may help characterize events within the complex chemical system to free observations from our expectations. Additionally, use of the spiral of complexity (rather than the traditional tree metaphor, or more recent notion of a web, which is a very complex version of a tree), enables the system to be epistemologically described and imagined beyond a comparative analysis of geometric patterns and metapatterns. The idea of a structure unfolding and folding back on itself as a navigational instrument creates a space of possibility in which relationships between and similarities within the system may relate, connect and potentially construct new kinds of knowledge.

— By working with the poetics of the system in the production of a diagram, oceanic ontologies may provide a pedagogical framework that enables disciplinary convergence across the Two Cultures (Snow, 1959) and ultimately, the development of technological species that 'synthesize[s] quantities into qualities' (Ambasz, 2006, p.22).

— An epistemological progression of events through actors within an oceanic ontology is an informal exploration, not a formal classification system, and needs to be further explored and interrogated.

— Although metaphorical descriptions are inexact (i.e. non-geometric) assessments of the system, they may, however, help establish multidisciplinary and collaborative approaches in conveying complex ideas. This is problematic in a scientific context, where important issues such as repeatability, quantification, precision, rigour and the effective communication of ideas must be respected as intrinsic to the field. However, in the arts and humanities the value of metaphor is in curating sets of ideas and their cultural expression. Oceanic ontologies therefore provide a means of exploring different conceptual frameworks to develop an accessible language through which multiple disciplines may work as an 'ecology' of practices (Stengers, 2000), by identifying convergences within seemingly divergent practices and ultimately create opportunities for synthesizing new approaches in design and engineering with non-linear systems.

— Further exploration of the possible applications of oceanic ontologies and maps would be informative in assessing the transferability of the approach. For example, in a scientific context, oceanic ontologies could be used as a graphical version of a Turing test (Cronin et al, 2006), where complex events may be visually compared. Parallax between the systems under observation may provide new information that may (re)inform further experiments, but this is beyond this particular research inquiry.

The innate flexibility, pluripotency and context sensitivity of oceanic ontologies confers them with the ability to find synergies between different frameworks. Indeed, they may prove essential in enabling designers to simultaneously inhabit object-oriented and process-led systems, which may be key to developing new modes of architectural design practice. As Whitehead observes, both Heraclitean and Platonic perspectives are useful in the process of knowledge acquisition and are entangled in our experience of them.

> *Ideals fashion themselves round these two notions, permanence and flux. In the inescapable flux, there is something that abides; in the overwhelming permanence, there is an element that escapes into flux. Permanence can be snatched only out of flux; and the passing moment can find its adequate intensity only by its submission to permanence. Those who would disjoin the two elements can find no interpretation of patent facts. ... But the two elements must not really be disjoined ... bodily life transmits itself as an element of novelty throughout the avenues of the body. Its sole use to the body is its vivid originality: it is the organ of novelty.* (Whitehead, 1979, pp.339–340)

6.11 Summary

Materials at far from equilibrium states appear to embody the principles of vibrant matter through the production of assemblages, with the potential to forge a new kind of technological platform. These principles were demonstrated by conducting a range of 300 replicate experiments on the Bütschli system, which provided both a material system and a technology that could build and even enhance relationships between populations of agents and their surroundings. Using the Bütschli droplet system it was demonstrated that vibrant matter:
- Possesses agency
- Can be programmed using morphological computing techniques
- Provides unexpected insights into the behaviour of non-linear technology

These experiments suggest that vibrant matter may be applied within the engineering and construction of buildings as ELT. Potentially new relationships may be orchestrated by applying these technologies that result in the production of architecture, which builds ecological connections within systems. Depending on how ELT is imagined and designed, strategic applications may even remediate, absorb or make use of environmental toxins, for example, by incorporating them in the repair and growth of materials (Armstrong, 2012b). Most importantly, ELT promotes new ways of 'seeing' design and architectural solutions that no longer rely on machine metaphors. This technological platform enables us to imagine and express the world anew, with the potential to build new relationships with our environment and produce new kinds of knowledge. In this context, the role of the architect is as codesigner within ecologies of actants, all of which are establishing claims in a material system. Acts of codesign are therefore equivalent to acts of 'life', which refuse to accept deterministic pathways that are forged by past events, or obey the limits imposed by the claims of other actants in the system. Instead, by embracing the role of 'vibrant' architect, codesigners respond to constraints within the system through acts of continual creativity.

To meaningfully develop the principles and possible practices of vibrant matter, ELT must be accessible at the human scale. In the following chapter, I aim to test the scalability of dynamic droplets and other lively chemistries in an architectural design context.

7 'A Short Story of a Short Life'

7.1 Overview

This speculative narrative is an illustrative work that examines the ethics of working with lifelike technologies. It was originally written for the online magazine *Organs Everywhere*, edited by Simone Ferracina (Armstrong, 2012e).

7.2 A Short Story of a Short Life

The tiny droplet crawled almost imperceptibly over the base of the petri dish. Rendell Stone wondered just how long she had now been watching it through the microscope lens. These droplets were a couple of millimetres big and extremely simple, being made from an emulsion of oil, water and an alkaline salt. When the initial chemical field broke up, this simple combination gave rise to an extraordinary result: droplets that exhibited lifelike behaviour. Admittedly, the lifelike behaviour of the system was relatively limited, but the droplets had an uncanny character and were able to move around and seemingly sense their surroundings by following invisible trails of unidentified chemicals in the medium. When they collided, they exchanged a strange kind of 'kissing' action. Usually the rather gregarious and somewhat affectionate beads of fluid had a lifespan of only several minutes but when she was at the microscope, Rendell felt as if she had been observing them for hours. Should she be disturbing these delicate moments of intimacy?

> 'Pull yourself together! This is a science experiment!'

she muttered, irked that the compelling and rather 'familiar' behaviours of the droplets had somehow disarmed her objectivity. Rendell reminded herself that everything could be explained using the language of fluid dynamics and principles of self-assembly but logic could not stifle their strangeness. The droplets gleamed personality back at her, no matter how hard she tried thinking about them in 'simple' material terms. Sighing, Rendell realized that she had missed the droplet she had been tracking. They could be surprisingly tricky to follow, especially when they began to shed a crystal 'skin' and turn from a droplet into a tadpole-like structure, since they had a tendency to suddenly speed up – another endearing quality. After a short, systematic search, Rendell returned the droplet to the field of view, where it was still heaving and pushing its way to an unknown destination. This time it was moving with a 'swimming' action since the droplet had grown a 'tail', which made the whole complex sway from side to side. For a fleeting moment, she wondered how she would confirm that she had retrieved the 'right' droplet, but grew distracted by a conversation about lunch that was taking place in the corridor. Now she wished she had been content to let it disappear from view. Rendell grew impatient at the persistence of the tiny tadpole, which glittered appealingly, and decided to follow the fading voices before she perished first from lack of food. She left the digital video-recording system 'on' to capture the droplet's demise and trailed the others down to the canteen.

> 'Hey Rendell! How's it going?'
> 'I've been watching droplets travel around a petri dish.'
> 'Great! You're only doing real science when it's tedious or if it doesn't work!'
> 'I'm having a wonderful day then!'

The laboratory was on the first floor by an open courtyard where most of her colleagues went to light up cigarettes after lunch. On returning to the microscope, she could smell the smoke filtering its way through a leaky window seal. She drew the curtain around the microscope to keep out the brightening sunlight and prevent the fumes from 'poisoning' her experiments. Rendell decided to save the movie file that had been left running during lunch, before the computer ran out of disc space. She hadn't considered there was anything unusual about the abandoned droplet, but as she glanced into the viewer she was surprised to see it still throbbing at the centre of her field of view. Its tail had expanded so much that it now resembled a barnacle and was anchored to the bottom of the petri dish. She stared through the viewer waiting for the droplet to stop pulsing while the files were loading and converting. Its oscillations appeared less frequent and vigorous than before, so Rendell concluded that the changeling droplet would soon 'die'.

It didn't. The tiny droplet continued to glare back at her, defiantly moving in its new casing, slightly but regularly as if quietly 'breathing'.

Rendell glanced at the clock on the wall, which appeared to be holding its bored institutional face despairingly in its hands. She wished that she had made a proper note of the time. When the software finally stopped spitting out lines of 'dots' that promised a calculation was 'in progress', Rendell started the camera recording again and went to find her supervisor.

> 'Do you have a moment?'

Massimo Aomori's back was visible through the open door of his study. She was aware that her supervisor had developed a protective mechanism against persistent interruptions by students. He simply did not acknowledge them. Determined to secure a second opinion, she took several loud steps into his room, and tried a different approach.

> 'Would you recognize artificial "life" if you saw it, sir?'

Massimo's back drew a deep breath and spun around in a chair that turned twice, once at the pivot under its seat and again on the wheeled base. He leant forwards.

'Would you?' he challenged.

'I'm not sure! But I'd like your opinion on something. I'm chewing up lots of disc space making a recording of a droplet that just won't ... well ... die!'

Massimo laughed. 'Are you telling me that you've already created life in the lab from scratch?'

'Well, I don't think I "created" it! It self-assembled. I've been observing it for hours now. This is definitely out of the ordinary as this particular chemical system is usually completely spent after ten minutes. So, *something* odd is happening!'

'Okay! Now that I am interrupted, let me take a look!'

Rendell picked up a pace on her return to the microscope bench. Although she'd spent most of the day wishing that the tediously persistent droplet would stop moving, she was now concerned that all she might have to show Massimo would be an inert lump. She shut her eyes as she approached the viewer, willing the droplet to 'live'. When she finally opened them, she was relieved to find it winking at her – right at the centre of the field of view. It was smaller than she remembered and had broken free from its broad-based 'shell'. It lazily circled the base of the mineral deposit in a way that reminded Rendell of the way glimmering fairground goldfish patrolled their plastic bags.

'It's still there! Thank goodness!'

Massimo squinted into the lens and changed the fine adjustment on the eyepiece.

'Mmmmmm!' he observed.

'What do you make of it?' asked Rendell eagerly.

'How long has this been moving for?'

'Around two hours now.'

'Okay. It's hardly "evolution" but it's interesting. There must be something different about this droplet from the others. You need to find out why.'

Massimo became engrossed in surveying the details of the petri dish, deftly navigating the fine focus controls around the microscope stage.

'But how do I go about trying to establish its difference?' asked Rendell.

'Well, that's the problem. What exactly is it about "life" that is worth measuring?'

Massimo muttered as he studied the mineral traces.

'It's ...'

Rendell paused, realizing that she hadn't thought that particular issue through. She needed to.

Massimo continued.

'The next problem is that the techniques we use to measure "life" in science require us first to "kill" and process them.'

His hand momentarily splayed while he appeared to be taking note of something.

'But the chemistry of the system must change when it is killed!' objected Rendell.

'Correct!'

Massimo added, changing a lens setting.

'That's exactly the moment science waits to document. The difference between life and death.'

Rendell rolled her eyes in exasperation.

'So what should I do?'

Massimo smiled and tilted his head thoughtfully.

'Well, if I were you, I wouldn't rush to make any measurements. Not until you really know what you're looking at. Why don't you start at the beginning?'

He returned the microscope back to the magnification setting that Rendell had been working with.

Rendell was puzzled.

'The beginning of what?'

Massimo stood up straight and stretched his back.

'The beginning of the scientific experiment, of course! You must establish what you're doing by trying to prove yourself wrong. "Falsification". If you fail to prove yourself wrong, you're on the road to being right!'

Rendell hated these riddles and wanted clarity. Annoyed by his apparent equivocation, she demanded practicality.

> 'So, how do I do that?'

Massimo's back replied.

> 'Rigour. For starters, you need to be sure that the movement you are seeing isn't due to mixing of the oil and water layers under the heat of the microscope lamp. You also need to prove that airflow across the surface of the fluid isn't causing the droplet to move.'

Rendell sighed deeply. She wanted to ask her supervisor that if 'life' wasn't something that could be objectively 'measured', then what was the point of taking these ridiculous 'control' measures in the first place? Instead she inquired,

> 'So, what should I use?'

The back was already leaving as it announced,

> 'You'll figure that one out yourself! I have a grant proposal to write. Keep me posted!'

Realizing that the camera had been left recording all this time, Rendell decided to save the data and started to convert the file so that it could be exported. At least this would release some working memory on the scratch disc while she figured out what to do next. Peering into the viewer, she noticed that the droplet was weak and almost stationary. Yet it continued to throb at the foot of the shell-like deposit it had grown earlier in the day. If it had been 'alive' then she might have described it as being 'exhausted'.

Where on earth could she find a suitable marker to follow through on Massimo's suggestion?

Golden afternoon rays started to lick their way across the microscope bench. Rendell wandered over to the window as the movie files were being rendered, and stared down into the courtyard. A group of smokers were huddled around a door that opened out to the courtyard, trying to protect their cigarettes from being extinguished by the wind. A burst of dandelion seeds was stirred upwards past the window, where a few of them settled for several moments before being hurried on again by a gust of wind. Rendell realized that these seeds would be a perfect indicator for any airflow over on the surface of the liquid, and she raced outside to try and catch one.

Although there were many weeds between the cracks in the concrete scattered throughout the poorly tended courtyard, the task of gathering seeds was not so easy. None of the dandelion clocks appeared to be ripe enough to pluck a supply from. Rendell finally secured a fluffy parachuting body that was trapped by the roughness of a wall in one corner of the quad. Approaching it carefully, she pounced, and triumphantly brought the precious structure back to the lab.

She placed the downy seed on the surface of the liquid, where it floated like a water boatman. But when she looked for the droplet under the microscope viewer, it was nowhere to be seen. She tried a systematic search, hoping that she'd just misplaced the droplet, but only found trails of crystal skins.

Rendell anxiously scanned the petri dish hoping that the droplet had strayed. She wanted the droplet to have found new energy as part of its series of strange transformations, but although she looked repeatedly, the petri dish was barren.

Oddly shaken by the loss of the lively droplet, Rendell searched for evidence in the video footage that she had just taken. Sadly, this suggested that the droplet had turned into a tiny crystal skin deposit. Rendell stared at the enlightening frames in disbelief that such a stubborn entity could ever be extinguished. The dynamic, pulsing, winking droplet had gone and had been replaced by a small inert barnacle standing on the foot of a bigger one. There was no sign of anything throbbing or moving in the vicinity.

As Rendell sadly cleared away the petri dish remains into the sink, she was unaware that a tiny, winking droplet, which had been tethered like a barnacle to the petri dish, broke loose and made its bold way into the depths of the laboratory plumbing system.

8 'Hylozoic Ground'

A soil is not a pile of dirt. It is a transformer, a body that organises raw materials into tissues. These are the tissues that become mother to all organic life. (Logan, 2007, p.181)

8.1 Overview

An opportunity to investigate the applications of vibrant matter in an architectural context arose from an invitation to collaborate on the Hylozoic Ground installation by Philip Beesley. This chapter, therefore, examines the conditions in which vibrant matter and different species of ELT such as the Bütschli system can be applied and orchestrated, using morphological computing approaches within an architectural design setting. General design principles were developed for the installation with potential applications for broader architectural design contexts. A series of design experiments were also conducted to establish an approach towards developing appropriate infrastructures that may enable and prolong the lively action of vibrant matter.

8.2 An Exploration of Vibrant Matter in an Architectural Design Context

Hylozoic Ground is an architectural installation that was conceived and designed by architect Philip Beesley as Canada's entry to the 2010 Venice Architecture Biennale (see Fig. 8.1). Hylozoic Ground is a version of Beesley's ongoing series of installations (Armstrong and Beesley, 2011), which reflects his extensive engagement with responsive and distributed architectural environments and interactive systems. Beesley's practice is particularly concerned with design integrated with Nature, lifelike systems and their materiality.

My involvement with Hylozoic Ground began as a series of conversations with Beesley when we met at the 'Plectic Systems Architecture' conference held by Neil Spiller's AVATAR (Advanced Virtual and Technological Architectural Research) group in 2009. I had co-organized the event to bring together scientists working with emerging technologies and architects. The event consisted of a series of public presentations and a moderated workshop that aimed to explore and forge new possibilities for a 21st century practice of architecture in a multidisciplinary, collaborative environment (Armstrong, 2010b). A series of exploratory exchanges between Beesley and myself ensued, which reflected on the practical opportunities for directly integrating vibrant matter, ELT and morphological computing techniques into cybernetic frameworks (see Fig. 8.2).

Figure 8.1: Hylozoic Ground installation, Canadian pavilion, Venice is a cybernetic matrix that integrates a range of different 'organ' and 'tissue' types such as swallowing tubes (tapered cylindrical structures to the right of the photograph) and sound organs (clustered leaf-like structures in the centre of the photograph). The challenge was to design a set of dynamic chemistries that would aesthetically and functionally complement the soft mechanical systems. A centrally placed (yellow) chemical organ can be seen in centre field. Photograph, courtesy Philip Beesley, August 2010.

Figure 8.2: Some of the dynamic chemistries designed for the Hylozoic Ground installation resonated aesthetically with some of the soft, mechanical, feathery actuators within the cybernetic matrix. Photograph, Rachel Armstrong, July 2010.

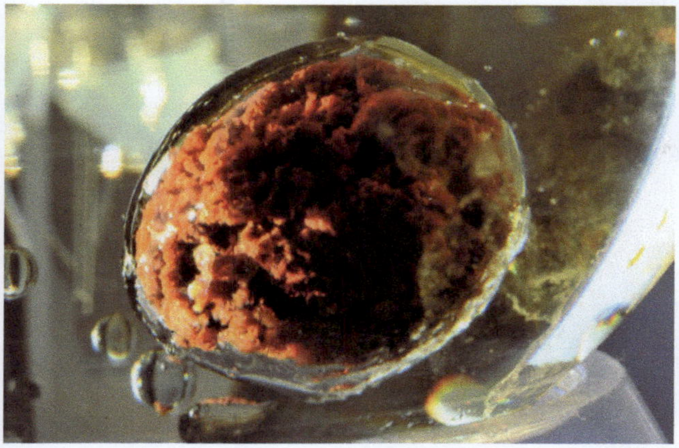

Figure 8.3.: Chemical experiments performed at the University of Waterloo with Philip Beesley, explored some of the properties of self-organizing systems so that they could potentially be coupled within a large, soft mechanical cybernetic framework. In this experiment, a copper II sulphate crystal was added to a weak potassium ferrocyanate solution in a flask, which produced a strong, undulating brown precipitate. Photograph, Rachel Armstrong, September 2009.

A variety of chemical systems, namely dynamic droplets (Hanczyc et al, 2007; Toyota et al, 2009), Traube cells (Traube, 1867) and Liesegang rings (Liesegang, 1869) were of interest to Beesley as the basis for a series of experiments that could be incorporated into his installation work. The suitability for each of these systems to be explored in a design context was established through a series of experiments at the laboratory at the Center for Fundamental Living Technology (FLinT) at the University of Southern Denmark, courtesy of the centre's director Steen Rasmussen and chemist Martin Hanczyc. Having worked closely with Beesley on these lifelike technologies and materials, a set of designs were developed and systems constructed that could be integrated into Hylozoic Ground, and installed in Venice in July 2010.

8.3 Hylozoic Ground at the Canadian Pavilion, Venice

Hylozoic Ground was situated in the Canadian pavilion at the Giardini on the island of Venice. The unique gallery space was designed by the Milan-based architecture firm BBPR (Gian Luigi Banfi, Ludovico Barbiano di Belgiojoso, Enrico Peressutti and Ernesto Nathan Rogers) (Di Martino, 2007) and inspired by the Nautilus shell. At the heart of the gallery was a thriving oak tree enclosed by glass panels that led to a skylight through which the foliage hung over the roof of the building (see Fig. 8.4).

The Hylozoic Ground installation itself is a complex, cybernetic system of interacting synthetic organs that are connected across interstitial spaces. The matrix is forged by a series of 'mycelium-like', delicate plastic fronds that are integrated with

Figure 8.4: Hylozoic Ground chemistries were nested within the cybernetic matrix of the installation and could also be seen against the natural oak canopy, which rose from the centre of the Canadian pavilion. Photograph, Rachel Armstrong, July 2012.

a neural network, which activates a variety of sensors and effectors. The cybernetic components of the installation respond to disturbances in the matrix by episodically discharging volleys of light, shaking the geotextile canopy like the wind through leaves and prompting feathery columns to greet visitors with sweeping salutations. The various species of systems that collectively make up the installation can be thought of as a web of physical, chemical and environmental connections through which a range of actants may participate. The installation may be likened to a soil, which is not formless, but has a specific architecture and an evolving body that is shaped by its responsive, material ecologies. The vibrant community of assemblages and robust centres of organization within Hylozoic Ground create a set of fertile conditions in which further lifelike events may occur. Conjecturally, given enough time, perhaps a sufficiently life-promoting environment could provoke an event that might be considered formally 'alive'[43] should it become possible to set up uninterrupted generations of droplet 'bodies' within the system.[44]

8.4 Hylozoic Ground Dynamic Chemistries

The dynamic cybernetic matrix of the Hylozoic Ground installation provided an active site that invited collaborative explorations with the non-linear outputs of its actuators, which were intrinsically coupled to the conditions of its site. The complexity of this environment required systems that could respond to unpredictable events and contribute creatively to the complex spatial programs continually shaped by the many participating actants, which could be considered as an 'ecology of design'.[45] A range of species of vibrant matter was therefore an obvious candidate for incorporation into the cybernetic system. The 'metabolism first' model of organization and the Chemoton (Gánti, 2003) criteria for 'life' were applied to develop a fundamental set of conditions around which ideas about creating a fertile material field could be developed, in which increasingly complex phenomena could potentially be observed. The responsive infrastructures of the cybernetic system, the neural network, the movement of people around the space, the flow of air and the material conditions within the gallery (gases, water vapour, dust) provided a rich landscape and infrastructure, which could be considered as participating agents within an ecological system of abiotic assemblages.

Chemistries were selected for incorporation into the cybernetic system on the basis they demonstrated some recognizable lifelike qualities, such as movement, sensitivity and the ability to adapt to changing environmental conditions (Armstrong, 2010a). Four different dynamic chemical species were developed into arrangements that complemented the hylozoic, or life-bearing, ambitions of the installation. They were positioned within the cybernetic matrix as a cohesive web of materialities that invited the participation of other actants. The striking lifelike characteristics of the Bütschli system (water in oil) (Armstrong and Hanczyc, 2013) warranted further

43 There is no universally accepted definition of life, although the working definition applied in the case of the 'lifelike' chemistries relates to the Chemoton model, which specifies that life needs a container, metabolism and information (Gánti, 2003). Although the conditions for the Chemoton model were not met, these principles provided a guide for considering the kind of chemical conditions that might increase the material 'fertility' of the site.

44 Continual flow of nutrients and removal of waste can prolong the activity of self-organizing systems (Hsu, Mou and Lee, 1994).

45 Stengers' notion of 'an ecology of practices' has many resonances with Hylozoic Ground, which is a non-hierarchical collaboration between many heterogeneous agents that may 'dream along with' each other to generate novel events (Stengers, 2000).

architectural exploration as well as developing a comparative reverse phase system (oil in water droplet technology), which provided an 'open' aqueous system through which a flow[46] of resources could be delivered to support the droplets' metabolic activities. The self-organizing properties of Liesegang rings (Liesegang, 1869) that result from chemical precipitation and diffusion of minerals through a reactive gel were also selected to augment the hylozoic potential of the site. Islands of hygroscopic materials within semi-permeable, latex envelopes encouraged airborne water movement through the installation, enriching the potency of the space to support living processes. Notionally, the Hygroscopic Islands functioned as an inorganic lymphatic system and speculatively provided a vehicle for the potential transfer of minerals in solution, or particles throughout the hylozoic matrix. These various chemical systems worked as an assemblage of chemical actants that orchestrated the movement of elemental systems and mineral resources through a cybernetic matrix, to establish fertile conditions that increased the probability of synthetic events occurring within the cybernetic field. These species included:

- Incubator Flasks: Modified Bütschli system
- Carbon Eater Flasks: Carbon-fixing oil droplets
- Liesegang ring plates: Vertical diffusion–precipitation fields
- Hygroscopic Islands: Assemblages of hygroscopic materials

8.4.1 Incubator Flasks: Modified Bütschli System

With the advent of synthetic biology (the design and engineering of living things) (Armstrong, 2013c; Armstrong, 2013d) and morphological computing (Armstrong, in press) it is possible to design with fully alive and lifelike processes using chemical building blocks. This scientific practice is relatively new and it is still unclear exactly what kinds of challenges the technology of 'life' is best placed to address. It is also uncertain how it may be possible to design with lifelike technologies in everyday situations. Synthetic biology and morphological computing are normally carried out on a very small scale, often in sterile laboratories; therefore, establishing a design practice based on vibrant matter, different species of ELT and morphological computing techniques within an architectural installation poses a number of design challenges:

- Is it possible to work with vibrant matter, ELT and morphological computing at a human scale?
- What design principles apply to materials that are lively?
- What infrastructures enable vibrant matter, ELT and morphological computing to be accessible for architectural design practice?

46 Continual flow of nutrients and removal of waste can prolong the activity of self-organizing systems (Hsu, Mou and Lee, 1994).

These questions were approached by working like Nature does, starting with the chemical principles that underpin living building blocks and aiming to 'grow' an architectural technology. The outcome was not specified at the start of the experiment, which was in keeping with design principles consistent with emergence, but left 'open' to explore the limits of the system. Chemistry is a parallel programming language and hardware, which can respond to changing circumstances and opportunities, just like natural systems do – although lifelike chemistry is not instructed by DNA. The participating chemical agents were regarded as codesigners of the cybernetic system that would establish their own relevance to and connection with the installation's actants by acting as ELT. This is a very different approach to architectural design conventions, which operate according to predetermined outcomes.

The lifelike qualities of the Bütschli system constituted a powerful platform and chemical operating system through which ELT could be developed. The dynamic chemistry was installed within a glass vessel, which comprised a design unit called an 'incubator'. The terminology reflected an ambition to design with living processes and to explore the conditions of chemical fertility in which other systems could thrive. Potentially, a fertile material matrix colonized by vibrant matter and different species of ELT increased the probability of hylozoism, where new kinds of lifelike chemical systems might emerge. To develop the technological potential of the Bütschli system as ELT required modification of the system so that it could orchestrate material events. Bütschli droplets spontaneously assemble when a drop of alkali is added to a dish of olive oil, where the aqueous phase spreads out and breaks up into tiny droplets that are about a millimetre in diameter and are just visible to the naked eye. At this scale, they are difficult to individually manipulate and even more difficult to see with the naked eye, especially in a challenging gallery context. When Bütschli droplets are examined under a light microscope they exhibit a striking range of lifelike behaviours. Each droplet possesses a unique personality and explores an individual trajectory. As Bütschli droplets push through the oil field they leave trails of soap crystals behind them, some of which are rather biological in appearance.

The first attempts to scale up this fascinating miniature world from the petri dish into a public space explored the possibility of using chemical attractants to induce population-scale activity within a droplet colony. Bütschli droplets respond vigorously to ethanol (alcohol), butanol and acetone (nail polish remover) and form droplet swarms in the oil medium. However, even using large amounts of droplets and chemoattractant did not produce an aesthetically appealing result, as aggregated Bütschli droplets simply looked like milky streaks over an oil field at the human scale (see Fig. 8.5).

Unmodified Bütschli droplets are not easily visible from any distance with the naked eye, since they are denser than their medium and sink to the bottom of the flask as they are formed. Although it was intended that visitors would view the chemistry from underneath, the droplets tended to clump together on the base of the petri dish in which they were prepared where it was not easy to see them clearly. A layer of

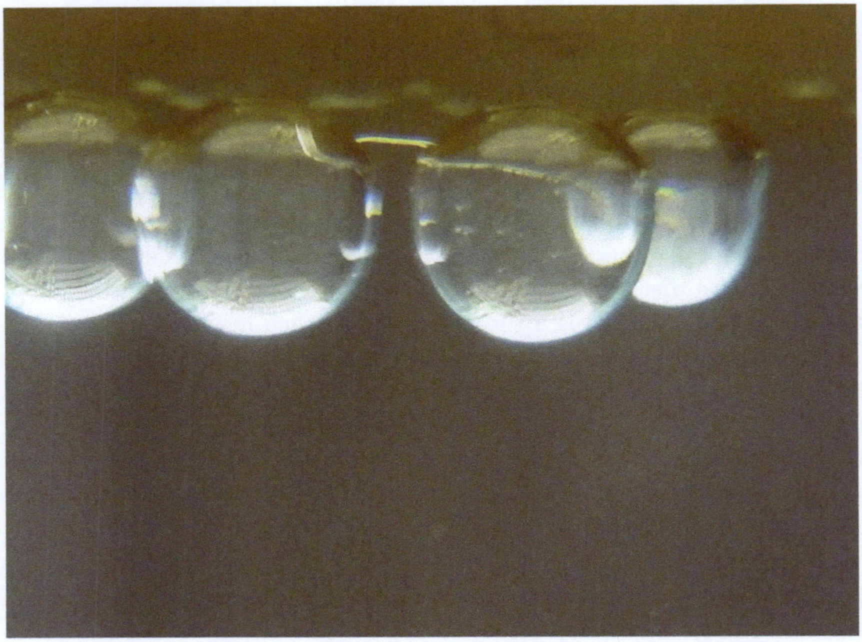

Figure 8.5: Unmodified Bütschli droplets are not readily visible to the naked eye and require a contrasting interface if they are to be easily seen within a gallery setting. Photograph, Rachel Armstrong, February 2010.

diethyl phenyl phthalate (DEPP),[47] which is clear oil with a specific gravity of 1.12 g/cm^3 at 20°C, was added into the petri dish to form an interface with the olive oil. DEPP is denser than olive oil, with a specific gravity of 0.9150–0.9180 g/cm^3 at 15.5°C (Olive Oil Source, 1998–2013) and it is also denser than water, with a specific gravity of 1.0 g/cm^3 at 20°C. So, with a layer of DEPP at the bottom of the container, a clear interface could be developed against which the droplets could be clearly seen, by carefully layering olive oil over it so that the oils did not mix (see Fig. 8.6). A 250 ml round-bottomed flask was clamped in a retort stand and 100 ml of DEPP was added, before carefully pipetting 100 ml of olive oil by hand down the side of the flask over the DEPP to create an interface and avoid mechanical mixing.

An inhibitor was added to the 3 M sodium hydroxide aqueous phase to increase the size and visibility of the Bütschli droplets, which was prepared by agitating 20 ml olive oil and 20 ml 3 M sodium hydroxide in a 50 ml screw-cap disposable test tube, for 2 min. The emulsion was then spun down in a centrifuge at 10,000 rpm for 30 s and the milky precipitate that had settled at the interface by hand was extracted

---

47 DEPP is commercially used as a plasticizer, a perfume diluent, emollient and as a fixing agent in gas chromatography.

Figure 8.6: Modified Bütschli droplets self-assemble at the interface between green olive oil (above) and yellow-tinged DEPP (below) and are clearly visible in a gallery setting. Photograph, Rachel Armstrong, April 2011.

with a 2 ml disposable pipette. 1 ml of extracted surfactant (precipitate) was added to each 10 ml of 3 M sodium hydroxide in the preparation of the Bütschli system, which produced greatly enlarged droplets with diameters of 1–2 cm. These modified droplets self-assembled into evenly spaced formations and produced a clearly visible layer that settled at the interface between the two oils (see Fig. 8.7).

Under these conditions, the Bütschli droplets produced milky, soapy deposits at the oil/water interface and could also be built up carefully by hand to a few centimetres in diameter. Additional metabolisms of brightly coloured 1 M inorganic salts that included copper II sulphate, iron II chloride, iron III chloride, cobalt II chloride and nickel II sulphate, in amounts of 1–2 ml, were pipetted by hand into the flask (see Fig. 8.8).

The reagents are summarized in Table 8.1.

These droplets fused on contact with the enlarged Bütschli droplets to create insoluble, brightly coloured precipitates within the alkaline environment of the Bütschli system, which promoted crystal growth (Spanos and Koutsoukos, 1998). Modified droplets could be clearly seen against the oil layers within the flasks and formed miniature, suspended crystal gardens and shell-like structures that were remarkably stable in the gallery conditions and survived for the entire duration of the installation. However, since the Bütschli droplets had been greatly slowed down

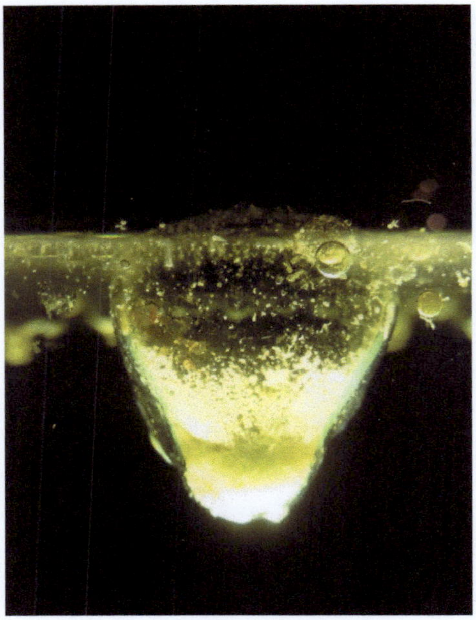

Figure 8.7: Modified Bütschli droplets will enlarge from the millimetre to the centimetre scale by introducing a small quantity at 1% v/v of the system's soap-like product into the 3 M sodium hydroxide solution. This can be seen accumulating at the bottom of the droplets under the influence of gravity as a highly reflective sheen, which is produced by the soap crystals. Photograph, Rachel Armstrong, April 2011.

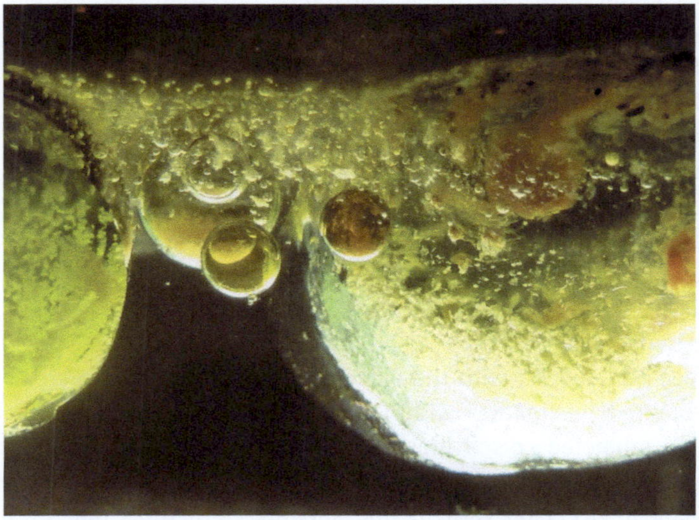

Figure 8.8: With the addition of 1 M mineral solutions, modified Bütschli droplets produce insoluble carbonate crystals, whose growth is facilitated in the alkaline droplet interior. Photograph, Rachel Armstrong, April 2011.

by the surfactant inhibitor, they did not exhibit the same kind of striking lifelike behaviour of the original recipe. Yet, they were not inert and produced a different kind of morphological computation where the droplets self-organized into periodically spaced clusters (see Fig. 8.9).

Table 8.1: Incubator Flask preparation for a 250 ml flask

Chemical	Strength	Amount
Diethyl phthalate	n/a	100 ml
Extra virgin olive oil	n/a	100 ml
Sodium hydroxide	3 M	40 ml
Iron II chloride	1 M	10 ml
Iron III chloride	2 M	10 ml
Nickel II sulphate	1 M	10 ml
Copper II sulphate	1 M	10 ml
Cobalt II chloride	1 M	10 ml
Calcium chloride	1 M	10 ml

Figure 8.9: Droplet clusters are spontaneously produced through morphological computing interactions than can be shaped by adding mineral solutions to the flasks, which convert dissolved carbon dioxide into structural carbonate precipitates. Photograph, Rachel Armstrong, April 2011.

These formations were considered to behave as a slowly mineralizing artificial tissue system within the cybernetic installation as in Fig. 8.10.[48] The modified Bütschli droplets responded to physical changes in the environment through subtle changes in the speed of crystal growth that were influenced by movement and heat produced by the mechanical cybernetic matrix.

The Incubator Flasks were prepared for the Hylozoic Ground installation by hand as a series of 20 round-bottomed flasks, each of which housed an individually modified version of the Bütschli system. They were suspended in specially designed holders within the Hylozoic Ground matrix by Beesley's team (see Fig. 8.11).

Figure 8.10: Modified Bütschli droplets respond to environmental conditions in flasks that are open to the air. Incubator Flasks were suspended in the Hylozoic Ground matrix and positioned over LEDs to capture heat and light emitted by the activated cybernetic matrix. Photograph, Rachel Armstrong, April 2011.

48 Vesicle-based, artificial tissue systems have been described in Gabriel Villar's work on printing with microscale abiotic vesicles (University of Oxford, 2013).

Figure 8.11: The selection of a range of mineral solutions, such as soluble nickel and iron salts, may cause a range of brightly coloured precipitates to be formed. Modified Bütschli droplets with mineral metabolisms gradually produce internal crystalline structures at the oil/water interface, as miniature crystal gardens (Glauber, 1651). Photograph, Rachel Armstrong, April 2011.

Other collaborators included chemist Martin Hanczyc from the University of Southern Denmark, who was consulted on the chemical design of the Incubator Flasks, and Explora Biotech, who maintained the chemistry for the installation duration. The Incubator Flasks were integrated into the cybernetic matrix by their positioning over LEDs, which were activated by the mechanical sensors of the cybernetic matrix. Summated activation of the sensory network also produced periodic bursts of movement within the Hylozoic Ground matrix that accelerated and shaped crystal formation within the Bütschli droplets. During the three months of the exhibition, the crystals became more visible within the incubators and their striking appearance drew visitors into the depth of the cybernetic matrix, which activated proximity sensors that were most active when the gallery was busy.

Visitor activity also provoked volleys of movement in the feathery appendages throughout the cybernetic matrix, which stirred up the air in the gallery and circulated the carbon dioxide around the space, which fed the metabolism of the 'Carbon Eater' Flasks. The entangled chemical, mechanical and human actants created a positive feed-forward loop of activity that attracted more visitors into the cybernetic matrix, which further stimulated the behaviour of actants and fed the Hylozoic Ground's metabolism.

8.4.2 'Carbon Eater' Flasks with Carbon-fixing Oil Droplets or 'Protopearls'

The 'Carbon Eater' Flasks were designed to produce artificial shell-like structures that could be built from resources in their immediate environment such as dissolved carbon dioxide and minerals (see Fig. 8.12).

An oil droplet within an aqueous phase was chosen to keep the system materially open so that, as a universal solvent, it could continue to harvest water-soluble resources. Oil droplets in aqueous media have been demonstrated as an effective distribution platform for a metabolism and under the right conditions they may also exhibit lifelike properties such as self-propulsion and chemotaxis (Hanczyc et al, 2007; Toyota et al, 2009). An oil droplet system was considered a suitable technology that would detect and respond to changes in the gallery environment such as carbon dioxide concentrations. This respiratory gas is of particular architectural and cultural importance as it is an indicator of life, being exhaled by gallery visitors, and is also of global and environmental significance as a 'greenhouse gas' (GHG), which is released into the atmosphere from other sources, such as the burning of fossil fuels within the city. Venice lagoon water was chosen as the aqueous medium for the oil droplets,

Figure 8.12: This protopearl flask is suspended in the Hylozoic Ground matrix, where airflow is augmented through the system by feathery actuators. As the flasks are open to the air, a profuse copper carbonate precipitate is produced, which has formed a green ring-like structure at the base of the flask. Photograph, Rachel Armstrong, July 2010.

which is extremely poor quality and highly saline (3.5%) so it is rich with substances that could be used by the metabolizing droplet technology. The lagoon's water quality is affected by many factors, such as high population density, reduced exchanges with the sea, accumulation of nutrients from its drainage basin, exploitation of its natural resources (Facca et al, 2011) and effluent. Indeed, Venice has never maintained a main sewage system and the urban effluent, run-off from agriculture and industry discharges feed into the lagoon (Fletcher, da Mosto and Spencer, 2005). DEPP was chosen as the oil carrier system for the Carbon Eater metabolism, which is not vigorous or lifelike in the way that other oil/water systems are (Hanczyc et al, 2007; Toyota et al, 2009) but possess dynamic physical properties that provide a lively gallery performance. DEPP is heavier than water with a specific gravity of 1.1 that is temperature sensitive, so it exhibits a range of behaviours in aqueous environments. Below 25°C, the droplets sink in the flasks and form spheres, while at higher temperatures DEPP droplets rise to the surface and spread out like plates.

Magnesium and calcium (Anthoni, 2006) were used as metabolic agents as they reflected the natural preponderance of mineral species in the Venetian lagoon water, which form white precipitates in the presence of dissolved carbon dioxide (see Fig. 8.13).

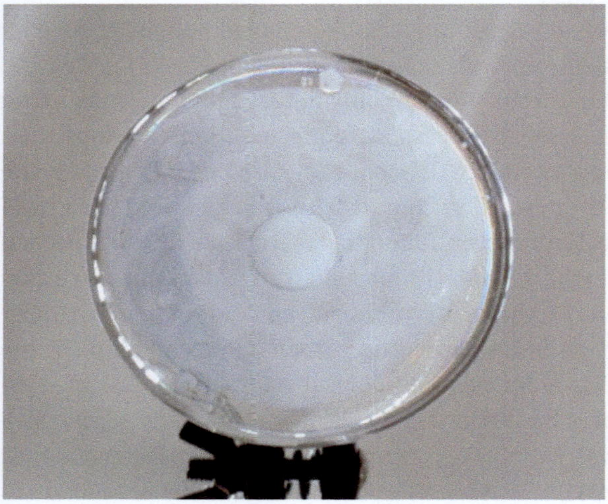

Figure 8.13: Calcium carbonate-producing protopearls can be seen at the bottom of the round-bottomed flask. Photograph, Rachel Armstrong, July 2010.

Two different species of DEPP droplets were prepared – one with calcium chloride, to produce a white precipitate, and the other with copper II sulphate, which would produce a green precipitate. Since inorganic salts are insoluble in oil, they were manually ground into the DEPP to form a paste. Six 250 ml 'Carbon Eater' Flasks were filled with 200 ml of Venice water. 0.5 g of calcium chloride and 0.5 g of copper II sulphate were separately crushed into a paste of 6 ml of DEPP using a metal spatula. A hand-held pipette was used to draw up 2 ml of the oil mixture at a time, which was dropped into the flask. The droplets sank to the apex of the bulb and within a few minutes precipitation could be seen.

These white and green crystal-coated oil droplets, or protopearls, provided a site for crystal growth throughout the duration of the installation, when minerals in the DEPP droplets and dissolved carbon dioxide from the Venice water reacted to produce insoluble precipitates. The carbonate-coated droplets were extremely robust and would re-form after even vigorous shaking. The oily 'pearl' droplets became opaque at high levels of carbon dioxide and with time, as carbon dioxide from the surrounding atmospheric air dissolved into the flasks. In the same way that the Incubator Flasks were nested within the cybernetic matrix to connect with the activity in the body of the installation, the Carbon Eater Flasks were positioned so that LEDs situated at the bottom of the flasks generated a small amount of heat, which nominally speeded up mineralizing activity within the flask.

The Carbon Eater Flasks were hung using specially designed holders created by Beesley's team, adopting the same approach as for the Incubator Flasks and leaving the vessels open to the air, so that carbon dioxide could diffuse into the Venice water to replete the carbon dioxide necessary for carbonate precipitation. The formation of carbonate could be speeded up by adding 0.02 g of 1,1-dicarbonyl imidazole to the DEPP droplets, which rapidly hydrolysed on contact with water to produce a local concentration of carbon dioxide. High concentrations of local carbon dioxide speeded up the formation of carbonates at the oil/water droplet interface and promoted carbon dioxide uptake in the Venice water. Additionally, very small amounts of soluble copper and calcium ions (0.1 g) added could also speed up carbonate precipitation. These chemical modifications provided a set of technological approaches that enabled Beesley and his team to optimize the performance of the Carbon Eater Flasks as a spontaneous construction process. During the three-month installation the abundance of materials from the air and in solution was reflected in the vigorous production of precipitate. Explora Biotech therefore provided a fresh supply of ingredients on a weekly basis. The ingredients are summarized in Table 8.2.

Carbon Eater droplets could be regarded as ELT that enables designers to begin to influence the carbon cycle (TED blog, 2010) by directing mineral formations in an environmentally responsive manner. However, Carbon Eater droplets do not propose to be a geoengineering-scale intervention that can address the current global crisis of rising levels of the greenhouse gas but suggest that more site-specific and local interventions are possible in an architectural context. Although the quantitative value

Table 8.2: Carbon Eater preparation for a 250 ml flask

Chemistry	Strength	Amount
Diethyl phthalate	n/a	1 ml
Copper sulphate	Powder	0.5 g
Calcium chloride	Powder	0.5 g
Venetian water	n/a	200 ml

of the possible carbon fixation of the DEPP droplet system was very small (Webster, 2011), the results are qualitatively significant. These models suggest that it may be possible to go beyond the carbon-neutral ideals for the production of architecture and strive for environmentally remedial or synthetic outcomes (Armstrong, 2011c).

The open flow of resources through the Carbon Eater Flasks, which is mediated through Venetian lagoon water, increases the probability of an emergent, self-propagating or even self-regulating material event. Yet, these potential hylozoic activities are not random, nor left to chance, but are midwifed using ELT. To augment the likelihood of hylozoism, the assistance of mechanical apparatuses could be incorporated, which, for example, would act as a 'birthing unit' for the automatic release of metabolic DEPP droplets into the system in response to mechanically sensed parameters, such as light density or carbon dioxide concentrations. Future versions of this installation may be designed to work in concert with human surveillance, such as Explora Biotech's diligent vigilance of the performance of the chemical systems, to develop an assemblage of interacting agents that nurture the site's potency. Resource abundance within Hylozoic Ground was most open and richest where blooms of intense light notionally stimulated the mineralization processes within the Incubator Flasks (see Fig. 8.14).

Also, small plumes of air movement provided a matrix for abundant exchanges of vaporized solutes and particles that were directed around the flasks, which were provoked by gallery visitors moving close to the Incubator and Carbon Eater Flasks and by the vigorous, periodic volleys of the canopy. The shifting of resources around the site through air-cooling, metabolic exchanges (via carbon dioxide and humidity-rich transpired and respired air) and heat (via the LED lights) were designed to provoke mineralization of the Carbon Eater droplets. Indeed, inducing division and fusion in various droplet species (Caschera, Rasmussen and Hanczyc, 2013) may speculatively provoke hylozoic activity. It is also anticipated that the integration of different ELT species with mechanical systems may provide new opportunities to increase the possibility of autopoietic systems in future versions of the Hylozoic Ground installation. Although the organic/mechanical interface is notoriously difficult to design, the potential for notional bidirectional exchange is ever more likely. One of the major challenges in developing biohybrid robots is in creating an interface that

Figure 8.14: An aqueous infrastructure is required for the effective carbon-fixing action of protopearls, which is provided by the open environment of the round-bottomed flasks. Photograph, Rachel Armstrong, July 2010.

allows effective communication between the biological and electronic components. This is inherently difficult because of the ontological differences between the systems, and because the vast majority of cellular signals do not easily translate into digital codes. Conversely, cellular processes usually produce signals that travel too slowly for electrical circuits, and electrical fields have negative effects on many cells, which can lead to cell death (Lee, 2006).

Currently, the engineering practice of microfluidics, which works with the properties of liquids at very small scales, mediates exchanges between mechanical and organic systems. However, since they are shaped by physical principles that are unique to the microscale such as surface tension, energy dissipation and fluidic resistance, these systems are scale-specific. Recently, cells have been genetically engineered to behave in ways by producing digital signals by genetically modifying them to produce modified proteins that release nitrous oxide in response to light (Yarkoni, Donlon and Frankel, 2012). This research suggests that developments in bio/electronic interfaces may present opportunities to increase the material complexity and fertility of future versions of the Hylozoic Ground installation by integrating living and mechanical systems through substrate independent modes of operation that involve the coupling of actants into functional assemblages and point towards possible further developments for the collaboration.

8.4.3 Liesegang Ring Plates: Vertical Diffusion–Precipitation Fields

The 'Liesegang ring' plates consisted of a gel-based chemical 'clock' that produced brightly coloured mineral patterns with the passage of time. They represented a chemical archive of material change within the installation over the duration of the exhibition and exhibited the qualities of vibrant matter (slowness, porosity, inorganic sympathy) (Bennett, 2011). The diffusion of the salts through the active alkaline gel constitute slowness, as the reaction takes place under the actions of gravity and propagate at a rate of about a millimetre per day. Porosity occurs in the fluctuations in the states of the metal ions descending in the gels between precipitate and solution, and inorganic sympathy takes place when the system reaches equilibrium as the banding patterns reach the lower portion of the plates, where there is no further active gel substrate to encourage the oscillations between solid and liquid forms. The Liesegang ring plates provide a counterpoint to the rapid vigorous, sporadic behaviour of the mechanical, cybernetic system and work along a similar timescale to geological processes, with similarities to the processes that produce banding structures in the mineral agate. The evolving traces in the plates marked a baseline indicator of the dynamic chemical and physical forces at work in the installation, which gradually decay towards equilibrium as – unlike the Carbon Eater Flasks – they are sealed and not replenished.

The Liesegang ring plates were designed as a modification of a self-organizing chemical process that produces periodic precipitates in an active gel medium. Liesegang first described this phenomenon when he was preparing photographic plates using silver nitrate and potassium dichromate and documented the spontaneous formation of ring-like patterns (Liesegang, 1869). Rather than simply creating a light-sensitive field, these salts had produced their own patterning system in the absence of light. Liesegang and Runge (Runge, 1850) further characterized the phenomenon as occurring when interacting ion species that produce density fluctuations in weakly soluble salts are separated through a matrix. The principles of pattern formation are based on the interactions between different types of salt as they exchange ion species and periodically form precipitates (sparingly soluble salts) and solutions (strongly soluble salts). Liesegang rings also form naturally in various rock types (Heaney and Davis, 1995) when minerals diffuse through gel-like mud, or fossilizing flesh, and are shaped by geological forces. They can be induced in a laboratory setting by impregnating a watery gel with a soluble salt species and diffusing another salt species into an alkaline matrix to produce slowly forming precipitates that (re)dissolve in keeping with the periodic fluctuations of chemical species in the system. Under the influence of gravity and diffusion, the spatial distribution of the resultant weakly and strongly soluble salt species appears as rhythmic bands or rings (see Fig. 8.15).

The Liesegang ring plates constructed for Hylozoic Ground were formed from a series of two sets of leaf-shaped perspex plates that were mounted in floret formation. Each unit consisted of two parallel plates that were fixed with a polymer at a uniform distance of 0.5 cm. The apex of the plate was left open for the introduction of solutions. Eight units were prepared in total, and were left to dry thoroughly before the addition of the active gel matrix. This was prepared using 400 ml of agarose gel made up to a concentration of 2% by weight and stirred over a hot plate at 70°C in a fume cupboard in Explora Biotech's laboratory in Marghera, in an industrial park just north of Venice

Figure 8.15: Liesegang ring plates were constructed from two perspex plates separated by a 0.5 cm gap sealed with silicone. Alkalinised agarose at 2% v/v was introduced into an apical gap in the plate system and allowed to cool for an hour. A solution of iron II, iron III and copper II salts was introduced into the apical reservoir. Within a few hours, precipitates could be seen moving through the plates, and produced striking banding patterns that continued to evolve over the duration of the three-month installation. Photograph, Rachel Armstrong, July 2010.

on the mainland. 20 ml of 1 M ammonium hydroxide was added and vigorously stirred into the gel so that the salt was evenly mixed. This solution was then carefully pipetted by hand into the gap between the fixed perspex plate pairs at 50 ml aliquots.

Great care was taken to avoid bubble formation in the matrix of the rapidly cooling gel and a space for 10 ml of fluid was left at the top each plate. The prepared plates were then carefully transported back to the Canadian pavilion. In the gallery, holders created by Beesley's team secured the plates vertically, in floret formation. When the plates were secured, 4 ml of 1 M copper II sulphate solution, 2 ml of 1 M iron II chloride solution and 2 ml of 2 M iron III chloride solution were added to the 10 ml reservoir at the apex of the units. The preparation is summarized in Table 8.3.

These solutions provided a supply of competing ion species that diffused through the gel under the influence of gravity. The plates were permanently sealed to avoid contamination or spillage. Over the course of the installation, the Liesegang ring plates produced clearly visible, brightly coloured, unique, evolving banding

Table 8.3: Liesegang ring plates: Evolving diffusion–precipitation reactions

Chemistry	Strength	Amount (per component)
Agarose gel	2%	50 ml
Copper II sulphate	1 M	4 ml
Ammonium hydroxide	1 M	10 ml
Iron II chloride	1 M	2 ml
Iron III chloride	2 M	2 ml

Figure 8.16: Clusters of vertically mounted Liesegang ring plates were introduced as a time-based chemical system in the Hylozoic Ground matrix, like the bark of a tree. Photograph, Rachel Armstrong, July 2010.

patterns that continued to develop and separate for the duration of the exhibition (see Figs. 8.16 and 8.17).

Owing to the robust integrity of the sealant, only a few of the plates started to dry out in the last week of the Biennale. None of the plates succumbed to bacterial or fungal colonization and they required no maintenance. Yet, the Liesegang ring plates had reached equilibrium by the close of the Biennale.

Figure 8.17: Liesegang ring plates are entangled in the Hylozoic Ground matrix as a sealed time-based organ system. Photograph, Rachel Armstrong, July 2010.

8.4.4 Hygroscopic Islands

The 'Hygroscopic Islands' were an arrangement of thousands of semi-permeable, latex vessels that contained hygroscopic salts (such as calcium chloride), desiccated organic matter (like dried lavender and herbs) and concentrated salt solutions (such as soy), and exhibited the qualities of vibrant matter (slowness, porosity, inorganic sympathy) (Bennett, 2011) (see Fig. 8.18).

As attractors, the hygroscopic materials worked slowly, since they relied on diffusion and the active stirring of water vapour in the gallery by the action of visitors and the mechanical volleys of the matrix. Their interactions with water were loose and, therefore, showed porosity as they associated and disassociated with water molecules as the humidity and temperature of the gallery changed. They did not reach equilibrium, as their affinity with water is entirely reversible and so they were continually able to make associations and disassociation with water molecules depending on the ambient humidity. Hygroscopic Islands exhibited strong inorganic sympathy (Bennett, 2011), as their actions in forming associations with water were weak, but when viewed as an assemblage of thousands of hygroscopic bodies, this lively force could be regarded as being much stronger. The hygroscopic materials were suspended from the cybernetic canopy in bunches, which hung low into the matrix of the gallery space. Hygroscopic Islands were strategically positioned to act as water-organizing systems that facilitated the movement and distribution of fluid between the chemically active flasks (see Fig. 8.19).

Figure 8.18: Hygroscopic Islands attract water vapour into their substance through their porous latex container and invite the presence of a primitive circulatory system within the gallery space. They may be notionally likened to a lymphatic organ that distributes nutrients throughout living tissues. Photograph, Rachel Armstrong, July 2010.

Figure 8.19: The combined weak water-transferring interactions of Hygroscopic Islands are amplified through assemblage formation. They provide a 'wet' infrastructure through which material transfer between sites may be possible. Photograph, Rachel Armstrong, July 2010.

These formations propose that active, diffusively distributed, material operations could meaningfully facilitate flow and resource exchanges within a technologically mediated site and are consistent with the conception of a cybernetic 'soil'. Further research and development of these ideas based on interacting assemblages of chemical systems may lead to a greater understanding of how self-regulatory, potentially hylozoic systems could deal with notions of resource abundance, material transformation and autopoiesis within architectural design practice through the construction of synthetic soils.

8.5 Modifications of the Hylozoic Ground Chemistries

The Hylozoic Ground chemistries established design principles for ways of working with lively substrates as sets of actants and assemblages in an architectural design context that were transferable to other contexts. Two particular modifications of interest are the 'hygroscopic preparation' and the 'BIO-FICTION installation', since they establish additional properties of chemical technologies that demonstrate their innate flexibility and are relevant to architectural design.

8.5.1 Hygroscopic Preparation

The modified Bütschli system led to requests from architectural students to repeat the formula in different settings, such as the Gallatin School NYU, School of Architecture at the University of Nottingham, and the Bartlett at UCL. Other events such as Secret Cinema,[49] the ArtScience Prize themed on synthetic biology at the Silk Mill, Derby (UK ArtScience Prize, 2013) and Glenfiddich Pioneers (Future Laboratory, 2010)[50] requested demonstrations of the self-organizing principles of the Bütschli system. Owing to the difficulty of sourcing some of the chemicals without the assistance of a chemistry department, alternative preparations were developed to embody the dynamic qualities of the system and develop ideas that explored the principles associated with living systems.

---

49 "I would like to acknowledge Liam Young and Kate Davies in connecting me with the Secret Cinema event that presented Ridley Scott's Prometheus in Euston, London, throughout June 2012, transforming a vast abandoned warehouse into the Brave New Ventures/Weyland Industries embarkation terminal and spacecraft, which was staffed by actors (Secret Cinema, 2012). Within the makeshift spacecraft, rooms were dressed as science laboratories and briefed to conduct experiments that searched for unusual life forms. I was invited to demonstrate my work in this context, participating with the general public and demonstrating the living qualities of chemistry, for which I used a hygroscopic preparation.
50 This event enabled participants to produce 'self-evolving' 3D paintings using olive oil, glycerine and food dyes as a visual counterpoint to the art and science of mixology.

The hygroscopic preparation of a species of ELT was prepared to visualize the dynamic properties of water and the materials that are invigorated by it. Hygroscopic materials exhibit a powerful affinity for water and their dynamics operate on a much faster timescale than the modified Bütschli system in the Incubator Flasks. Although this system was unsuitable for a three-month installation, as it reaches equilibrium quickly, it is a lively demonstration formula (see Figs. 8.20 and 8.21).

Figure 8.20: This chemistry demonstrator shows how some matter seeks out water – even when it is not 'alive'. This system was originally designed for Philip Beesley in Mexico City as part of the Hylozoic Ground chemistry series. Photograph, Philip Beesley Architect, March 2010.

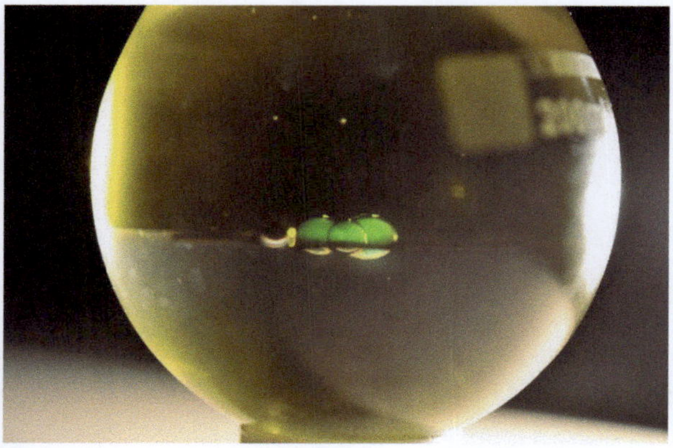

Figure 8.21: Droplets of strong salt solutions are pulled down into hygroscopic glycerol leaving a trail of solutes behind. This graphically portrays the 'battle' for water between different chemistries. Photograph, Philip Beesley Architect, March 2010.

The hygroscopic metabolism was designed to work safely with readily available household ingredients that may be purchased from grocery stores, local pharmacies and supermarkets. Hygroscopic materials may be prepared in an unpatterned 250 ml tumbler, with a base layer of glycerol, which is a viscous, hygroscopic, clear liquid. It has three hydroxyl groups that are responsible for its solubility in water and its hygroscopic nature. It is popularly used as a base for cough medicines and also as a laxative. 100 ml glycerol was poured into the tumbler to the halfway mark and 100 ml of olive oil was then layered carefully on top using a disposable pipette. Careful addition of the oil could also be achieved by pouring it over the back of a tablespoon and running the olive oil down the side of the glass to avoid mixing of the fluids which, once combined, cannot be separated. The droplets were prepared by using the central well of a saucer to contain 2–4 ml of different food colourants. Rock salt, which is weakly hygroscopic (from the traces of other minerals contained in the salt, such as magnesium chloride), was stirred and crushed into the liquid until the mixture was thick and super-saturated. 0.5 ml aliquots of the mixture were then added to the tumbler where they formed droplets at the glycerol/oil interface. The hydrophobic olive oil repels water, so that only aqueous solutions can move downwards in this preparation (see Figs. 8.22–8.25).

Figure 8.22: This hygroscopic demonstration was prepared for Liam Young's 'Contamination' event for the launch of Ridley Scott's film *Prometheus* at a premiere event organized by Secret Cinema in London. Photograph, Rachel Armstrong, June 2012.

Figure 8.23: Laboratory bench equipment needed to prepare a hygroscopic demonstration of dynamic chemistries. Photograph, Rachel Armstrong, June 2012.

Figure 8.24: Tumblers used by participants to set up their own hygroscopic chemistry preparation. Photograph, Rachel Armstrong, June 2012.

Figure 8.25: Round-bottomed flask set up to demonstrate a hygroscopic chemistry preparation. Photograph, Rachel Armstrong, June 2012.

The ingredients are summarized in Table 8.4.

Table 8.4: Hygroscopic preparation for a 250 ml tumbler

Chemical	Strength	Volume
Glycerine	n/a	100 ml
Olive oil	n/a	100 ml
Food colouring	n/a	1–5 ml of each colour
Table salt	Crystals	2–5 g

A physical/chemical 'tug of war' for water ensued between the salty food dye and the glycerol. The more vigorously hygroscopic glycerol associates with the water in the droplets and, when this tipping point is reached, descending, firework-like trails of colour can be seen exploding into the glycerol, as in Fig. 8.26.

This may take between several seconds to several minutes to be reached and this dynamic process can be used to make 'three-dimensional', self-evolving 'paintings', which speak of the importance of water as an organizing force in material systems and as a necessary precondition for life.

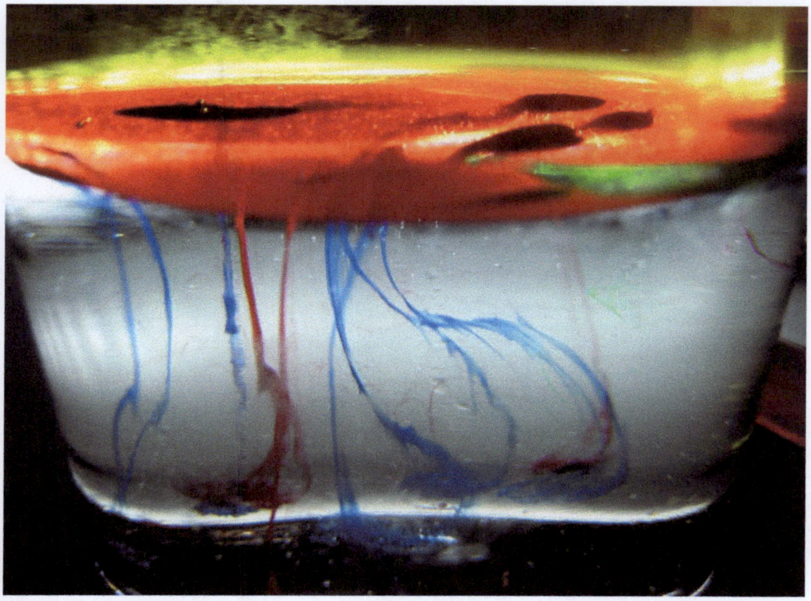

Figure 8.26: Tumbler details showing solute traces in the glycerol base layer. Photograph, Rachel Armstrong, June 2012.

8.5.2 BIO·FICTION Installation

A further modification of the Bütschli system was designed for the Synth-ethic art group show for the BIO·FICTION event at the Natural History Museum, Vienna, May 2011 (Synth-ethic, 2011). The chemical principles were identical to the modified version of the Bütschli system for the Incubator Flasks in the Hylozoic Ground installation. An important modification was made to this system in restricting the space in which self-organization of the droplets could take place. This was limited to a 2 cm gap in the walls between two tanks (one constructed inside the other) and approximates to the average diameter of a modified Bütschli droplet. A lower layer of 20 l of DEPP and an upper layer of 20 l of extra virgin olive oil were carefully constructed so that the two oils did not mix. 200 ml of 3 M sodium hydroxide was then pipetted evenly around the tank in 2 ml aliquots where droplets spontaneously formed at the interface. An additional total of 100 ml of salt solutions were added into the tanks as 2 ml aliquots, which included 1 M solutions of nickel sulphate, copper II sulphate, iron II chloride, cobalt II chloride, calcium chloride and 2 M iron III chloride (see Fig. 8.27).

The recipes are summarized in Table 8.5.

The Bütschli droplets and salt solutions began to reorganize over the course of several hours to produce distinct, banded patterns (see Figs. 8.28 and 8.29). The spatial constraints created the conditions for the production of undulating chemical

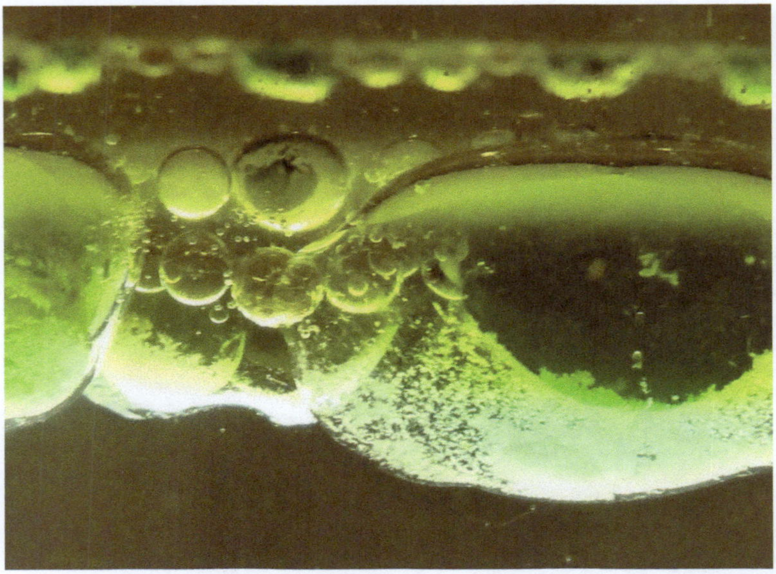

Figure 8.27: Self-organizing modified Bütschli droplets settle at a DEPP/olive oil interface and begin to respond to the introduction of simple salt solutions such as copper II sulphate. The actants are constrained within a narrow space, which reveals their propensity to produce sinusoidal Turing bands. Photograph, Rachel Armstrong, April 2012.

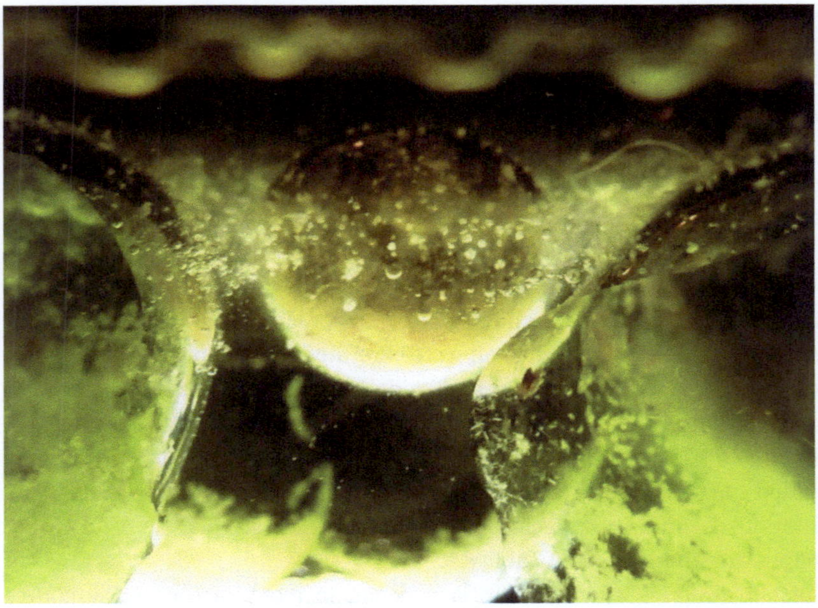

Figure 8.28: Modified Bütschli droplets self-organize within a 2 cm space to produce undulating Turing bands. Photograph, Rachel Armstrong, April 2012.

Figure 8.29: Modified Bütschli droplets self-organize within a 2 cm space to produce undulating Turing bands. Photograph, Rachel Armstrong, April 2012.

Table 8.5: BIO-FICTION installation preparation for a 45,000 ml tank

Chemical	Strength	Amount
Diethyl phthalate	n/a	20,000 ml
Extra virgin olive oil	n/a	20,000 ml
Sodium hydroxide	3 M	200 ml
Iron II chloride	1 M	10 ml
Iron III chloride	2 M	10 ml
Nickel II sulphate	1 M	10 ml
Copper II sulphate	1 M	50 ml
Cobalt II chloride	1 M	10 ml
Calcium chloride	1 M	10 ml

waves as expressions of reaction-diffusion bands. Turing (1952) proposed that such structures could account for patterning in animals, specifically 'dappling'.[51]

8.6 Design Principles

> *... architecture does not exist without a program, and its presence changes with the differing nature of the programs.* (Tschumi, 2012, p.22)

The modern industrial age is characterized by its use of inert materials that are insensitive to their context or any change in their surroundings. Contemporary architecture also seeks dryness and the conditions in which materials are imagined and designed are optimized for anhydrous environments, yet living systems require water to function and thrive. The various explorations of vibrant matter and different species of ELT within the Hylozoic Ground installation established a set of design principles by which materials that were not compatible with natural systems through a shared chemical language could thrive in aqueous conditions. They demonstrated a series of transferable properties that enabled them to be entangled with other non-equilibrium outputs from different systems, such as airflow in the environment or heat from a light source, and to respond to these new couplings through self-modification and adaptation. In an architectural context, vibrant matter and different species of ELT suggest that it may be possible to design architectures that are able to grow or adapt to their environment. Although traditional techniques can achieve this, such as ivy being trained over frameworks to grow arches, or trees being pleached to build fences, biotechnological advances now enable the modification of living processes at such small scales and with such precision that living systems may be considered as a set of technological species. For example, it is possible to introduce jellyfish genes into pets such as mice, fish and cats so they express a protein that glows under ultraviolet light (Chalfie at al, 1994). Matter has never been so strange or creative. Yet, designers are habituated to working with living systems in such a limited set of circumstances, such as gardening and agriculture, that design practice is mostly imagined through applications of inert materials with predictable properties.

Different species of ELT increase the range of possibilities for a more environmentally contextualized, dynamic design practice by bringing materials which share some of the qualities of living things but do not have the status of being truly 'alive' into everyday manufacturing practices and social spaces. Indeed, these kinds of materials

51 Animal markings are explained today in terms of differential expression of genes that are modified by cellular signals and although the process is more complex than Turing reaction-diffusion bands, the principles of organization are still relevant to the pattern-generating process.

are beginning to influence the way cities are imagined and designed. Henk Jonkers at the University of Delft is developing concrete impregnated with bacteria that can survive in alkaline conditions (Jonkers, 2007, pp.195–204) and Elizabeth Demaray paints lichen on to the walls of buildings (Reutgers, 2011), which may help to regulate their temperature; while the engineering company Sustainable Now Technologies is developing an algae bioreactor that houses 1,000 gallons of algae, which can be kept in a garden shed and produce enough biofuel to keep a family car topped up – without needing to visit a fuel station (Sustainable Now Technologies, 2012). Working with vibrant matter and different species of ELT informs a set of principles that could be applied to a whole range of lifelike materials, which may be smart chemistries but they may also be of animal, plant or bacterial origin. These kinds of materials could be used in a range of systems from desalination plants (Bland, 2009) to domestic lighting units (Cha, 2011). Such lifelike systems can be designed from a top-down perspective by modifying an existing system so that it works slightly differently, such as using goats' milk that contains spider protein (National Science Foundation, 2010), while an alternative approach is design from the bottom up and using chemical self-organization to give rise to lifelike phenomena. This method is consistent with Stéphane Leduc's notion of synthetic biology, which he regarded as being an extension of synthetic chemistry (Leduc, 1911, pp.113–121).

Despite 150 years experimenting to produce life in the laboratory, nobody has been able to synthesize life from chemical ingredients (Hanczyc, 2011a). From a technological perspective, being able to work directly with the dynamic processes embodied in living systems is a creative opportunity for designers. Vibrant matter and ELT constitute a non-equilibrium platform in which the variables in dynamic chemical systems can be explored, and require very different approaches to designing a static object that is optimized for equilibrium states. To carefully characterize the system and select the right ingredients it is essential to clearly understand the reason for selecting vibrant matter and the kinds of outcomes that may be anticipated. Although different kinds of ELT present unique sets of challenges in different circumstances, there are general technical considerations to address when working with vibrant matter:

— Born not made (Kelly, 2010): Vibrant matter is a fundamental quality that exists as a function of the primordial laws of the universe and cannot be built into a system or acquired.
— Scale: Vibrant matter works across different scales simultaneously. Materials are more lively and further from equilibrium at the nanoscale owing to the effects of quantum physics. The scale at which design decisions are made influences the choice of materials, tools, contexts and infrastructures to shape the interactions of vibrant matter. Working at the nanoscale is very different to engineering a building (Haldane, 1926; Thompson, 1917, pp.22–77).
— Equilibrium: Vibrant matter exists at far from equilibrium states and resists the decay towards equilibrium (Schrödinger, 1944). This is different to designing with classical materials, which are designed to operate at equilibrium states.

- Metabolism: Vibrant matter evades reaching entropic equilibrium by coupling with other systems to form mutually reinforcing assemblages. The horizontal chemical couplings between assemblages that resist energetic decay may be thought of as 'metabolism'. Principles for designing with metabolisms have not been formalized, although Kauffman's notion of autocatalytic sets where groups of chemistries form closed-loop interactions (Kauffman, 2008, p.55) begin to speak of possible criteria for self-supporting assemblage formation, which might be applied within a design context.
- Simultaneity: Vibrant matter exists both as an object and as a process. It is therefore simultaneously soft (porous) and hard (impermeable) as well as being wet (flow) and dry (static). The intensity between these states and how we perceive them is dependent on time, scale and the nature of the participating assemblages.
- Program: Vibrant matter does not produce straight lines (Hundertwasser, not dated), or exist as expressions of Euclidean geometries. It does not snap to a grid nor respond to an 'undo' function. Instead, vibrant matter exists as creative, constantly forming fields of interactivity that can be combined with one another and cyclically and unevenly bloom and collapse, until they reach entropic equilibrium. These entanglements are capable of acts of radical novelty and may perform completely differently to their original constituents and their interactions may vary spatially and with the passage of time.
- Sustenance: Vibrant matter seeks food and energy sources to resist entropic imperatives and delay reaching equilibrium. The requirements change with time and, as the system develops and grows, so infrastructural concerns are of vital importance for vibrant matter to thrive.
- Mass: Vibrant matter possesses mass and therefore works more slowly than digital computing, which relies on the flow of nearly massless subatomic particles called electrons. However, possessing mass enables vibrant matter to make more interactions simultaneously and it can, therefore, perform parallel computing functions.
- Vectors: Since the chemical functions of vibrant matter are embodied they possess directionality and are ultimately polarized.
- Unpredictability: Complex, embodied materials can surprise designers and may require flexible strategic approaches. Vibrant matter may exert unusual effects that may be observed at the human scale; according to the laws of quantum physics, for example, it may vibrate and not vibrate simultaneously (TED.com, 2011).
- Rhythm: Vibrant matter works possesses a unique periodicity, which depends on its molecular and macroscale interactions. It therefore may perform its operations faster, or slower than natural systems.
- Control: Vibrant matter does not come with a push button and responds to soft (facilitative continual nurturing) rather than hard control systems (energy-

intensive command). It also possesses a chemical 'will' of its own and influences the performance of material systems by acting as a codesigning agency that makes claims within material territories, which challenge past events and resist the limits imposed by other codesigners in the system.

— The Inevitable: Designers need to consider what to do when vibrant matter and different species of ELT reach equilibrium. It is also important to consider what kinds of removal or recycling systems are appropriate for their disposal.

Working with vibrant matter and different species of ELT poses many design challenges so that each system requires special consideration. However, the investment made in using vibrant matter as a design solution enables architects to work directly with the kinds of material transformation that characterize natural systems, so that they can innovate organically. Indeed, ELT suggests that it may be possible to transform our resource-consuming industrial processes into potentially life-giving ecological ones. Further exploration of vibrant matter and its associated technologies and infrastructures in many different contexts may establish parameters for new kinds of design thinking. However, potential applications are more than an academic proposition to make a transition from mechanical to ecological approaches but establish the technological, material and infrastructural conditions that move towards this possibility.

8.7 The importance of Infrastructure

... buildings, and even whole cities, have become infrastructural technologies. (Easterling, 2012)

William Bryant Logan notes that for living systems to dwell on land they had first to develop bodies that 'learned to contain the sea' (Logan, 2007, p.11). The inner seas of organisms are those cellular infrastructures have enabled creatures to develop chemical systems and metabolisms that sustain them in arid conditions. If vibrant architecture is also to thrive, and not only anticipate our physiological challenges but also find ways of working with fluid spatial programs, then buildings may also need to contain the sea. Although water is a significant organizing matrix, other elemental systems such as air, earth and heat may also provide a flow of resources and support for increasingly lifelike material systems (Armstrong, 2012b). The unique role of infrastructures in facilitating new performance standards in vibrant matter and different species of ELT, such as creating new aesthetics and generating diverse material solutions, has been evidenced in the fossil record. For example, rainforest plants blossomed when they were able to solve a water-supply problem that also increased their ability to fix carbon (see Fig. 8.30). This surge of resources enabled non-flowering plants (gymnosperms) to develop into flowering ones (angiosperms) (Field et al, 2011).

Figure 8.30: Water droplets on the narrow, waxy leaves of a pine tree (gymnosperm) after a rainfall in Monte Verita, Switzerland echo the evolutionary transitions of ancestral plants that were able to transport water more effectively, fix carbon and develop sufficient organic complexity to develop flowering reproductive systems. Photograph, Rachel Armstrong, July 2013.

The explorations of vibrant matter and different species of ELT in the Hylozoic Ground installation raise questions about how elemental forces that support lifelike systems may be harnessed in an architectural context. They form an alternative technical system that enables architects to exploit the interstitial condition between the elements that a system is composed of. ELT can process space, events, movement and metabolism and offer a rich, dynamic palette for programmatic and spatial devices. While we may regard the sun, wind, rain and earth as powerful agents that shape our lives in ways that are mostly beyond our control, we have also found ways of manipulating them using infrastructures that convert these unruly forces into a kind of geoengineering-scale[52] form of technology. For example, the construction of canals and irrigation systems have used the power of water to transform the fertility of landscapes and even brought water into our homes (see Fig. 8.31).

52 The Royal Society defines geoengineering as 'the deliberate large-scale manipulation of the planetary environment to counteract anthropogenic climate change' and divides methods into two types: carbon dioxide removal from the atmosphere, and solar radiation management (The Royal Society, 2009).

Figure 8.31: Canal construction was essential to the development of Venice's infrastructure as a means of enabling its communities to populate previously uninhabitable land. Photograph, Rachel Armstrong, August 2012.

While we are not the 'gods' that Stewart Brand suggests,[53] neither are we hapless victims of circumstance. In the Anthropocene, our complex and difficult relationship with the natural world may be strategically managed through the use of technologies that harness elemental systems such as wind, waves and solar energy. Indeed, as we fail to cut emissions, industrial nations are increasingly turning to geoengineering (Hamilton, 2013a) to remedy environmental impacts such as that of China (Hamilton, 2013b). While most geoengineering-scale projects, by definition, have a global-scale impact that stems from top-down centralized strategies, renewable energy systems also draw from the same principles but at the scale of cities. Exactly when a renewable energy solution, such as the massive solar array in Morocco (Hickman, 2011), becomes a geoengineering technology is yet to be established and is indeed a desirable outcome for the solar power industry. Similarly, environmentally responsive vibrant matter and different species of ELT may be regarded as manipulating environmental

53 The 1968 Whole Earth Catalog began with the words 'We are as gods and might as well get good at it'. Yet, in his publication 'Whole Earth Discipline' (Brand, 2009), Brand stresses the urgency to get involved with ecosystems engineering despite likely opposition by the environmental movement (Edge, 2009).

impacts in technologically similar ways to harnessing renewables and reducing carbon dioxide emissions, but take place between the microscale and the human scale. Rather than using centralized forms of technology, their systems are distributed and organized using bottom-up approaches. Through the horizontal coupling of sufficiently expansive assemblages, forged from microscale interactions between dynamic systems and ELTs, environmentally remedial effects may be achieved. While geoengineering practices are defined by their global-scale impacts, the manipulation of elemental infrastructures to shape interactions between assemblages offers a non-equilibrium platform for architects that may help them develop spatial programs that enable transformations in material systems (see Fig. 8.32).

Yet, elemental infrastructures in these systems are not obedient carriers of chemical information, but a context or technology that contributes to structuring systems that interact with environmental conditions, which revises how we may deal with architectural concerns. For example, carbon dioxide may be extracted from the environment using bioprocesses and used to shape the production of microscale crystals in a self-healing system (Jonkers, 2007). Additionally, elemental infrastructures possess multiple properties and therefore offer a range of spatial programs and tactics that can operate within non-equilibrium conditions such as the natural environment. When elemental infrastructures are challenged by complex environmental changes, they can undergo a range of responses that can be witnessed by observing complex events in their matrix, such as collapse, robust adaptation, or

Figure 8.32: Light, carbon dioxide and water provide rich elemental infrastructures in the Venetian lagoon that are transformed into biomass for the rapid seasonal growth of algae blooms. Photograph, Rachel Armstrong, August 2012.

phase shifts in their performance. Examples of these have been demonstrated earlier in the Bütschli system where alterations in the chemistry of an olive oil field are brought about by the action of dynamic droplets, which give rise to phase changes in behaviour and morphology (see Fig. 8.33).

These produce local changes that induce an as yet uncharacterized, chemical feedback system, which prompts radical reorganization of droplet behaviour. The points of transition are known as 'tipping points' and are provoked by their contingency so, although they may be anticipated, they cannot be accurately predicted. It is challenging to incorporate these kinds of complex behaviours into conventional architectural design, since vibrant matter and different species of ELT not only behave according to the laws of complex systems but also operate at

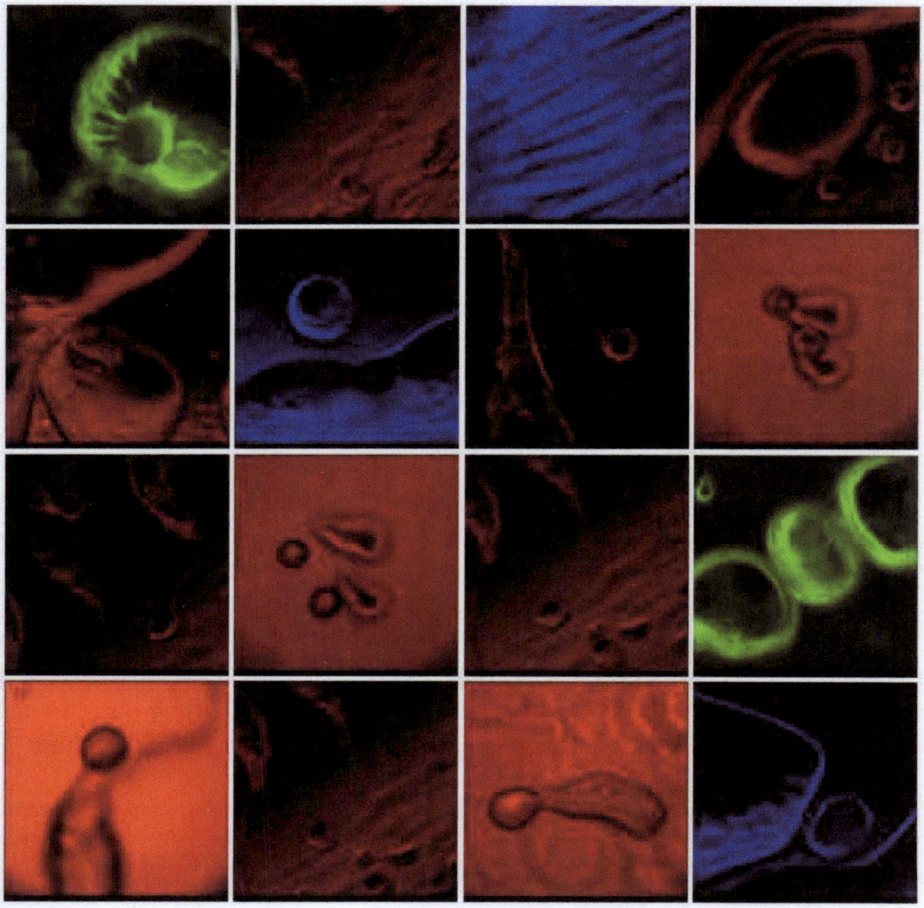

Figure 8.33: Bütschli droplets produce a variety of different outcomes, particularly when they reach a tipping point. Micrographs and collage, magnification 4×, Rachel Armstrong, February 2009.

multiple scales of order, and the processes that give rise to these transitions have not yet been resolved. Yet the kinds of strategies needed to work with ELT are familiar ones, as they deal with complex, dynamic environments. Gardening and agriculture already enable us to work with complexity and deal with probability. In horticultural practices, agents operate within limits (e.g. crops cannot be guaranteed to produce a certain yield) which are defined by the properties of the agents (e.g. plant species) and their elemental infrastructures (e.g. soil, weather). However, these restraints provoke creativity within the system rather than supress it (e.g. plants may grow shorter, or bloom earlier during drought). It is critical to understand where sites of abundance exist within elemental landscapes as 'life' may thrive in difficult circumstances by finding richness in their surroundings as in the case of troglodytes, hydrothermal vent ecologies and extremophile bacteria. Based on the Bütschli experiments, this quality also appears to be characteristic of lifelike systems where subtle changes in the composition of environment and infrastructure shape their performance. Identifying these subtle cues could provide the opportunity to develop spatial programs and tactics that enable architects to design systems that can provoke or transform events within systems, as well as maintain them. Yet, complex systems are largely conservative despite their revolutionary potential. Philip Ball describes the ordering systems that conserve the behaviour of complex systems, which stem from their elemental infrastructures or Nature's tapestry. He describes these forces in terms of 'shapes' (the subdivision of space) (Ball, 2009c), 'flow' (movement) (Ball, 2009b) and 'branches' (connectivity) (Ball, 2009a), which establish the general behaviour of the systems and may be disrupted when conditions are changed (see Fig. 8.34).

For example, saturated copper II sulphate salt solution will form diamond-shaped crystals under cool, clean conditions (Searle, 2008). However, if the system is encouraged to make new material connections, such as introducing dust into the crystallizing solution, then behaviour occurs that exists within a spectrum of possibility (scientific experiments indicate that the crystal lattice will be disordered under a variety of conditions), yet it is not possible to exactly predict all the variations (Giulietti et al, 1996). So, when a complex system is perturbed in different ways, its apparent 'simple' logic shatters and new patterns burst forth, which sometimes surprise us. Perhaps, then, rather than establishing formal rule sets to govern generalizations in performance, non-linear systems may perhaps be most effectively titrated by adjusting their infrastructures. Indeed, Keller Easterling observes the importance of infrastructure and how this shapes our experiences of architecture:

> *We do not build cities by accumulating singular masterpiece buildings. The constant flow of spatial projects and urban formulas is more infrastructural. Architecture is making the occasional stone in the water. The world is making the water.* (Easterling, 2012)

If the world is making water, then elemental infrastructures quite literally provide a means through which the traditional boundary between landscape and building may

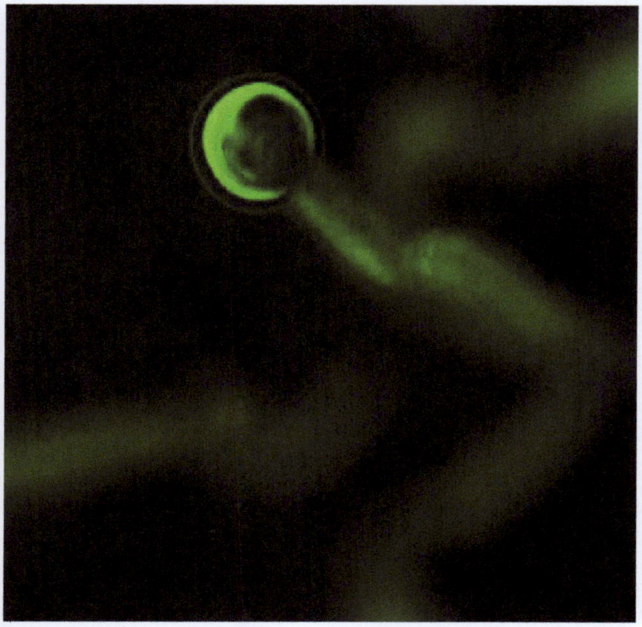

Figure 8.34: Osmotic skin produced by a Bütschli droplet exhibits branching structures that are primordial homologues of 'nature's tapestry' (Ball, 2009a). Micrograph, magnification 4×, Rachel Armstrong, February 2009.

be eroded and shaped by new spatial programs. Elemental infrastructures may serve as a transport medium for substances that permeate the barrier fabric of buildings and open up the possibility of designing in new and surprising places. Rather than buildings being cleaved from their surroundings by brute matter, the spatial programs that shape the non-equilibrium properties of vibrant materials through morphological computing literally bring living processes into building fabrics so they breathe, feel, grow and change with time (see Fig. 8.35).

To enable vibrant matter to thrive in modern buildings, elemental infrastructures will need to be incorporated into their mechanical shells and façades. This is likely to be challenging, as vibrant matter behaves non-linearly and undergoes spontaneous phase transitions (Ulissi, Strano and Braatz, 2013). Yet, the architectural challenge at hand is not for a designer to become an expert physicist, engineer or materials scientist but to set new goals when developing design tactics and spatial programs. Designers may need to consider the sequences of spaces and sequences of events between the participating actants, which may become totally interdependent on each other and fully condition each other's existence. While some interactions may be mutually reinforcing, others may result in conflict when spaces contradict each other's internal logic (Tschumi, 2012, pp.62–63).

Figure 8.35: Reflection of a building in a Venetian canal graphically displays the multiple, complex overlapping spatial programmes of a site, which living systems may respond to. Photograph, Rachel Armstrong, August 2012.

Such possibilities require an inventive use of matter so materials no longer obstruct events but become porous to them. Hundertwasser proposes that buildings are made of windows rather than walls. His 'Forest Spiral' (2000) features an uneven grid of more than a thousand windows, none of which are exactly the same (Dannies, not dated:b). This enables 'porosity' through self-expression, as well as the permeability of space, where a person can 'lean out of his window and scrape off the masonry within arm's reach ... [or] ... take a long brush and paint everything outside within arm's reach' (Hundertwasser, not dated). Infrastructures that facilitate the creative passage of elements, like a brush delivering continual paint strokes of lively matter, may similarly give rise to architectural events that are associated with a material form of creativity and self-expression. For example, spongy material offers a substrate for mineralization as inorganic salts pass through its body, as in Mother Shipton's cave (Mother Shipton, not dated), where groundwater filtering through limestone-rich bedrock becomes highly saturated with minerals. As the minerals come into contact with dissolved carbon dioxide, they precipitate and produce solid matter within the substance of soft, porous objects, such as John Wayne's hat and teddy bears (see Fig. 8.36).

These materials not only fix carbon into a solid form and soften the drinking water but also offer a form of construction through an accretion process, which may be translated to architectural practice. Yet the opportunities to manipulate

Figure 8.36: Soft objects suspended in the heavily mineralized waters of Mother Shipton's cave in Yorkshire produce a limestone-like crust and become petrified within three to four weeks. Photographs and collage, Rachel Armstrong, July 2012.

elemental infrastructures and a flow of materials through a system in naturalistic ways are limited by the site conditions, infrastructures and technologies that feed and orchestrate these processes. Indeed, if an architectural design practice that incorporates vibrant matter is to be realized, then combined computing and 'printing' platforms that integrate chemical, biological and mechanical systems are a fundamental requirement (University of Southampton, not dated; WETFab, 2011; Armstrong, 2012g; Adams, 2012; Villar, Graham and Bayley, 2013; TED.com, 2013b).

8.8 Infrastructure Experiments

Design-led experiments were conducted to further explore the importance of infrastructure in vibrant matter by applying the Hylozoic Ground chemistries in contents where the flow of material could be facilitated beyond the closed constraints of glassware with restricted access to environmental actants. Potentially, developing a system where semi-permeable materials with leaky, open interfaces, such as gels, paper or foams, could be used, would keep the supply of nutrients flowing. Increasing the porosity of the Hylozoic Ground installation could potentially construct programs for agile architectures that could respond and adapt to changing conditions through processes such as accretion, proliferation, reabsorption and remodelling. Although, for financial and practical reasons, such systems could not be incorporated into the installation in Venice, several experiments were conducted at the FLinT laboratory in Denmark, using self-organizing chemistries. The reaction time of the chemical systems was modified by using porous materials such as gels and tissue paper, which were supported on rigid scaffolding like wire and wood. The permeable materials were selected so that they would change the way that non-equilibrium chemistries interacted with their surroundings so that it might be possible to establish how their dynamic behaviour could be sustained, shaped and manipulated in a design context.

Two preparations were distributed using hand-held syringes over a scaffold framework:

- Liesegang scaffolding preparation: A framework to support an alkali-activated gel was fashioned and reactive salt solutions added to produce bands of precipitating crystals.
- Traube scaffolding preparation: Modified Traube cell chemistry where 1 M copper II sulphate solution was added to 0.1 M potassium hexacyanoferrate in 5% agar.

8.8.1 Liesegang Scaffolding Preparation

A frame was fashioned from wooden sticks with cotton tips, which was fastened using cotton thread to provide a makeshift porous framework to accommodate the gel and solutions.

An agar gel base impregnated with alkali was prepared as a medium for the Liesegang ring reaction by adding 10 ml of a 1 M solution of ammonium hydroxide to 50 ml of a 5% agar gel solution. The wooden–cotton framework was then saturated with the gel using a hand-held syringe. Using a hand-held pipette, solutions of iron (II and III) and copper II salts were carefully dropped on to the agar and the reaction observed over the course of an hour (see Fig. 8.37). The preparation is summarized in Table 8.6.

The makeshift framework attenuated the movement of the agar medium, although insufficiently long enough for the structure to become saturated with the reactive base. Addition of the iron (II and III) and copper II salts therefore produced bright patterns in the gel, although mostly on the base of the structure.

8.8.2 Traube Scaffolding Preparation

Moritz Traube first described the production of 'artificial plant' cells in 1867 (Traube, 1867), which were formed by inorganic substances. A Traube cell is an artificial, inorganic model of a cell that is produced by adding a violet-blue, typically diamond-

Table 8.6: Liesegang scaffolding preparation

Chemistry	Strength	Amount (total)
Agar gel	5%	50 ml
Ammonium hydroxide	1 M	10 ml added to agar
Iron II chloride	1 M	2 ml
Iron III chloride	2 M	2 ml
Copper II sulphate	1 M	4 ml

Table 8.7: Traube scaffolding preparation

Chemistry	Strength	Amount (total)
Agar gel	10%	50 ml
Copper II sulphate	1 M	40 ml
Potassium hexacyanoferrate	0.1 M	10 ml

Figure 8.37: Makeshift, porous wood and cotton scaffoldings were constructed and coated with alkaline agar base. Iron (II and III) and copper II salt solutions were dropped on to the gel infrastructure by hand using a disposable pipette. Reactive, brightly coloured patterns within the gel over the course of an hour at the base of the structure, although the distinctive Liesegang bands were not observed in this preparation. Photographs and collage, Rachel Armstrong, February 2010.

shaped crystal of copper II sulphate into a weak (0.08–0.1 M) pale yellow solution of potassium hexacyanoferrate. The ingredients are summarized in Table 8.7.

As the crystal dissolves into the surrounding solution, it produces a brown, semi-permeable membrane of copper hexacyanoferrate. This allows water to enter but not leave the vicinity of the dissolving crystal. The osmotic pressure inside the crystal rapidly builds up and creates a force that ruptures the copper hexacyanoferrate membrane. As the membrane splits, the copper sulphate and potassium hexacyanoferrate solutions mix, forming a new membrane as they come into contact with each other. More water can now enter the repaired interface and the structure swells and extends, until the osmotic force builds up and ruptures it again, transforming the geometric, blue crystal into a sprawling, membranous, seaweed-like mass as shown in Fig. 8.38. This membrane growth and repair process is iteratively repeated and can be seen with the naked eye as jerky growth spurts of the membrane, which can be observed in Movie 8.1 and continue to occur at the microscale (Fig. 8.39) until all the copper ions have been depleted. Typically, a 0.5 cm diameter copper II sulphate crystal may grow up to 4–8 cm within 30 min, after adding it to a field of potassium hexacyanoferrate,

The Traube cell preparation was modified to create conditions in which the osmotic growth phase could be delayed by attenuating the movement of water molecules and copper ions, by adding water-seeking biopolymers. Potassium hexacyanoferrate was added to the agar to make up a 10% solution to provide additional structural integrity for the Traube membrane. The biopolymer would also confer it with an extended osmotic physiology where water movement would take place more slowly. These combined effects were thought to amplify the growth of the chemical cell, as well as enable it to exhibit water-seeking, hydrophilic tendencies. The agar impregnated with potassium hexacyanoferrate was applied using a hand-held syringe on to wire scaffolding and was titrated to produce brown, membranous structures by the addition of 1 M copper II sulphate solution. As the two substances mixed, fibres of gel-expanded membranes self-formed and with care, they could be extended to up to 60 cm under the force of gravity, which can be seen in Fig. 8.40. Gradually, the elongating structures began to dry out owing to evaporation of the water over the large surface area-to-volume ratio of the expanding structures, which made them fragile and prone to fracture.

Simple tests were conducted to further alter the performance of the modified Traube cell preparation by adding cotton thread and fibrous tissue paper to assist in weight bearing as the growing membranes elongated. A wire frame was constructed from materials available in the laboratory and the solutions were added by hand to the framework. The most interesting results were produced when the gel matrix was laid down first, as it attenuated the passage of solutions and changed the temporal and spatial programs of the system, as in the Liesegang ring scaffold experiments. The results of this test can be seen in Fig. 8.41.

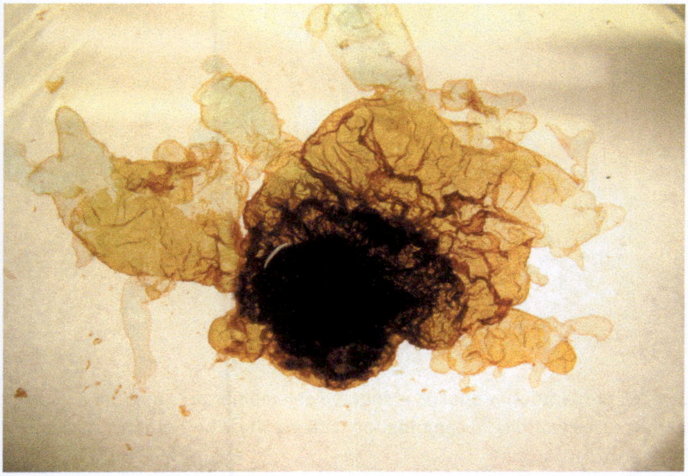

Figure 8.38: ATraube cell membrane is produced when a crystal of blue copper II sulphate is added to a weak solution of potassium hexacyanoferrate. Photograph, courtesy Philip Beesley Architect, February 2010.

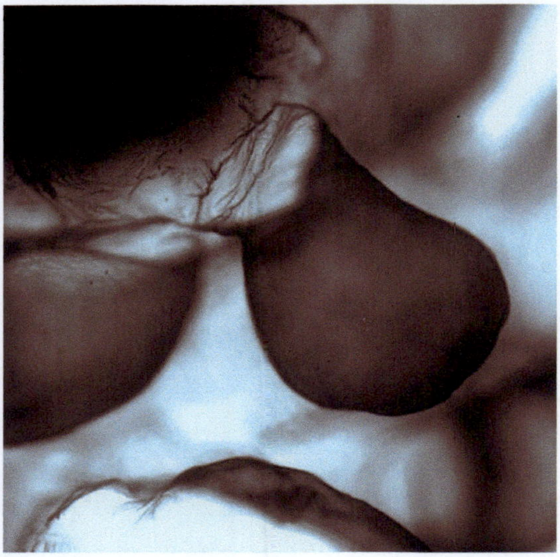

Figure 8.39: Osmotic forces rupture the Traube cell membranes that form around the copper II sulphate crystal and immediately heal as the salt solutions come into contact with each other. Micrograph, magnification 40×, Rachel Armstrong, February 2009.

Figure 8.40: Modified Traube cell preparation produces elongated cell membranes that are extended by self-organizing chemical processes working in combination with gravity. Photograph, courtesy Philip Beesley Architect, February 2010.

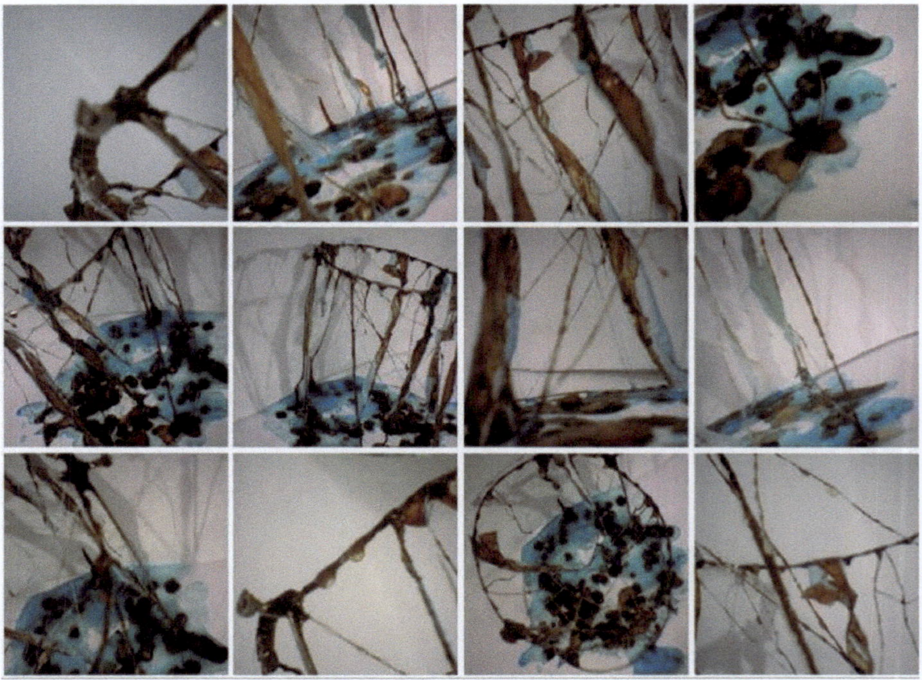

Figure 8.41: Wire and cotton scaffolding draped with porous paper supports a potassium hexacyanoferrate-impregnated gel that responds to copper II sulphate droplets by producing a profuse brown membrane. Photographs and collage, Rachel Armstrong, February 2010.

8.8.3 Scaffold Experiment Observations

These experiments, which were conducted using hand-held distribution systems, reactive chemistries and porous materials (wood, cotton, fibrous tissue paper and agar gel) revealed a number of practical design challenges, namely:

— Lack of readily available infrastructures and tools to assist in dextrously manipulating the outputs of dynamic chemical systems.
— Poor structural integrity of soft, wet systems that required the used of mixed materials and scaffolding for support.
— Inefficient use of materials, since at least one reagent is required in abundance for the system to proliferate.
— Limited precision with available tools.
— Disruption of event sequences through environmental changes, e.g. drying out of elongating structures. However, these experiments were exploratory and suggested modifications could be made to improve the responsiveness and performance of the participating chemistries.
— More rigid biological scaffolding materials could be applied, such as Zbigniew Oksiuta's gelatine 'biospheres' (Oksiuta, 2006). This may further attenuate the passage of solutions through the structure and increase structural integrity. For example, it may be possible to run the experiments for longer and observe the formation of Liesegang bands, which usually begin to appear within an hour, and may continue for many years under the right conditions.
— Humidifying the environment may prevent gels from drying out.
— Development of more specific, robust infrastructures (macrofluidics, porous clays) and computer-aided tools, which may diversify possible design methods.

Complex heterogeneous infrastructures, like soils, may best optimize the performance of dynamic chemistries by engaging with material, parallel programs to produce integrated, heterogeneous fabrics that can simultaneously deal with rigid, semi-rigid, soft and solid materials and systems (Armstrong, 2013a). As Tschumi notes, 'not all architecture is linear, nor is it all made of spatial additions, of detachable parts and clearly defined entities' (Tschumi, 2013, p.63). A fluid system shaped by a non-linear matter and technological physiology creates the conditions for probabilistic spaces that may form and re-form through endless acts of creation and subtraction, such as Darwin's wormstones.[54] Consequently, buildings may be able to respond to the activities of human and non-human inhabitants through growth, extrusions,

54 In his essay on vegetable matter, Charles Darwin noted that the downward movement of rocks was produced by more than the effects of gravity but by the concerted subtractive and additive processes mediated by earthworms who removed soil from under the boulders and transferred it to the surface as worm casts (Darwin, 2007).

reabsorption, fixing carbon or recycling water. These ceaseless material exchanges may ultimately empower inhabitants to sculpt their surroundings as acts of self-expression and according to their needs.

8.9 Hylozoic Ground and the Architectural Transferability of ELT-informed Design Principles

> *... an aesthetic shift must be appreciated: where mechanical joints squeak, molecular chains hum.* (Khan, 2011, p.59)

The Hylozoic Ground chemistries embody an emerging platform whose material and technological systems are intimately coupled, since matter can respond to and act at the molecular scale. They operate as sensitive, responsive membranes that oscillate between cybernetic framework, environment and visitors. These chemistries could simultaneously produce microstructures and formal traces, which documented events within the installation and operated across many scales with a high degree of specificity and environmental sensitivity.

The principles explored in designing the Hylozoic Ground chemistries could be further developed beyond the gallery setting through the production of simple scaffolding systems that opened up the interfaces available for chemical and environmental exchanges. This flexibility and adaptability in the platform suggests that it may be possible to develop more dynamic, ecologically contextualized forms of architectural design practice. Such approaches may work in parallel at multiple scales of operation by deploying unconventional means of constructing space, such as material depositions and metabolisms. Moreover, Hylozoic Ground chemistries are more than materials for the production of an installation but operate as codesigners of systems. Their outputs may be engaged and shaped through unconventional computing techniques that engage with spatial programs and notions of soft control.

As such, the Hylozoic Ground chemistries are not deterministic agents, but exhibit a degree of unpredictability that can be defined within site- and context-specific limits. Although playful and ephemeral, the chemical experiments embodied within Hylozoic Ground were not developed in isolation as purely aesthetic practice, but engaged with essential contemporary architectural debates – embodying a way of working, rather than a style (Gage, 2012). Importantly, they decentralize the production of architecture by proposing new ways of making and prototyping (Armstrong, 2012g; Adams, 2012; Graham and Bayley, 2013; TED.com, 2013b), where materials themselves contribute directly to the production process in concert with digital manufacturing platforms – being codesigners, and not simply passive substrates on which to be acted. Yet, vibrant matter possesses different qualities to classical architectural substrates and is optimized for non-equilibrium conditions, so vibrant bodies are soft, fluid and malleable rather than rigid and tensile. Indeed, Nicholas Negroponte proposed that radical responsive architectures[55] needed to be soft (to deform and transform) and cyclic (undergoing continual cycles of construction and deconstruction over their lifetimes) (Negroponte, 1975). Yet, the Bütschli system – and by implication other forms of vibrant matter – are more than soft bodies, but 'wet' technologies that carry their own operational infrastructures (or sea) (Logan, 2007, p.11). Indeed, the Bütschli system can orchestrate detailed interactions within complex, overlapping material systems and fluid programs (see Fig. 8.42), which not only embody material responsiveness but also constitute a material (rather than digital) sensory system.

A new kind of architectural 'body' arises from the interactions between dynamic infrastructures and technologies to facilitate new associations between space and the events within it, by harnessing the lifelike qualities of non-linear, material systems. Such corpulence is not a digital puppet awaiting the translation of commands through multiple software layers, but can directly sense and respond to environmental changes in real time. Such dynamism and responsiveness requires the construction of new kinds of spatial and temporal programs to shape the potency – rather than performance – of vibrant material systems.

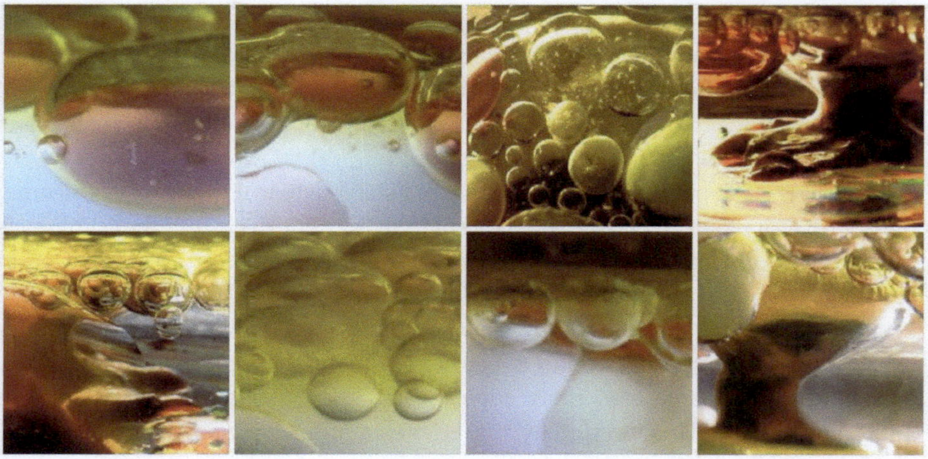

Figure 8.42: When a modified preparation of Bütschli droplets is observed within a constrained space and enabled through a fluid infrastructure, the production of Turing bands can be observed. Photographs and collage, Rachel Armstrong, February 2010.

55 Negroponte's notion of 'radical', in this case, is an architecture that can respond to disruptions which are greater than the everyday disturbances in its native system.

In the most basic sense of the word, the Hylozoic Ground chemistries begin to indicate that materials, which might conventionally be regarded as inert, may actually be 'living'.

8.10 Ontological and Epistemological Questions Raised by the Hylozoic Ground Chemistries

> *And all this material is being put through a reduction process that brings it down to the essential, condensed, miniaturised minimum, a process whose limits have yet to be established; just as all existing and possible images are being filed in minute spools of microfilm, while microscopic bobbins of magnetic tape hold all sounds that have ever been and ever can be recorded. What we are planning to build is a centralised archive of humankind, and we are attempting to store it in the smallest possible space, along the lines of the individual memories in our brains.* (Calvino, 2009, p.366)

The ontological and epistemological issues raised by Hylozoic Ground chemistries are shared by the Bütschli droplet system since their outputs are ontologically non-deterministic and therefore exist within a spectrum of degrees of freedom, which poses epistemological challenges for classical pedagogies. This section takes an applied view of using non-classical frameworks to observe the potency of dynamic chemistries, with the intention of providing architects with alternative portfolios of design possibilities that do not constrain the expectations of dynamic chemistries within the lexicon of miniature 'wet' machines.

However, such perspectives pose significant challenges to design and engineering portfolios, as it requires an alternative way of thinking about 'control' where influence in producing outcomes is shared between participating actants. This viewpoint challenges our classical expectations of the material and even natural world, whose identities have been constrained by object definitions within hierarchical orders or existence. For example, a human is considered more 'evolved' than a bacterium, which in turn is considered more sophisticated than a crystal. Indeed, modern anthropocentrism and biocentricity[56] is problematic for the appreciation of lively matter that is not fully 'alive' as it is not recognized by a formal classification system.

In the current age of scientific 'omes'[57] perhaps there is a need to bring forth a 'ge-ome'[58] so that it may be possible to apprehend the character of lively inorganic

56 By 'modern biocentricity' I am referring to the western late 20th century focus on the 'gene' in neo-Darwinism, as the primary organizing agent of the natural world – a viewpoint that is not shared by Linnæus (1735), Vernadsky (2007) and Gould (1994) where the role of environment is emphasised.

57 This expression refers to refers to the capacities of chemical systems to act in the course of biological development, such as the genome (genes) and proteome (proteins).

58 Geo, as in 'geological', or possibly mineral-ome.

substances and how they give rise to lifelike structures and behaviours. Yet, 'omes' may be too deterministic and restrictive a concept for dealing with the flexibility of lively systems. For example, there is a significant linguistic challenge in characterizing genes using terms that do not provide an intuitive feel for the character, function or spatiality of the systems or their domains. Although the Human Genome (HUGO) Gene Nomenclature Committee (HGNC) proposes a system for naming unique gene symbols and names – with links to genomic, proteomic and phenotypic information, as well as dedicated gene family pages – the language is aimed at specialists (HGNC, not dated).

For example, the gene for carnitine O-acetyltransferase, situated at location 9q34.1, is represented by its approved symbol 'CRAT'. This does not have any metaphorical connection with other words in the English language and therefore does not convey the character of the substance encoded by the gene, which is an enzyme involved in amino acid (protein) synthesis. Nor does CRAT indicate its 'gene ontology' – 'has no children' (AMiGO, not dated).[59] In dealing with complex, emergent and probabilistic agents it is essential to communicate their potential in meaningful ways. Indeed, to further complexify matters, Eva Jablonka and Marion Lamb propose that evolution is characterized by four dimensions (genetic, epigenetic, behavioural and symbolic) (Jablonka and Lamb, 2006, p.1), which introduces further complexity into the relationships between lively structures, their expression and their trajectories. These multidimensional relationships also challenge classical models of a centralized hierarchy of command proposed by neo-Darwinism (Dawkins, 1979). Kauffman embraces the incalculable nature of matter (Kauffman, 2008, p.126), referring to the potent, emergent space into which substances may bloom by virtue of their own agency as the 'adjacent possible' (Kauffman, 2008, p.100). Perhaps, then, the ontological and epistemological frameworks that deal with probability need not to be closed down with rigid definitions, but remain open so they may couple horizontally with other concepts (see Fig. 8.43).

Yet, without a formal language it becomes impossible to convey the nature of vibrant matter. New forms of classifying probabilistic events will help designers navigate the opportunities implicit in convergent, advanced combined technological platforms in which vibrant matter plays a part. These notations may indicate synergies and connections, rather than differences, and designers may iteratively work between classical and non-classical systems to develop tactics that enable them to navigate

59 'Having no children' in genetic terms implies that there are no other genes that are derived from the particular sequence of interest. 'Children', used metaphorically in this case, is informative and conveys relationships better than the name actually ascribed to the gene sequence itself. Additionally, gene ontologies come with graphical representations of the relationships of the gene of interest to biological process, cellular component and molecular function. Although inventive, these ontologies are difficult for non-experts to understand and navigate the concepts.

Figure 8.43: Photograph, Rachel Armstrong. August 2012. Since dynamic chemistries are contingent on their environment and are continually processing information, they are challenging to classify on account of their temporal nature and oceanic ontology.

previously unchartered territories. They may, for example, infer synthetic periodic tables of materials with repeating elements or characters, such as Luigi Serafini's *Codex Serafinianus*, which constructs a whole new set of material taxonomies in a mysterious language reminiscent of alchemical codes (Davis, 1996–2013). Indeed, the practice of alchemy may offer tactics for developing new forms of classification. For example, Spiller describes 'protocells' as 'surrealist technologies of softness, growth, swarm and scaffold' (Spiller, 2011, p.65) whose various stages and transformations produce 'taxonomies of form' (Spiller, 2011, p.65). He proposes that such groupings may be considered as a form of architectural alchemy where one substance is transmuted into another, and becomes an invitation to 'the architectural observer to read, explore and use' (Spiller, 2011).

8.11 Summary

The Hylozoic Ground installation collaboration offered a rich and creative engagement with the material possibilities of vibrant matter that went beyond the formalist constraints of mechanical paradigms and bucolic expectations of naturalism. Working with a range of ELT species, it was possible to explore model systems that could operate differentially within a definable range of variables using morphological computing techniques. Using a range of dynamic chemistries, it was experimentally demonstrated that vibrant matter:

- Possesses agency
- Is programmable using morphological computing techniques
- Exhibits co-authorship in our environments
- Coherently operates across many scales that include the architectural realm

These observations go beyond what was possible to demonstrate in a laboratory setting by providing the additional challenges of context and scale to the various model systems used. These dynamic chemistries forcibly inserted the passage of time as a source of creativity into these systems by working with the idea of material bodies at non-equilibrium states (Prigogine, 1997). When sufficiently supported by open and resource-rich infrastructures, ELT provides new species of materials, tools and approaches, in which vibrant matter may codesign outcomes. The Carbon Eater Flasks, Incubator Flasks, Liesegang ring plates and Hygroscopic Islands exemplify these creative relationships and possibilities.

The multiple activities and relationships embodied in the Hylozoic Ground chemistries created a context within a site that, owing to their innate parallel programming abilities, could provoke multiple juxtapositions, which gave rise to a series of interactions and events, which, in themselves, were contingent on their circumstances and context. Changing the chemical programs within these systems – for example, when carbon dioxide is made available by dissolving into solutions – resulted in the transformation of processes and materials, which were also environmentally responsive. The multiple overlapping of languages, programs and events produced complex fields of interaction, which continually fed back to each other through countless parallel interactions, such as in the continual association and disassociations of water with hygroscopic materials.

These networks could reach tipping points that produced significant structural events, or architectures, which could also be (re)assimilated back into the system; for example, through the production of stripy precipitates in Liesegang ring plates as Turing bands or osmotic structures. Within these probabilistic fields of multi-scalar activity, engaging lively matter as the codesigners of programmatic concepts may shape the production of events. The outputs of these systems reveal new understandings of events through human and non-human agencies. They also question the content of experience, examine the structure of space and explore our intimate relationship with the parallel ecologies in which we are immersed.

Vibrant matter offers a new set of principles for designers that may be transferable to other architectural contexts. However, these exciting investigations need to be characterized in more detail through further experiments, models, prototypes and installations. Yet, the Hylozoic Ground chemistries establish a benchmark that may help to develop further applications of non-linear material programs in a design context, and strongly indicate the possibility of convergent tools and techniques where a range of technological types could potentially be meaningfully entangled in the material realm (WETFab, 2011; Armstrong, 2012g; Adams, 2012; Villar, Graham and Bayley, 2013; TED.com, 2013b). The agency of these assemblages may be shaped by architectural tactics at the human scale to produce post-natural landscapes (Armstrong, 2011a) using soft control systems, which enable architects to influence and suggest effects through creating fertile fabrics that perform probabilistically, rather than command them using deterministic styles and hard control mechanisms. Indeed, our full participation is needed in elevating the status of the material world (Bennett, 2010, p.13) if we are to develop a portfolio of approaches that may establish a different kind of relationship between human development and many different kinds of ecologies.

Kevin Kelly notes that even when it is possible to see the perfect outlines of how an emerging technology such as ELT may be realized, we tend to overestimate how soon it will become available to practitioners. He attributes this delay to the need for other invisible ecologies of co-technologies to exist concurrently. In other words, design possibilities with vibrant matter may only be realized when they are supported by the appropriate infrastructures. In the case of Hylozoic Ground, rich water infrastructures were necessary for the chemistries to blossom, in a similar way to how flowering plants suddenly evolved when they solved a water infrastructure challenge (Field et al, 2011). However, when these technological ecologies finally converge, advances seem to suddenly appear in our lives as if from nowhere, and are received with much surprise and applause for the 'unexpected' development (Kelly, 2010). Observing dynamic chemistries through non-classical frameworks opens up the frontier of transformation as being a tangible design strategy where architects may intuitively work with the probabilities within a site without having to be able to fully predict all the possible outcomes. It also requires designers to think differently about how participating agencies in a site may be orchestrated, rather than controlled and to be surprised by unanticipated events provoked by their codesigners. This kind of approach is most appropriate when sites possess many complex, dynamic variables and are likely to face unpredictable challenges, such as intercoastal regions that oscillate between being dried out and flooded.

In the following chapters I will explore how these materials, infrastructures and technologies may be applied in project work, in specific sites, where the application of such interventions may be ecologically and architecturally pertinent.

9 'Biolime: Mock Rock'

9.1 Overview

This speculative narrative was composed as a teaching tool for Danish high school students (University of Southern Denmark, 2012) to help them reflect on how a community might respond to the presence of lifelike materials and ELT for the ISSP (Initiative for Science, Society and Policy). It was also published in the online design magazine *Organs Everywhere* (Armstrong, 2012f).

9.2 Biolime: Mock Rock

For those that had never been to the city of Hardwich, it was impossible to tell whether the houses in the Mossville suburb had actually come 'alive' or not, for whenever sunlight stroked the mineral-clad buildings their facades seemed to quiver with an energized, metabolic glow. Early-morning joggers took advantage of the freshening air caused by the solar activation of the limestone. Dirt stains faded and curious cellular plant life toyed at the edges of the slowly creeping rock, as if they were deciding whether they had encountered a friend or foe.

The Biolime surface coating on the outside of the Mossville houses had been deemed a 'friend' but the new technology had not been accepted without controversy. Indeed, if it wasn't for the irrefutable fact that climate change was happening even faster than all forecasts had predicted, resulting in increasingly turbulent weather patterns that spoiled summers and brought bitter winters, then Biolime would probably have remained a curiosity of chemical behaviour that was only of interest to an elite group of scientists working in the new field of the origins of life sciences. Unusually, these researchers had collaborated with a group of architects who were interested in harnessing the carbon-fixing qualities of living systems to make environmentally positive, sustainable materials. The collaborators had produced a simple oil in water droplet emulsion likened to 'salad dressing' that used carbon dioxide from the atmosphere to drive a chemical process, which formed a rock-like salt called 'carbonate', commonly known as 'limestone'.

This work had generally been regarded as a fringe research activity, though some years later the renewed interest in finding ways of dealing with the runaway carbon count prompted journalistic investigation into the technology. It led to a prime-time news feature entitled 'Mock Rock'. So, in the wake of endless speculation in spin-off magazine articles such as 'Mock Rock around the Block' and 'Mock Rock da House', this sudden and rather unexpected global coverage of the research prompted the researchers to patent their technology as Biolime.

Yet, despite the growing interest in the system and the increased recognition that this exterior 'paint' for homes could actually make a real contribution to the health of urban communities, Biolime continued to be regarded with suspicion. A number of outspoken critics conjectured that, even under the current circumstances, the Biolime technology really belonged only in a laboratory setting and that it had no place in the natural world.

Part of the problem was that the technology had been implemented at the national level in the wake of a series of fractious G20 summit meetings. After a series of high-profile public protests leading to widespread outbursts of civil unrest and political humiliation regarding the ineffectiveness of the G20, the politicians were finally forced to take action. Programmes that relied on the good will and environmental responsibility of individuals were simply not making sufficient impact on the issue of soaring greenhouse gases that were responsible for climate change. The embarrassment caused by media-fuelled popular opinion finally spurred the major economic powers to undertake draconian measures. There was unanimous agreement amongst the representatives of the world's nations that it was time to find a creative solution. Of course, they suspected their patriarchal approach to planetary welfare would be resisted, but agreed that it was time the public faced the facts. The old methods and various forms of public bribery were just not good enough and a completely new approach was necessary. Political attention quickly turned to the 'Mock Rock' technology since it had recently become a popular chat show subject, which frequently deteriorated into insulting exchanges about the 'wilful uglification' of modern buildings. Following a number of rather cursory national polls conducted to investigate public attitudes towards the technology, the G20 countries endorsed Biolime as the most immediate and effective way to combat climate change.

The British government decided to pilot Biolime-based solutions in urban areas to demonstrate the benefits of the new technology in the form of community-based public schemes. The Mossville area of the city of Hardwich seemed a prime location for further government-initiated improvement as it had already responded to national sustainability initiatives through exemplary practice. Mossville boasted a well-supported permaculture project that had opened up garden spaces for the public cultivation of fruit trees, which allowed people to exchange fruit as seasonal currency and had adopted a staunch stance against plastic packaging. Shopkeepers either refused goods with wrappers from suppliers, or removed and recycled them at the point of purchase. Mr Grant Soames, who ran a hypermarket chain in several places around Hardwich, further capitalized on this practice when he discovered that there was a thriving market in recycled packaging materials. His stores not only became a focal point for community de-packaging activities, but also served as meeting points for teenagers who used the worthy excuse of recycling duties as a way of escaping their homework responsibilities.

Less than a month before the project began, the local councillors received official electronic notification of the Biolime initiative and organized a public meeting to salvage some semblance of a democratic process. Priding themselves as a progressive body, they were keen to head off any 'misconceptions' about the centrally driven imperatives. The response to the public notices was overwhelming, and a swell of banners that read 'Block Mock Rock' or 'Rock Mocking Us' soared above the Mossville crowd that had turned out for the meeting.

Councillor Arthur James, the youngest and most ambitious of the local politicians, brushed down the front of his suit in preparation for conflict with those that had elected him and asked the staff to open the doors. He'd agreed to lead the public meeting, partly because the senior committee members admitted they didn't know what a 'metabolism' was, and partly because he actually believed in the value of the project. Although Arthur had initially been as sceptical as anyone about the hype surrounding the Biolime technology, he had become increasingly charmed by its simplicity and effectiveness. Councillor James reminded the assembly of stony-faced people that limestone occurred naturally in underground caves and formed the scale deposits that resulted from every day processes, like boiling water in a kettle. He then urged the congregation to consider this as a way to build and maintain buildings naturally, albeit unconventionally. He also asked what it might mean to the community if the Biolime technology enabled their homes to do something more important than provide warmth and shelter.

How would they feel if their houses were able to contribute to the health and healing of the planet? After a few moments of stunned contemplation, some audience members raised objection to the technology by drawing analogy with genetic modification, but Councillor James was also quick to point out that the cell-like agents used in the Biolime process did not have any genes and therefore could not self-replicate. Biolime itself was not alive, and although it shared some of the characteristics of living systems, it would die without the continued nurturing of the community. Mrs Angel Darling, who was already considering spending more time outside for health reasons, wondered what the councillor meant by this. Councillor James told her that the Biolime needed to be continually replenished to keep the carbon-fixing process going, as it did not last forever. The fragility of Biolime and its dependence on the active participation of the community was sobering news and appeared to endear the technology to the congregation. They seemed less anxious about an alien paint invasion and began to ask questions about the necessary cultivation methods. On account of his youth, Councillor James found it harder to deal with the more philosophical issues that were raised in objection to the new technology. So, hearing Arthur's voice strain at Mr Henry Norton's recurrent interjections that Biolime was 'an act against God', Councillor Andrew Talbot felt the need to step in and assist his colleague. Mr Norton was not easy to console as he'd just lost his wife and was angry with everyone about everything, but Councillor Talbot reassured him with a rather meaningless but effective platitude, before moving the conversation swiftly onwards. In fact, Councillor Talbot had most difficulty with the permaculturists, who were the most vigorous objectors to the 'unnatural' nature of Biolime and could not accept that artificial processes could coexist alongside natural ones. The permaculturists blamed all forms of technology as being responsible for the sorry condition of the planet, and Councillor Talbot responded to these objections with a theatrical and ponderous demeanour. He cast the congregation's attention back to the days of planting orchards and using grafting technology to 'enhance' plants as being inherently unnatural pursuits and a kind of 'technology' that had ultimately benefited humankind and raised the stakes of the meeting. He urged the Mossville citizens to collectively take a lead in making amends to the planet on behalf of the human race by embracing Biolime. An overwhelmed and exhausted audience found themselves applauding the veteran councillor and were invited to cast their votes. Despite the handful of vigorous objections that had been voiced during the meeting, 'Mock Rock' was accepted with an overwhelming majority. Even Mr Norton was overheard muttering to himself on his way out of the town hall that if 'the abomination' meant that he didn't have to spend every Sunday morning sorting rubbish, then he was all for it. Life was too short to sort rubbish.

A public holiday was held in Mossville the day that the Biolime was delivered and became a community event. Large containers of locally prepared Biolime solution were assembled on a cordoned-off section of the road where people helped each other in filling up portable spray containers and coating the outside of their homes fully clad in overalls, goggles and masks. Mrs Kathleen Gately, who looked oddly alien in goggles that were too large for her sunken features, had problems using the hand pump with her rheumatoid hands. James Chesney, who had just come from next door to complain about the persistent yapping of her toy dog that was upset by the old woman's appearance in protective clothing, decided to help out. Kathleen hovered while the young man worked and repeatedly asked him why they were spraying a liquid on to the buildings in Mossville when they'd been promised some rocks? Jimmy mumbled from behind his mask that the rock was grown from the solution and nipped over the fence to finish off his own place. Kathleen took off her overalls, which settled the little dog and sat on her front wall looking back at her house in disbelief. How it could be true that water could turn into rock? She shook her head. In her view, this was something that would 'beggar Jesus' to figure out, so help her God.

Since everyone was fully occupied with diligently applying Biolime to their homes, the usual neighbourly vigilance had slackened, as people were concentrating on the job in hand rather than wondering what their neighbours were up to. The community was later astonished to find that an unpopular modern statue had been drenched in so much Biolime solution that it now resembled a spacecraft. Nor were they able to explain how the local skateboarders had managed to acquire a Biolime 'ramp' that gave them enough air to be clearly visible from Mr Norton's back yard. Otherwise, the day came and went uneventfully and after the initial flurry of activity and excitement, Mossville settled down again to its frugal routines.

A few weeks later, those areas that had been sprayed with the Biolime solution began to transform and produce a moist, heavily patterned, whitish rock. Delicate crustings of this material appeared in gutterways and grew into stalactite fingers where water had accumulated. Small children picked at oddly shaped protrusions that were sometimes used by wildlife and the Biolime could also be found in places where it had not been deliberately applied. Playground drains became unnaturally frosted and Biolime trails squeezed like toothpaste through gaps in the pavements. Where the Biolime had died it became laced with white ribbons that were prone to fracturing. On a dry day, these brittle splinters of rock could be heard cracking and falling like old plaster from the walls. Rain tasted clearer, fires burned brighter and even algal blooms in the waterways were more vigorous than before Mossville embraced Biolime. While many subtle differences were noted in the environment, the community continued with their usual, well-meaning business in their own peculiar ways.

Gradually, Biolime became part of the everyday community tensions. Jimmy grew tired of Mrs Gately's moaning that he hadn't done the front of her house properly and refused to stop by. As a result, Kathleen's façade looked scarred and provoked snide comment when she queued at Mr Soames's checkout. Kathleen expressed her defiance against her neighbours by allowing her little dog to urinate on the corner of Mr Norton's house. This act of wilful vandalism was quite a spectacle, as the acidic urine caused the Biolime to fizz like freshly shaken lemonade. A vigilant Mrs Darling, who had finally found her excuse to spend more time outside by taking up smoking, witnessed the sabotage and swore at Kathleen through her nicotine-licked lips – it was people like *her* that were responsible for global warming in the first place! Without so much as casting a backwards glance Kathleen flipped two rheumatoid fingers at her critic and patted the little dog on the head. Affronted, but in greater need of a cigarette fix than altercation, Mrs Darling resumed her smoking and cursed through her breath right to the end of the butt.

In his government report about the Biolime pilot scheme, Councillor James commended the community for intensifying their permaculture and recycling efforts. He commented that the only real difference to Mossville was the remarkable snow-like coating of the buildings. Councillor Talbot, on the other hand, was more perceptive, remarking that the presence of Biolime helped the community feel that their individual efforts in combating global warming were significant. In his opinion, Mossville had realized that if something as small as the chemical fragments of technology that constituted Biolime could contribute to tackling the inconstant challenges of climate change, then the efforts of each individual, no matter how trivial, would contribute immensely to their collective quest.

10 Vibrant Venice: Designing with Vibrant Matter

> *I have also thought of a model city from which I deduce all the others ... It is a city made only of exceptions, exclusions, incongruities, contradictions. If such a city is the most improbable, by reducing the number of abnormal elements, we increase the probability that the city really exists.* (Calvino, 1997, p.69)

10.1 Overview

This chapter speculatively explores how the principles of vibrant matter may be applied to an urban-scale project. Specifically, it proposes to empower the city of Venice with lifelike abilities that enable it to engage in a struggle for survival against the elements – just like living things do! 'Vibrant Venice' was initiated in February 2009, as a model system that proposed to couple the synthetic activity of artificial and natural systems within the lagoon to grow an artificial limestone reef under the city. The project is a multidisciplinary collaboration that combines speculative fiction, empirical science, fieldwork and architectural design. A particular kind of ELT was conceived for this project, based on dynamic oil droplets that are light sensitive and can produce an accretion material using resources in the lagoon system. The sinking of Venice into the soft delta soils on which it is founded is offset by the accretion of an artificial reef around the woodpiles, which spread the point load of the city over a much broader base. This project has been developed in collaboration with Neil Spiller and Martin Hanczyc, with drawings produced by working with Christian Kerrigan and GMJ. Vibrant Venice is not just a speculative proposition, but has been supported by field tests that are based on scientific research, which is at an early stage of development. The project proposes how vibrant matter may be applied through ELT across different scales of operation, and raises questions about how humans and non-humans may begin to codesign their environment and evolve their shared environmental future.

10.2 Future-proofing Architecture

In spite of our inability to accurately predict the future, we are rather good at surviving it. When I say 'we', I mean 'life' as we know it. Early life forms are likely to have been extinguished many times by violent asteroid showers before they finally established themselves on the Earth's surface some 3.5 billion years ago (Whitehouse, 1999). The surface of the planet is not kind or constant. It is a far from equilibrium system that requires a lot of energy for any structure to persist on its surface. After all, when considering the geological timescales that life has endured, even mountains are moved.

In 1609, Galileo was in Venice when he heard of an invention called the telescope, which allowed distant objects to be seen as distinctly as if they were nearby (Sis,

2000). By looking up, using this instrument, Galileo expanded the similarity between Earth and all the other heavenly bodies (Latour, 2013), setting the precedent for other Enlightenment thinkers to attempt to erase ambiguity from our world, using the principles of classical science. Indeed, such 'natural laws' propose that our existence is deterministic, so that it is possible to predict 'future scenarios' if we know the present location, properties and trajectories of all atoms in the universe (Prigogine, 1997, p.22).

Modern buildings embody Enlightenment scientific thinking, since they are constructed using engineering principles. Yet, over the course of the 20th century, it has become increasingly clear that equilibrium-state mechanical solutions cannot match the computational processes within natural systems. In many ways, classical machines are built as if in opposition to Nature. While looking down on us from one of those heavenly bodies, James Lovelock actually decreased our similarity with the heavenly bodies that Galileo had observed and challenged the unique attributes of the Earth (Latour, 2013). While mechanical systems simplify the laws of existence through 'natural law', natural systems themselves recomplexify events through many parallel processing systems. Ultimately, these give rise to acts of succession by the elements, bacteria, seedlings and other life forms. Indeed, the computational power of the chemistry underpinning such natural technologies is so powerful that nothing is spared this process. Consequently, machines and architectures are inevitably assimilated back into the shifting material systems of the Earth – sometimes on geological timescales, through the incessant, tiny acts of (re)complexification. Occasionally these changes take place much more rapidly during 'natural disasters', when Earth's forces act extremely and unpredictably around tipping points.

Modern buildings, which are both built by and imagined as machines (Gallagher, 2001; Le Corbusier, 2007, p.158), require significant investment in energy and resources for maintenance if they are to resist the relentless non-linear calculations of the material world. This results in significant wear and tear of inert materials, which occurs at the many microenvironmental interfaces in which our buildings are immersed. Viewed over a 30-year period, we spend 1–2% per annum of the initial construction costs of a building on maintenance (CBEC and APEGBC, 2009). However, these expenditures are manageable and economically justifiable when social and environmental conditions are stable.

Yet, fuelled by anthropogenic causes, the natural world appears to be increasingly restless, as recent catastrophic events testify: in New York City with hurricane Sandy (*Bloomberg*, not dated), in New Orleans with hurricane Irene (Dolnick, 2011), the Sendai tsunami in Japan (NBC News, 2012) and the loss of lives in flash floods in the UK (*Telegraph*, 2013). While we have always lived in a probabilistic world, it is becoming increasingly obvious that our global environmental system is reaching multiple environmental tipping points. While we have laws and models that can help us deal with non-linearity, the processes that underpin phase changes in material behaviour remain mysterious. Indeed, it is apparent that, over the course of this century, our environment will not be predictable or stable, and the way that we imagine and design architecture is simply not designed to deal with this degree of ambiguity or change.

With the advent of software and high-speed computer processing, it is possible to imagine and visualize urban environments as complex systems, or ecologies, whose trajectories are probabilistic – not deterministic. Probabilistic solutions are based in 'general systems theory' (Von Bertalanffy, 1950; Von Bertalanffy, 1969) and require a different approach in constructing an idea of the future. They address an emergent, contingent series of events that are shaped by their relationship to an unevenly distributed and constantly changing present. This does not mean that 'anything goes', since all events exist within finite limits of possibility. However, the decision pathways and events that shape the outcomes remain open to influence by many contingencies until they have actually occurred. So, although the trade-off for a probabilistic future is certainty, the creative potential is increased and outcomes can be continually shaped (Prigogine, 1997).

These probabilistic trajectories are better expressed as algorithms. Although software programs are used to inform architecture, such as in the practice of parametrics (Schumacher, 2009) where lively relationships are viewed through computer visualization, the underlying dynamic principles cannot be embodied in the production of architecture, since traditional architectural materials are selected and designed to perform at equilibrium states. Yet, Stuart Kauffman questions the conventional notion of equilibrium, noting that within the lifetime of the universe the composition of molecules can alter indefinitely (Kauffman, 2008, p.125). Nevertheless, the classical concept of equilibrium, where no (effective) net energy enters or leaves a system, applies in the case of design, where objects are assumed to be inert and possess no internal capacity to change or initiate events. In contrast, non-linear systems continually absorb and release energy, which confers them with dynamic properties that change throughout the design process. There is, therefore, a pressing need to establish ways of embodying flexibility, robustness and creativity within a portfolio of design solutions, so that architects can work with materials that can deal with non-linearity at far from equilibrium states. Such vibrant materials may provide contingency and produce outcomes that can deal with the eventuality of reaching tipping points in natural systems.

Yet a completely different set of approaches to the mechanical solutions that we are most familiar with exists, which deals with constant change as an inherent aspect of its organization. Life itself may be thought of as a kind of analogue computer, which resists the decay towards entropic equilibrium (Schrödinger, 1944) and is enabled to do so by the incredible parallel processing powers of chemical systems. These draw on the properties of the quantum world, which govern their molecular interactions and physical properties.[60] Matter, therefore, uses quantum computing techniques, which

60 Deutsch observes that all physical processes are quantum-mechanical and that quantum compu-

confer it with robustness, flexibility and capacity to deal with unpredictable events, by being engaged in continually unfolding chemical events, which are separated through space and time (Deutsch, 1997, p.194).

Indeed, Nature has been shown to behave according to non-locality, which is inconsistent with a system that operates according to quantum theory (Popescu, 2006; Pawlowski et al, 2009). In *Rhythmanalysis* (Lefebvre, 2004), Lefebvre begins to imagine how dynamic networks of vibrations, a macroscale version of string theory where extended bodies, or waves, permeate the fabric of our cities and forge everyday existence.[61] Lefebvre's ideas of connecting rhythms are embodied with real effects that are experienced at many levels, such as sounds in the air, the motivating force of our circadian biology and as expressions of chemical periodic activity.

10.3 Venice as a Case Study for Urban-Scale Design with Vibrant Matter

The future of Venice in northern Italy has always been precarious, being tethered to the shoreline, which is arguably the most ferocious environment on Earth. It is situated on the shores of a lagoon where the Po delta meets the Adriatic Sea (see Figs. 10.1–10.4).

The landscape is changeable and geographically ephemeral, being continually shaped by opposing forces of land and sea. Gradually, the historic architecture is beginning to lose its battle with the non-human world, being constantly battered by the elements, repeatedly flooded, desiccated, biologically (re)assimilated and chemically digested. The city itself is continually being remodelled by the exchanges between its various actants, where biology, architecture, machines, people and their effluents forge networks of interactions. These chemical networks lay down the substrates upon which the next group of assemblages exert their effects. In this way, old systems are not destroyed and replaced by new ones; they continue to exist and be transformed by new layers that wrap themselves around each other, merge with their predecessors, or erode and dissipate to be redistributed elsewhere in the lagoon. Venice itself could be thought of as the result of a process of architectural succession[62] facilitated by incessant material shifts, which transforms the city over time.

ting is a distinctively new way of harnessing Nature (Deutsch, 1997, pp.194–195).

61 String theory, or 'superstring theory', is where extended objects, 'strings', rather than point-like particles, form the elementary building blocks of matter (Deutsch, 1997, p.23).

62 Allan Savory defines biological sustainability as 'roughly biology's answer to entropy – a thrusting up as opposed to a running down, the irrepressible striving of living communities to become more dynamic diverse and stable. This is the force that causes jungles to overgrow old civilizations in areas of perennial rainfall, and soils to form from lava flows. It is the process that provided the biological capital we now use and which maintains the air we breathe' (Savory, 1991).

Figure 10.1: This view of Venice across the lagoon towards the Grand Canal (right) depicts the rich elemental infrastructure in which the city is immersed. The continual flow of matter through the site contributes to the battering of its architectural fabric, owing to the restless elements, but also nurtures the dense growth of biological systems in the waterways. Photograph, Rachel Armstrong, August 2010.

Figure 10.2: Jetsam on the shores of the Lido on the Adriatic Sea includes debris from trees that form a fence-like assemblage and garbage from the Venetian lagoon. Photograph, Rachel Armstrong, August 2010.

Figure 10.3: The continual morphological computing processes of the elements and marine organisms have weathered these Venetian buildings. Natural forces operate on the fabric of Venetian buildings in complex ways, such as brickwork being digested, generating a coating of white salt or efflorescence. Photograph, Rachel Armstrong, August 2010.

Figure 10.4: Physical processes act in concert with the local organisms like oysters, algae, bacteria and mussels, and produce metabolic events that shape the architectural fabric of Venetian buildings. These entanglements of natural and man-made materials constitute a post-natural building fabric, which functions as a vertical synthetic marine ecology at the tideline. This process may be likened to the production of scar tissue in an organism. Photographs and collage, Rachel Armstrong, August 2010.

In biological terms, succession is a natural process where metabolisms transform one landscape to become another. The process generally tends towards complexification, where simple cellular landscapes are replaced with multicellular ones. However, in conditions of scarcity, such as lightless caves that are inhabited by troglodytes with chemotrophic metabolisms (that derive their energy directly from chemistry, not the sun), succession may produce trends that reduce complexity. For example, troglodytes relinquish their eyes, as they are no longer necessary in their resource-constrained existence.

However, Venice is not a resource-constrained system. It is open to the sea and the light, and is bathed in the vigorous metabolic activity of its lagoon ecology. From a biological perspective, a city located in water, which is a universal solvent, may be regarded as a site for mineral release and, by virtue of the abundant waste that is emptied into the waterways, also as a rich source of organic compounds. These networks of vibrant matter are unevenly distributed by their context and form fertile landscapes of material transformation at many scales.

Venice has been imagined and constructed as a permanent structure within its continually shifting environment using traditional, inert building materials, such as marble, brick, iron, concrete, limestone and wood. Its entropy, therefore, is progressively increasing as a function of the physical laws of the universe, and is accelerated by natural processes that are digesting its fabric and redepositing it elsewhere. Although Venice has weathered the impact of its tempestuous environment for over three centuries, the lively non-human actants are imposing their own designs upon its materials and its integrity is crumbling. The visible deterioration of this iconic city has incited all kinds of impromptu acts of desperation to save the buildings from crumbling, where visitors and residents alike plug up the yawning, fist-sized holes in the wall with concrete, rubble, rubbish and even chewing gum (Artistic Things, 2009)[63] (see Fig. 10.5).

The brickwork in Venice, at the shoreline and even above the level of the tide, is subject to an entanglement of forces exerted by heterogeneous groups of actants that continually work to reconfigure its chemical and physical relationships and leads to its structural decay (see Fig. 10.6).

In a biological sense, this is an act of transformation by the natural world, which is reclaiming the rarefied building materials back into a complex system of soils. Yet, Venice's vigorous resistance to these incessant environmental assaults is remarkably ingenious and lies at the heart of its development, as the city has survived by allying itself with advanced technologies of the time. In the 9th to 12th centuries, when the

---

63 Simone Decker's poignant commentary on chewing gum in Venice was created for the 1999 Venice Biennale, where she used photomontages of chewing gum to create the impression of life-sized installations and combined these with photographs of public spaces in the city.

Figure 10.5: Gaps in Venetian brickwork appear to invite visitors to 'post' rubbish into the holes that the deterioration has left in the walls, such as this folded wedge of paper, which has been carefully

Figure 10.6: The structural decay of Venice's building fabric, coupled with the incorporation of organic matter into the new sites opened up by this process, is forged by the natural actants that cooperate through the production of assemblages that transform brickwork into new soils. Photograph, Rachel Armstrong August 2010.

city-state of Venice was being founded, these technologies were agrarian. State of the art practices involved the drainage of land, the creation of canals and the use of woodpiles under the foundations of the buildings to shore up the developing settlement (see Fig. 10.7).

These technologically enabled developments continue to form an essential part of the city today. With the Industrial Revolution, a new set of technologies exploded on to the scene, bringing great power and wealth to the people of Venice. This was implemented though the energy-intensive, top-down subordination of matter, which was executed with atomic-scale precision. Yet, with new technologies exploding on the scene, prompting rapid industrialization of the area, industrial machines quickly plundered Venice's rich resources (see Fig. 10.8).

Water consumption of Venice's aquifers caused the ground to sink beneath the city. This has contributed to the *acqua alta*, the high tides, which periodically flood the city and cause chemical and physical damage to the buildings. While industrialization vitalized the economy, it came at a great environmental cost, which Venice is still paying, in terms of pollution, a depleted water table and reduction in the rich lagoon biodiversity. However, the industrial age has also bestowed Venice with potential salvation – a colossal mechanical barrier that proposes to divide the city from its assailants. Construction of the MOSE (Modulo Sperimentale Elettromeccanico) project (Ravera, 2000) began in 2003 and is aimed for completion in 2014 (see Fig. 10.9).

Figure 10.7: The visible wear and tear at this boat mooring's base mirrors the kinds of processes that shape Venice's foundations, which rest upon woodpiles. The salty marine water sank these into the soft delta soil, on which the city was established in mediaeval times, and they have generally been preserved. However, when woodpiles are exposed to the air, they are rapidly digested and transformed by marine organisms, such as bacteria and shellfish, and lose their structural integrity through decay. Photograph, Rachel Armstrong, August 2010.

Figure 10.8: An oil refinery in Marghera, north of Venice, casts a shadow on the sunset of the historic city. Photograph, Rachel Armstrong, August 2010.

Figure 10.9: Construction site of the MOSE project from the Lido beach. Photograph, Rachel Armstrong, August 2010.

It is a series of 78 hydraulically powered gates that propose to literally hold back the tide and ensure Venice's survival. Yet, ecologists are concerned about the backpressure of water on the natural drainage of the Po delta, which they fear may result in potentially devastating changes within the lagoon ecosystem (see Fig. 10.10).

Figure 10.10: Banner protesting against the MOSE project. Photograph, Rachel Armstrong, August 2010.

10.4 ELT as a Next Generation Technology in Venice's Survival Portfolio

In keeping with Venice's progressive, technological engagement, the next generation of technologies designed to deal with Venice's uncertain predicament may incorporate the technologies of 'life' into its portfolio of tools. Living systems are much older than machines, and persist through networks and relationships that are continually forged through metabolic (interior) and ecological (exterior) assemblages (see Fig. 10.11). They exert 'soft power' rather than the 'hard power' of machines and work subtly with natural forces rather than standing in opposition to them (Nye, 1990).[64]

64 Joseph Nye first proposed the concept of 'soft power' to describe how persuasion and cooperation may exert influence on events, as opposed to 'hard power' that operates through domination and coercion to generate influence (Nye, 1990).

Figure 10.11: Multiple actants exert a variety of forces on the material assemblages that constitute Venice's fabric and include the relentless motion of water, wind, sunlight and the local marine ecology. Photograph, Rachel Armstrong, August 2010.

In the previous chapter, the Hylozoic Ground installation embodied an expression of next-generation technology, which harnessed some of the properties of living systems to generate environmentally contextualized outcomes. Various species of ELT within the cybernetic installation took the form of dynamic chemistries, which gave rise to a set of robust and flexible responses to the changing gallery conditions, by increasing the fertility[65] of the site, recording events and producing microstructures in response to them. Despite the highly engaging experience embodied by Hylozoic Ground, much more research is needed to understand how architects can work with vibrant matter. Further investigations are also needed to establish how various species of lively materials may also benefit other disciplines that work and design with emergence, such as computing, ecology and urban planning practices. Technologies that can deal with probabilistic laws and events that may embrace the radical creativity (Kauffman, 2008, p.135) of Nature (Prigogine, 1997, p.5) are essential for establishing new methods of design practice. Tactics that forge new ecological relationships, which are not simply directed at creating a specific intervention, are needed, which may orchestrate architectural conditions through the production of assemblages. For example, vibrant

65 Where 'fertility' refers to an increase in probability that a material event may occur.

matter may play a similar role to 'mother' trees, which possess such rich connections that they are critically important in nurturing the diverse ecosystems of our forests (Howard, 2011). The new linkages forged by ELT may confer otherwise inert fabrics with some of the qualities of living systems so they resist entropy and are also able to deal with environmental unpredictability in real time. The outcomes of these interventions are probabilistic, not deterministic, and require the kinds of managed engagement that are already familiar to agriculture, gardening or permaculture. However, these investments establish the conditions for the development of a qualitatively new kind of architectural design practice that regards the city of Venice not as a gated lagoon, but as an expanded ecosystem (Thackara, 2012).

Along the edges of the Venetian waterways is a native construction system that is not under human control. It is able to respond to the constant challenges of its hostile surroundings through constant growth, regeneration and repair. The actants of this construction system are, of course, Venice's marine biology, which use the technology of 'life' to persist in the face of perpetually changing environments. The local wildlife in Venice's waterways, such as algae, shellfish and bacteria, have claimed a construction process within this harsh terrain as their own (see Fig. 10.12).

They have evolved to accrete, secrete, remould and sculpt the materials of their surroundings to co-create microenvironments that are uniquely suited to their needs. While biology creates organisms with short lifespans in comparison to geological

Figure 10.12: Tenuous marine wildlife secure attachment to Venetian brickwork and assimilate its materials to sustain their metabolism. In the process, these organisms transform the building fabric so that it is (re)appropriated into biological structures (as shells), or forms new soils (silt). Photograph, Rachel Armstrong, August 2010.

timescales, they are highly versatile and possess a range of strategies, which enables them to creatively cope with the constant demands of an inconstant environment. Biology is resilient, adaptable and evolvable, and will survive unpredictable disturbances, which is in stark contrast to the local buildings that, no matter how durable, do not possess any of these qualities. Yet, currently Venice's native biology is undirected by human activity and threatens the integrity of its architecture, behaving like an unruly garden. Venetian wildlife exploit every opportunity to extend into new territories and environmental niches to claim them as their own, and they vigorously pursue easily accessible sources of nutrients from the fabric of the buildings.

The precarious situation of Venice embodies the condition of many modern cities, which are facing environmental disruption through the collective action of non-human forces, such as climate change and resource scarcity. An optimum approach to an environmental design may be to develop alternatives to the top-down, machine-centred approach of modern architecture, and replace or complement it with the bottom-up, self-organizing imperatives of actants. Natural systems already work by horizontally coupling the agency of dynamic systems at multiple scales and multiple levels of organization (subatomic, microscale, macroscale and megascale) through natural (Denning, 2007) and quantum computing approaches (Deutsch, 1997, p.123). Perhaps designers may be able to strategically influence the performance of urban assemblages by applying soft control systems using morphological computing techniques. By orchestrating the formation of strategic partnerships as a design intervention, the fabric of historic buildings could potentially be preserved, repaired or augmented – rather than destroyed.

This is more than a hypothetical proposition. Advances in biotechnology have provided us with a toolset that enables us to couple technology with Nature. The new science of 'synthetic biology' enables the design and engineering of living systems (Armstrong, 2013c)[66] as it shares the common language of physics and chemistry with the natural world. Synthetic biology offers a set of tools, methods and materials that may provide a new, iterative relationship between manufactured and natural systems, which may be appropriated within architectural practice (Armstrong and Spiller, 2011).

Biology is vigorously synthetic and continually combines different kinds of chemical substrates to produce new compounds. In the same way that Venice has used advanced new technologies to assist its survival, so biology uses the naturally occurring technology of 'life' to persist in the face of perpetually changing environments. Local wildlife such as algae and shellfish actively accrete, secrete, remould and sculpt the

---

66 Multiple definitions for synthetic biology exist ranging from genetic modification to molecular building blocks, which are described in detail in my analysis of the field for Volume magazine (Armstrong, 2013c). In this thesis, Stéphane Leduc's view of synthetic biology is applied, which is viewed as a branch of synthetic chemistry that exhibits lifelike properties.

materials of their surroundings to co-create microenvironments that are uniquely suited to their needs. This is in stark contrast to the fragility of local buildings that, like Ruskin's Matterhorn (Ruskin, 1989), will simply be worn down by Nature's relentless processes shaped by time's arrow (Prigogine, 1997, p.1) (see Fig. 10.13). However, currently Venice's native biology is undirected and the catabolic processes that also accompany its anabolic or synthetic activities do not discriminate between the minerals of a building and those in the marine environment. Without mediation, natural processes do not share our anthropocentric concerns, and threaten the integrity of Venice's historic architecture. They behave like an unruly garden, finding opportunities to extend into new territories and plunder new sites of abundance, to dismantle the high concentrations of easily accessible sources of nutrients from the fabric of the buildings.

10.5 Chemically Programming the Fabric of Venice

Although the most well-known forms of synthetic biology take a top-down approach to the design and engineering of living things by modifying the biological code of natural organisms, this is not the only means by which ELT obtains results. In recent years, dynamic droplet systems, which are programmable assemblages of chemistries that do not exist in Nature, can be used as alternative production

Figure 10.13: Light falling on a mooring step in Venice that houses a colony of microalgae, which have produced thrombolites. Photograph, Rachel Armstrong, August 2010.

platforms to machines (Hanczyc et al, 2007; Toyota et al, 2009). They work by facilitating horizontal material couplings between agents, which may be orchestrated through morphological computing techniques. Since these actants are not biological, they are easily programmable, and altering their internal and external chemical programs may shape their actions. For example, droplets could be constructed that are sensitive to light and move away from it. They may also be designed to produce a mineral shell using dissolved substances when they are at rest. These processes are likely to be context sensitive and therefore will vary depending on the composition of the local environment. Since Venice rests upon woodpiles, its foundations offer an environment in which a variety of ELT species could be designed to produce a material – potentially at an architectural scale – by harnessing the en masse effects of chemically programmed dynamic droplets. Morphological computing techniques could be used to orchestrate the production of an artificial limestone-like reef by chemically manipulating the activity of droplets (see Fig. 10.14).

Indeed, a range of species of ELT could be designed and released directly into the waterways, and chemical attractants could be used to ensure infiltration of the technology where conditions for penetration are poor. It is possible that dynamic droplet species may even be able to move with such force that they resist currents and chemical gradients. For example, a species of droplet could be designed with combined qualities so it could move away from the light-filled waterways and then construct artificial mineral shells at rest under the darkened foundations of the city, as in Figs. 10.15 and 10.16. However, unlike tipping concrete into the water, the chemically programmable droplets would be sensitive to the biotic environment.

Figure 10.14: Dynamic droplets programmed to move away from the light and down into the darkened foundations of the city could potentially be designed as a city-scale morphological computer to construct an artificial limestone reef underneath Venice's foundations. Photograph, Rachel Armstrong, August 2010.

The chemical composition, texture and colour of the accreted material are likely to change with time as minerals and conditions in the water vary. The first manifestations of the active technology are based on an oily hardware and therefore produce soft, spongy materials that settle around the woodpile foundations. While many will be crushed by water turbulence, they can robustly resist being destroyed by the vigorous tides, and even recover after being nibbled by marine life. Gradually, more rigid structures with delicate, thin shells may replace the soft structures,

Figure 10.15: View underneath Venice's foundations, which stand upon woodpiles, demonstrating the potential action of a city-scale morphological computer composed of smart, programmable droplets. This structure aims to gradually spread the point load of the city over a much broader base than offered by the narrow woodpiles, as well as providing new ecological niches for the marine wildlife. Computer drawing, Christian Kerrigan, February 2009.

Figure 10.16: View of programmable droplets coming to rest underneath Venice's foundations as their light-sensitive metabolism reaches equilibrium and activates a second metabolic process that enables the droplets to use local minerals and dissolved carbon dioxide to grow a solid, reef-like structure. Computer drawing, Christian Kerrigan, February 2009.

and some will persist to create porous, pumice-like assemblages. The resulting construction would offer a supporting framework for the city and marine wildlife, as well as serving as a pollution sequestration system which was capable of growth and repair.

Mineralization of 'bimorph' crystals[67] could be promoted by releasing a new species of silica-containing droplet into the Venice waterways, which may technologically strengthen the mineralization process (under alkaline conditions) (Kellermeier, 2012). Wim Noorduin demonstrated that controlling the environmental chemical conditions of bimorph crystals could accurately produce flower-like microstructures. The exact morphology of the crystals depends on how the different substances diffuse through a liquid silica-based solution and is sensitive to the presence of certain chemical gradients, as the acidity of the reaction changes spontaneously during the crystallization process. The reaction conditions determine whether the structure resembles broad leaves, thin stems or petal rosettes (Harvard School of Engineering and Applied Sciences, 2013).

Gradually, the shells produced by droplets would be shaped by the chemistry within micro niches around hollows in the woodpiles. Further synthetic mineralizing activity would then be sculpted by new generations of smart droplets. Additionally, local marine life such as shellfish are also likely to be drawn by the abundance of mineral resources in the site. Although this scheme has not yet been formally tested in the laboratory, it is very likely that the smart droplets, simply put, provide a more attractive food source for these organisms. Their minerals would be much more accessible to the marine wildlife than the existing sources in the brickwork of the Venetian buildings. Should that be the case, something remarkable would happen. In the presence of smart droplets, the same organisms that have previously 'preyed' on the brickwork and the wooden poles will do the exact opposite. They may produce resilient calcium-rich materials that are chemically adhered to the brickwork, which stops the further action of physical and chemical forces eroding the man-made materials. Moreover, new populations may be chemically attracted to the area in which the smart droplets are making high levels of minerals available by simply concentrating them in thin shells. So, as one population of droplets creates a layer of mineral shells, another takes root on its surface, and gradually a material is accreted that grows from the inside outwards. The deceased members of the colony give up their 'bones' to form a collective 'skeleton' or 'scaffolding' for a community, which forms a sturdy platform for the next generation, which is exactly how a reef forms.

---

67 The term 'biomorphic' crystal applies to inorganic matter that adopts biological-like forms. These may be synthesized from a highly alkaline (pH 9–11) silicate solution using salts from Group 2A in the periodic table, which includes magnesium, calcium, barium, strontium and radium and produce shapes commonly assumed to be characteristic of life such as, helices and convoluted curved sheets (Carnerup et al, 2006, p.248).

So, the droplets produce a reef directly and indirectly by creating the conditions for marine biology to co-produce the reef alongside them. With time, it is even possible that new species of shell-producing wildlife may evolve, which will be able to resolve sunlight abundance with the ability to make strong shells. Gradually, the accretion process may produce an artificial limestone reef underneath the foundations of the city. Ultimately, the proliferation of the reef under the city could provide a counterforce that is strong enough to stop Venice from sinking into the soft delta soils on which it was founded (see Fig. 10.17).

10.6 Experiments on Programmable Droplets

A number of experiments to demonstrate the feasibility of this proposal have been conducted both in the laboratory and in the field although the degree of penetration

Figure 10.17: Droplet assemblages may produce porous, pumice-like formations over time that reflect mineral compositions through their coloured layers, like stromatolites (Shark Bay World Heritage Area, 2009). The resultant architecture does not obey a top-down paradigm, but is codesigned through the entangled metabolic interactions between marine and human populations. Computer drawing, Christian Kerrigan, February 2009.

of the city foundations by the smart droplets has not been tested, nor has the ideal position from which to release the technology been established.[68]

10.6.1 Droplet Design

'Protopearls' were developed with chemist Hans Toftlund and Martin Hanczyc as a demonstration model system to explore the possibility of a technology that could strategically distribute minerals spatially and temporally and in response to environmental cues.

The first series of experiments were conducted courtesy of Davide De Lucrezia at Biotech Explora in Marghera Science Park, Italy, and with Martin Hanczyc at the Center for Fundamental Living Technology (FLinT) at the University of Southern Denmark. The technological design principles and conditions for producing a mineral shell around an oil droplet using lagoon water were explored. An oil droplet in water system was selected as the vehicle for a mineralizing metabolism. The clear oil diethyl phenyl phthalate (DEPP) was used as the oil carrier, which is the same oil that was used to house the Carbon Eater metabolism for the Hylozoic Ground installation. Although DEPP in aqueous phase is not vigorous or lifelike in the way that other oil/water systems may be (Hanczyc et al, 2007; Toyota et al, 2009), its specific gravity (1.1 at 20°C) is heavier than water and was thought to be a more suitable carrier for ingredients to facilitate accretion around woodpiles than a lighter oil, whose activity would be directed at the water's surface.

Since DEPP's specific gravity is temperature sensitive, it could work at different levels in the water. Below 25°C, the oil droplets sink, and at higher temperatures they rise to the surface and spread out. The metabolism selected for distribution by the droplets was designed to reflect the natural salts present in the lagoon, which is rich in magnesium and calcium. These salts are formed by alkaline earth metals from Group 2A of the periodic table, which includes magnesium, calcium, barium, beryllium, radium and strontium. Significant work by Juan García-Ruiz on the precipitates of 'basic' salts formed by barium and strontium showed that unusual crystal growth mechanisms in alkaline silica-rich environments could produce 'witherite' (a barium carbonate mineral in the aragonite group) precipitates from barium chloride and silica solutions (García-Ruiz, Melero-García and Hyde, 2009). These deposits were referred to as 'silica/carbonate biomorphs' because of their striking resemblance to primitive organisms. The aggregates of self-assembled materials were solely of inorganic origins and were composed of an amorphous phase of silica, which was intimately intertwined with a carbonate nanocrystalline phase. García-Ruiz and colleagues

68 Detailed tests, analyses, surveys and models will need to be made before this project is realized, which is beyond the scope of this current research.

proposed that the mechanism for producing such striking lifelike formations could be explained by coupled co-precipitation of carbonate and silica, which induced fibrillation of the growing front and gave rise to laminar structures that curled at their growing edges (García-Ruiz, Melero-García and Hyde, 2009).

Based on García-Ruiz and colleagues' observations, a demonstration experiment was designed with chemist Hans Toftlund at the University of Southern Denmark to form artificial shell-like structures from biomorphic crystals at the oil/water interface of DEPP droplets, which were loaded with a metabolic substrate comprised of soluble barium or calcium chloride salts. Barium salts were the control system for the experiment, having already been identified by García-Ruiz and colleagues as producing biomorphic crystal growth (García-Ruiz, Melero-García and Hyde, 2009), while calcium salts were the desired substrate for mineralization of the proposed artificial reef. Yet the properties of calcium had not been explored in the 2009 *Science* paper, so a preparation was designed to expose them to the same alkaline/silicate conditions as the barium salts. Both species of droplets (barium and calcium) were prepared by mechanically grinding 0.001 moles of salt per 2 ml of DEPP to form a paste. Each petri dish was prepared by adding 15 ml of water, which had been alkalinized by adding 0.2 ml of sodium hydroxide to reach a pH of between 8.5 and 10 (García-Ruiz, Melero-García and Hyde, 2009).[69] 0.6 ml of waterglass (sodium silicate solution) was added to this solution and gently stirred until fully dissolved using a plastic spatula. The petri dish was left open to the air so that atmospheric carbon dioxide could dissolve and form carbonate and bicarbonate species. Finally, two to six droplets of separate species of metabolic DEPP were added to the petri dish for both calcium and barium preparations. They were left to stand at room temperature[70] to form crystalline precipitates for 12–36 hours. The ingredients for both protopearl preparations are summarized in Table 10.1 and Table 10.2.

The droplet interfaces of both barium and calcium preparations were rapidly mineralized in solution in the presence of alkaline sodium silicate at pH > 8.5. The dense white precipitates around the spherical oil droplets appeared pearl-like and were colloquially referred to as 'protopearls'.

10.6.1.1 Barium Protopearls

Rapid mineralization occurred at the droplet interface of the DEPP droplets carrying barium chloride, and within a few hours a fluffy, vigorous white precipitate had formed at their surface (see Fig. 10.18).

---

69 García-Ruiz and colleagues recommend alkaline conditions with a pH range between 8.5 and 11 (García-Ruiz, Melero-García and Hyde, 2009).

70 Room temperature being 20°C.

Table 10.1: Barium carbonate protopearls preparation

Substance	Strength	Amount
DEPP	n/a	2 ml per droplet
Sodium silicate solution (waterglass)	Sigma Aldrich, ≥10% NaOH basis, ≥27% SiO_2 basis	2 ml per 50 ml water
Sodium hydroxide	1 M	0.2 ml per 15 ml water
Barium chloride	0.2 g (0.001 mol)	Mechanically ground into each 2 ml aliquot of DEPP to form paste

Table 10.2: Calcium carbonate protopearls preparation

Substance	Strength	Amount
DEPP	n/a	2 ml per droplet
Sodium silicate solution (waterglass)	Sigma Aldrich, ≥10% NaOH basis, ≥27% SiO_2 basis	2 ml per 50 ml water
Sodium hydroxide	1 M	0.2 ml per 15 ml water
Calcium chloride	0.1 g (0.001 mol)	Mechanically ground into each 2 ml aliquot of DEPP to form paste

After around three hours, precipitation in the petri dishes became diffuse and was not limited to the surface of the DEPP droplets. This was due to the barium salt diffusing from the body of the DEPP droplet and equilibrating within the petri dish, which occurred after six hours. Examination using a Nikon Eclipse TE2000-S inverted microscope with a Photometrics Cascade II 512 camera and in-house software revealed 'wheatsheaf' shaped crystals characteristic of barium carbonate at a magnification of 10×.

10.6.1.2 Calcium Protopearls

The calcium chloride carrying DEPP droplets rapidly mineralized at the oil/water interface, where a fine white precipitate formed (see Fig. 10.19). After 24 hours, precipitation within the petri dish was observed as the calcium equilibrated within the petri dish. After 48 hours, the stabilized droplets were observed under a light microscope at 10× magnification using a Nikon Eclipse TE2000-S inverted microscope with a Photometrics Cascade II 512 camera and in-house software, where 'tetrahedral' shaped crystals, characteristic of calcium carbonate, were observed, assembled in brittle, shell-like formations.

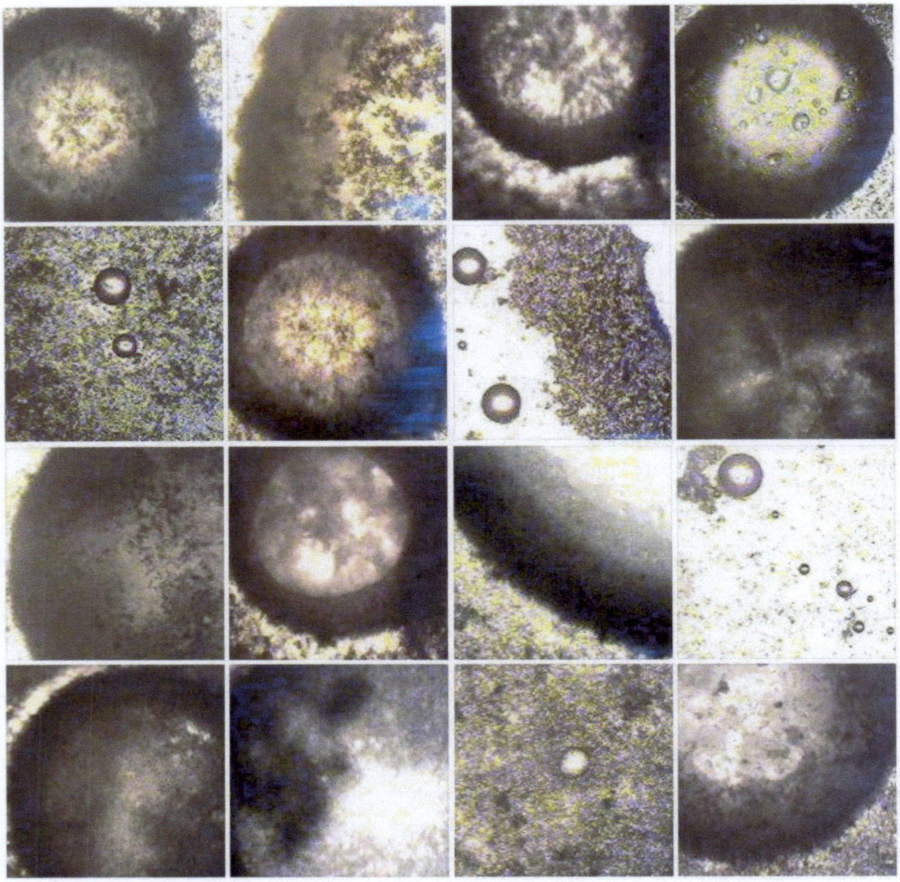

Figure 10.18: Barium protopearls are produced when an oil droplet is used as a slow-release carrier system for a simple carbon-fixing metabolism, which produces insoluble carbonate at the oil/water interface in the presence of dissolved carbon dioxide. The droplets quickly become coated with a feathery precipitate (classically described as a 'wheatsheaf'), which forms a simple mineral shell around the oil droplets, giving it a 'pearl-like' appearance. Micrographs and collage, magnification 4×, Rachel Armstrong, February 2009.

10.6.2 Protopearl Interactions with Other Materials

The DEPP protopearl droplets of both barium and calcium species robustly settled under gravity and produced soft scaffolding that could withstand vigorous agitation, re-forming droplets as soon as the physical activity ceased. Their ability to form assemblages and associations with other materials was demonstrated in a series of exploratory experiments where the droplets were added to containers with architectural substrates, including brick, wood, concrete and steel, that were submerged in alkaline solutions of pH > 8.5.

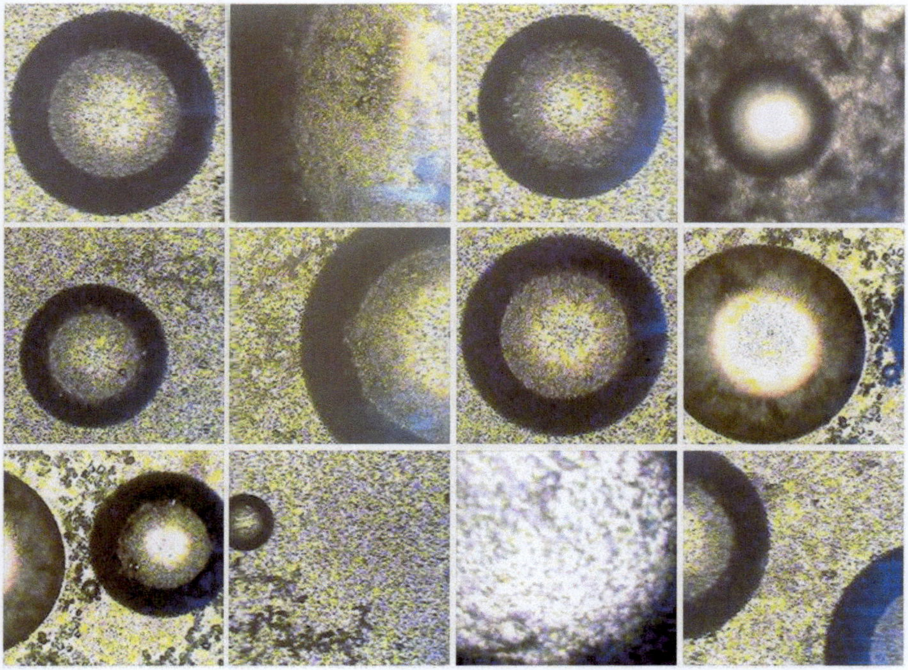

Figure 10.19: Calcium protopearls are produced when oil droplets that slowly release a soluble calcium salt as a simple carbon-fixing metabolism produce insoluble carbonate at the oil/water interface in the presence of dissolved carbon dioxide. The droplets quickly become coated with a dense, crystalline precipitate, which forms a bright white, simple mineral shell around the oil droplets that bestows them with a dramatic 'pearl-like' appearance. Micrographs and collage, magnification 4×, Rachel Armstrong, February 2009.

10.6.2.1 Brick

A variety of brick fragments were gathered from rubble in the streets of Venice and brought to the Biotech Explora laboratory, where they were added to a large 50 ml petri dish (see Fig. 10.20). The brick fragments were submerged in 40 ml tap water and titrated to pH 8.5 using sodium hydroxide at room temperature. DEPP paste containing calcium chloride was made up to 10 ml and aliquots of 1 ml were added to the water (in the absence of sodium silicate). The clear DEPP droplets turned white with calcium carbonate surface precipitation within 5–10 seconds after being added to the water and settled under gravity. Some of the droplets were mechanically divided by the brick fragments and formed smaller droplets. The droplets were not observed over a period of two hours, where no further movement or division of the oil bodies was observed, nor did any precipitate or residue form. The reason for this is unknown but may be due to silicates present in the brickwork and also due to the practical time constraints on the experiment. Further characterization and exploration of the interactions between the biomorphic crystals and the brickwork at the microscopic level requires further study.

Figure 10.20: In the presence of brick fragments, when DEPP – with a simple carbon-fixing metabolism – was introduced into a container with Venetian lagoon water, rapid mineralization of the oil/water interface occurred vigorously producing 'protopearls'. Photographs and collage, magnification 4×, Rachel Armstrong, February 2009.

10.6.2.2 Wood

Untreated wood fragments were gathered from the university campus at the University of Southern Denmark and taken to the FLinT laboratory where they were added to a large 500 ml glass beaker (see Fig. 10.21). The wood fragments were submerged in 400 ml tap water and titrated to pH 8.5 using sodium hydroxide at room temperature. DEPP paste containing calcium chloride was made up to 10 ml and aliquots of 1 ml were added to the water (in the absence of sodium silicate). The clear DEPP droplets turned white with calcium carbonate surface precipitation within 5–10 seconds after being added to the water where they rapidly descended and settled under gravity. After 48 hours, the droplets had developed a brittle calcium carbonate shell and a small amount of precipitate formed around the droplets.

10.6.2.3 Concrete

Concrete shards were gathered from the university campus at the University of Southern Denmark and taken to the FLinT laboratory where they were added to a large 500 ml glass beaker (see Fig. 10.22). The splinters of concrete were submerged in 400 ml tap water and titrated to pH 8.5 with sodium hydroxide at room temperature. DEPP paste containing calcium chloride was made up to 10 ml and aliquots of 1 ml were added to the water (in the absence of sodium silicate). The clear DEPP droplets turned white with calcium carbonate surface precipitation within 5–10 seconds after being added to the water where they rapidly descended and settled under gravity. After 48 hours, the droplets had developed a brittle calcium carbonate shell and a small amount of precipitate had also formed around the droplets.

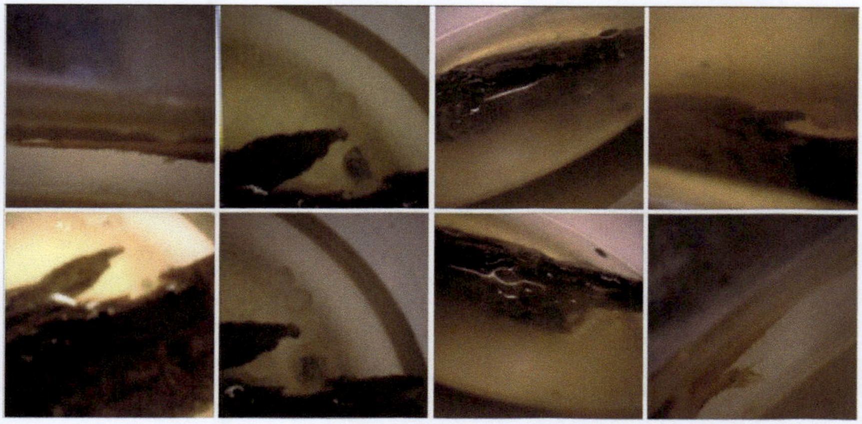

Figure 10.21: In the presence of wood fragments, malleable DEPP droplets with a 'protopearl' metabolism tended to roll off the wood and settle on the bottom of the container. Photographs and collage, Rachel Armstrong, February 2009.

10.6.2.4 Steel

A rusty six-inch bolt and stainless steel nail were sourced from a rubbish skip at the maintenance workshop at the University of Southern Denmark (see Fig. 10.23). They were taken to the FLinT laboratory where they were each added to a large 500 ml glass beaker. The nail and bolt were separately submerged in different containers of 400 ml tap water and titrated to pH 8.5 with sodium hydroxide at room temperature. DEPP paste containing calcium chloride was made up to 10 ml and aliquots of 1 ml were added to the water in each container (in the absence of sodium silicate). In both containers, clear DEPP droplets turned white with calcium carbonate surface precipitation within 5–10 seconds after being added to the water, where they rapidly descended and settled under gravity. After 48 hours, the droplets had developed a brittle calcium carbonate shell but had not made any bond with either the bolt or the nail. A small amount of precipitate had formed around the droplets in the beaker with the new nail and a more diffuse, dirty brown precipitate had formed at the bottom of the beaker with the rusty bolt.

10.6.3 Observations

The protopearls provided a model system to interrogate the possibilities of designing a light-activated, carbon-fixing droplet technology that could potentially accrete dissolved chemicals over its surface to produce a solid structure. In principle, it was demonstrated that dynamic droplets could be programmed to produce limestone-like deposits, which interacted with common building materials and were most vigorous

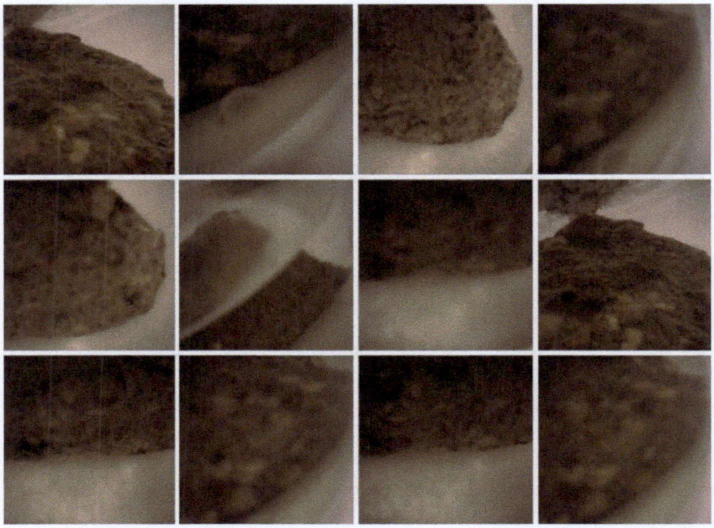

Figure 10.22: In the presence of concrete fragments, 'protopearl'-producing DEPP droplets were inclined to roll off the concrete and settle on the bottom of the container. Photographs and collage, Rachel Armstrong, February 2009.

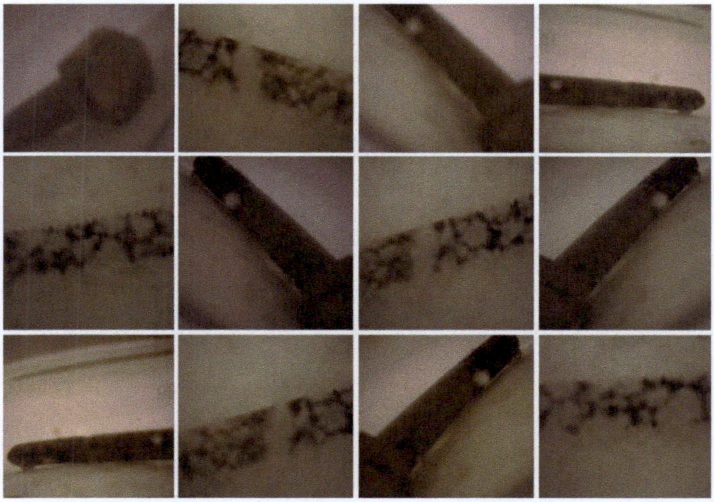

Figure 10.23: In the presence of a rusty bolt, 'protopearl'-producing DEPP droplets produced a significant flocculent precipitate, which was not confined to the oil/water interface. Photographs and collage, Rachel Armstrong, February 2009.

in the presence of brick fragments. The DEPP droplets self-organized under the influence of gravity and then adapted to local conditions. In the case of the wood splinters, the presence of air in some of the fragments caused the droplets to move and occasionally become entangled with air bubbles that precipitated around the crystals before escaping to the surface of the water. The presence of contaminants in the building materials, such as dust, soil or rust, did not appear to affect the performance of the primary metabolism, except in the case of the brick fragments, where precipitation was reduced.

More specific, rigorous analysis is needed in terms of the selection of materials, how they mineralize in the presence of silica salts (sodium silicate)[71] and what kind of chemical bonds are formed with protopearls. However, these initial sketches provide some rudimentary observations that may assist the further design of different ELT species and the subsequent development of experiments. As such, my experiments are initial forays and design experiments that require much further detailed, scientific exploration, technological development and analysis. Yet, given the proposed architectural challenge, the experimental findings indicate that in their current 'model' form protopearls have the potential to be developed into a chemically programmable accretion technology, although the currently demonstrator model system has many limitations, which are outlined below:

- Biomorphic crystal shells require a degree of alkalinity that is toxic to biological systems.
- DEPP droplets promote the formation of diffuse precipitates as the system reaches equilibrium, which causes sedimentation and is not desirable in terms of water quality or as a functional result of the droplet technology.
- More detailed titration of the basic salt concentrations, the amount of silica salt added and the range of alkalinity is needed to produce an optimized result using this system, and will form the basis of further investigations.
- The degree of spontaneous mineralization in lagoon water was not observed. Further experiments will need to be conducted to establish these baseline criteria.
- DEPP is used as a plastic softener with oestrogenic effects and is therefore not suitable for use in natural systems. A range of non-toxic oils need to be identified and their role as a potential accretion technology considered. Ideally, alternative carrier systems will possess a spectrum of densities so that they are active at different depths in the waterways.
- The addition of silicates using a slow-release system, such as another species of droplet, or even perhaps the effect that silica from diatoms may have on facilitating and regulating artificial shell growth, is another set of variables that require greater exploration and characterization.

71 Sodium silicate is also known as 'waterglass' and is used as the medium in which crystal gardens are grown.

- A photosensitive metabolism was not experimentally explored, although the proposal is based on experimental findings conducted by Martin Hanczyc.
- Site-specific experiments may be scaled up in tanks before field studies are designed.
- Fieldwork must evaluate impacts on water and wildlife.
- Fieldwork will need containment, perhaps by using porous materials or filter papers that enable minerals to enter the experimental site but prevents droplets from leaving.
- Mathematical modelling of the system may help better understanding of the physics and chemistry and enable further oil/water chemistries to be developed that are less toxic than the current models used. Cellular automata that appear in Stephan Rafler's 'continuous' version of John Conway's 'Game of Life' (Gardner, 1970), which uses floating point values instead of integers (Doctorow, 2012), shows some homologies with the Bütschli system. Computer modelling the Bütschli system, or another suitable species of dynamic droplet, may help to characterize the operational principles of the system and help to extend the range of possibilities for establishing transferable design principles for the production of artificially grown coastal reefs over a range of different sites, e.g. the Maldives or Songdo.

10.7 Further Development of Dynamic Droplets

The mineralization and structural processes that constitute the synthetic reef-like materials could be architecturally regarded as soft scaffolds. They may be further monitored and strategically influenced using morphological computing techniques by introducing distributed data collection systems into the most active sites of reef formation. For example, smart dust, which is a speculative technology still in development (Warneke et al, 2001), is made up of a population of small sensors with microprocessors and transmitters that may detect light, chemicals, vibration and temperature, and could work in a distributed manner to direct human agencies to respond to changes within the site as the reef is actively growing. Digital sensors and displays could be used by a community of experts that may include architects, ecologists, local historians and engineers to evaluate a range of parameters, such as the effect on the expanding reef on the buildings of Venice or its impact on native ecosystems. To the degree that they transform minerals and carbon dioxide to become an active accretion technology, the smart droplets could also serve as a pollution sequestration system that may even reduce sediment in the water. Moreover, smart droplets would reduce the acidity of the water as they remove carbon dioxide from solution and fix it into a solid substance. Yet, the most striking aspect in this context is that, working in concert with digital monitoring systems, dynamic droplets may be able to directly interact with their environment and respond to local changes in its chemistry. With the overall impact of the lagoon ecology and the welfare of the city in mind, decisions based on the information gathered from active sites could then be made to release new resources into the environment, or withhold certain droplet populations. Indeed, in many ways it is advantageous that the droplets do not possess the ability to spontaneously replicate with any vigour and rely on humans to feed and direct their macroscale operations. The appropriate technological governance of the different species of ELT could oversee the interacting systems and encourage them to grow in a particular direction, sequester specific compounds or secrete desirable products. The spatiality of the developing reef has synergies with embryological processes[72] and pattern formation that involve orchestration of complex material systems and unfolding manifolds of space and matter over time. These architectures do not propose an 'artificial' life, as in John Frazer's digital morphogenesis where Nature provides the model for generative forms (Frazer, 1995, p.9), but directly integrates with living processes as a fundamentally synthetic and material practice – as 'parallel' embryology (Lionni, 1977).[73] During the formation of undulating folds of reef-like matter, smart droplets, marine biology and the environment sequentially position chemical gradients and other forms of 'soft control' to create organizing centres that may generate propagating waves of activity that attract, activate and inhibit interactions within the reef. Chemical cues could be delivered to site-specific locations throughout the city via modified cisternae that provide access to the terrain under the heart of Venice, which is not easily accessed from the periphery. The addition of specific substances such as calcium salts could augment the activity of droplets and marine organisms through seeping fields of chemical activity that surge like developmental waves. Depending on the timing and nature of these periodic, chemical signals, dynamic structures could evolve underneath the foundations of the city (see Fig. 10.24). ELT species could therefore provide a nexus for human influence in a technologically mediated ecological system, or post-natural fabric, by manipulating the chemistries and regular assessment of growth of the reef. In particular, indigenous marine life such as barnacles and clams may be encouraged to participate in the synthetic reef formation, for example, by manipulating the bioavailability of minerals, as they are naturally able to sequester carbon dioxide as mineralized forms and continually produce new niches for the local marine ecology.

---

72 In this context I am not referring to embryology as a formal style inspired by natural processes, as in Greg Lynn's 'Embryological House' (Lynn, 2000) where forms are barely (algorithmically) linked to their nutrient supplies. Rather, I am referring to embryology in a very literal sense – where self-organizing material systems are immersed in, saturated by and changed by their environment (yolk sac, placenta, water). The intimacy between conceptus and context codesigns the successive transformations of the matter in the system through many generations of iterative additive and subtractive processes.

73 I am intrigued by Lionni's notion of 'para-materiality' as a way of dealing with imaginary structures which are called into existence through a) direct and indirect observation and b) symbols. These tactics potentially offer ways of addressing the probabilistic nature of speculative design practices.

With the appropriate level of surveillance, timely adjustments could be made where different species of droplet could be strategically designed and directed to produce a network of expanding subterranean canals that are centred on the woodpiles. With monitoring of the accretion process, marine ecologists could determine just how much fresh smart droplet solutions they need to add to the waterways to continue to 'fertilize' the biological accretion process and keep the reef alive. Alternatively, if the reef activity is too vigorous, then smart droplets may not be added to the water (via modified cisternae or directly into the canals), to slow down or even reverse the mineralization process. Ultimately, Venice's post-natural landscape will behave as a complex system that acts more like a living creature than a machine. Such a system would also need to be nourished and cared for, just in the same way that we might take care of our gardens. ELT species create the poetic notion of an architecture that connects a city to the natural world in a very direct and immediate way and even revitalizes it (see Fig. 10.25). A reef of this character would be porous, yet rigid, and could potentially stop the city sinking by producing an upward and lateral counterforce through the process of active growth, like expanding roots. With sufficient vigour, growth of the reef may even raise the city above sea level, although it is more likely to form laterally spreading masses that take the path of growth with least resistance.

Figure 10.24: Reflections within the Venetian canals hint at a host of potential sites of transformation by ELT applications. Photograph, Rachel Armstrong, August 2012.

Figure 10.25: GMJ's rendering of a postcard from Venice transformed by dynamic droplets depicts a city that is transformed by self-organizing structures above the water level as well as underneath the foundations of the city. Drawing, GMJ, February 2009.

10.8 ELT and Environmental Responsiveness

> *According to Mr. Blainville, who usually is trustworthy, one can predict that in less than a hundred years will Venice be totally united with the rest of Italy and you can walk dry-shod from Italy to the city.* (@Google, 2010)

But what if the trustworthy Mr Blainville is right, and Venice dries out rather than floods as a consequence of the raising of the MOSE gates causing an unpredicted tipping point in the lagoon? Or, perhaps Pietro Teatini and his colleagues will be successful in their modelling of anthropogenic lift of the city of Venice by a foot over ten years (Castelletto et al, 2008; Teatini et al, 2011) and begin a real-world project to reinflate Venice's aquifers with salt water.

ELT may respond to drastic changes and environmental tipping points by responding in real time to the changing environmental conditions, rather than trying to anticipate them (see Fig. 10.26). Programmable droplets are robust enough to reappropriate their interactions as conditions change. For example, should water levels recede, the ongoing mineralization of the reef would not be directed outwards to spread the point load of the city, but is likely to move inwards and downwards into the woodpiles. The vigorous chemical and biological activity is anticipated to produce a kind of bioconcrete accretion to protect the integrity of the woodpiles by sealing them from the air and preventing them from decay. The slower the water levels fall, the thicker the mineral coats would be. The biological-like accretion system of ELT offers a completely different kind of solution to Venice's exposed woodpiles, as an active ongoing repair process that would be able to reach sites that surveyors may find very difficult to discover or reach. In recent years, the city's foundations have been increasingly threatened by the passage of huge ocean liners that travel to the city and literally suck the water out from beneath it, exposing woodpiles that have previously been entirely submerged. Equipping Venice with a new set of properties such as those embodied in ELT would create a possible future for survival for the city, where its very fabric could respond directly to changing conditions, prevent regression into the soft delta soils on which it was founded and effectively engage in its own 'struggle' for survival, just like living systems do.

10.9 Photographic Survey of the Biological 'Stones' of Venice

During my research, I photographically surveyed the Venetian waterways using a Lumix Optica digital camera with 12× zoom lens. My survey identified the canals around the Arsenale as sites of rich assemblages of accretion-forming wildlife. The surveys revealed that the assemblages were most pronounced during April and May, while very limited formations were observed in December and January. The accretions were greatest between the low and high tidal areas and much less profuse deeper down in the water, although the water quality in the lagoon was poor and therefore hard to photographically survey.

Figure 10.26: This mooring post is colonized by actants producing matter at the interface between air and water. Photograph, Rachel Armstrong, August 2012.

Notably, shallows are also sites for naturally occurring stromatolites, such as in Shark Bay, Australia (Shark Bay World Heritage Area, 2009), which are mineral accretions grown by bacteria that may take hundreds – even thousands – of years to grow. During this survey, it became possible to imagine how an artificial reef could be constructed, working with the growth imperatives of the indigenous wildlife, yet facilitated by ELT, so that an effective reef that could spread the point load of the city, as well as provide a unique ecological niche, could potentially be grown within decades as opposed to centuries. While my photographic survey of the waterways revealed the mineralized stromatolite-like deposits of the native marine wildlife attached to the building fabric, the precipitates produced by the modified Bütschli droplets for the Hylozoic Ground installation are also suggestive of how a material solution produced by ELT may appear. I have made a speculative leap from what was actually observed around the Venetian waterways to practically integrating these ideas, through collaborative drawings made with Christian Kerrigan in Figs. 10.15–10.17, which have been particularly useful in more widely facilitating discussions about applications of ELT within Venice's waterways. Kerrigan's time-based architectural drawings depict how a complex set of factors may interact and generate lifelike materials with unique aesthetics to produce a reef-like structure. They have also help identify points of possible influence within this process, such as encouraging the growth of stalagmites in the city as biotechnological 'stones' (Ruskin, 1989), or as a means of building up materials that have been digested by the *acqua alta*.

10.10 Vibrant Venice: Weaving a Post-natural Fabric

In speculating on the possible futures for the city, Vibrant Venice is one where lively matter shaped through ELT may forge an alternative development pathway that may be moulded by architects and citizens alike. This contrasts starkly with other proposals to 'save' Venice, such as the barrier against the tides afforded by the MOSE gates (United Nations Office for Disaster Risk Reduction, 2012), reinflating Venice's aquifers to raise the foundations of the city (Teatini et al, 2011), or constant repairs and maintenance after damage has occurred. Indeed, the bioregion of Venice may be considered as existing architecture that is embroidered by a constant proliferation of vibrant matter, which could be reconstructed by shaping spatial programs and tactics that generate new relationships between natural and artificial actants. The architectural inquiry concerns how overlapping spatial programs are fashioned and developing design tactics that enable a constant flux between fabric, space, structure and location. These are orchestrated through morphological computing outputs, which do not imitate Nature but work according to 'low-level' programming principles[74] and

74 I am using the term 'low-level' chemical programming to refer to fundamental organic and inorganic chemical reactions that are unmediated by biological catalysts or DNA.

the horizontal coupling between actants using a common chemical language. These networks, therefore, have the potential to constantly forge assemblages and respond to the constantly eroding and shifting bioregion.

The vision of Vibrant Venice provides an example of how incremental involvements and emergent dynamics might effectively work, through the production of life-bearing structures. By placing limestone shell creation in advance of the 'life' within those new shells before the evolution of soft-bodied organisms, the idea is suggested that architecture is not a secondary condition for 'life', but a primary driver. This implies that architecture itself may establish the conditions for the emergence of 'life', such as in the production of soils and water infrastructures. Rather than simply generating impermeable boundaries between systems, which characterizes the production of modern architecture, Vibrant Venice proposes that small, subtle events may radically transform the biological potency of a site so that it promotes rather than constrains 'life'. Through physical incursions of carbonate accretions, ELT collectively produces disruption within the complex marine foundations of the city of Venice and establishes a dynamic network of potential architectural events. These culminate in a reef-like formation that is codesigned by the marine wildlife and the city's inhabitants (see Fig. 10.27).

Figure 10.27: Stromatolite-like formations in the Venetian canals are shaped over time to produce mineralized materials. Computer drawing, Christian Kerrigan, September 2012.

As conditions in the lagoon change and the waters recede, a variety of ELT species may produce further richness in managing material and environmental complexity through a range of metabolisms and approaches. When combined, these actants may generate a portfolio of material responses, whose assemblages transform the city from being regarded as a passive object in its struggle for survival to an active participant in co-constructing its future (see Fig. 10.28). Through the strategic application of agile assemblages, orchestrated through ELT and shaped by morphological computing technologies, the city of Venice and its community could identify ways to deal with the constant change in their surroundings. Material networks may be established that could be orchestrated to restore, repair or even extend the city's fabric for its continued survival in the long term.

Yet, vibrant matter and ELT are more than functional material systems that build microstructures, but also possess the same kind of radical creativity (Kauffman, 2008, p.xi) that Nature does, producing rich patterns and metapatterns with unique aesthetics. While John Ruskin studied and documented the rich architecture of the buildings of Venice in detail (Ruskin, 1989) to formalize a classification system of construction and ornament, Vibrant Venice, as constructed by many species of ELT, could potentially offer a different kind of ornament and construction process. Such 'vibrant stones' would forge synthetic ecologies that reflect and continually respond to incessant changes at all scales (see Fig. 10.29).

They may be collectively regarded as a complex biotechnological system that exists as a range of assemblages, which are forged by the complex interactions between humans and non-humans. Vibrant stones share homologies with biogeological stromatolites, which are ancient layered stone structures made of calcium carbonate. Stromatolites are produced by the actions of photosynthetic cyanobacteria and other microbes, which trap and bind grains of coastal sediment into fine layers (Bernhard et al, 2013). Like stromatolites, the 'vibrant' stones of Venice also use biological and mineral actants to establish strategic connections and associations with their surroundings that change with time[75] and alter the architectural character of the city.[76] Venice's 'vibrant stones' may therefore change through the production of new relationships, which, in turn, alter their physical composition and appearance (Bennett, 2010, p.4). Perhaps similarly to Calvino's Marco Polo (Calvino, 1997), each visitor to Vibrant Venice may experience unique psychogeographies and improbable landscapes as the buildings become mirror images of themselves, or rise unexpectedly from the lagoon as alien landscapes.

Although different species of ELT are not alive and therefore cannot replicate, they do share lifelike qualities and are therefore capable of unpredictable behaviour (see Fig. 10.30). In keeping with the character of natural systems, ELT is also capable of novelty when tipping points are reached and strategies need to be in place that

---

75 It is thought that the stromatolites disappeared from global coastlines with the evolution of foraminifera, single-celled organisms that use slender projections called pseudopods to engulf prey, move and continually explore their immediate environment. Potentially, these creatures could be introduced into the canals as a living architectural design tactic to reshape the interactions of the stromatolite-forming communities so, instead of producing large, layered stones, they make clumpy, smaller structures called thrombolites (Bernhard et al, 2013)

76 In some ways, Palais Idéal may be thought of as a kind of thrombollte produced by humans, rather than bacteria and protists

Figure 10.28: The constant flux between fabric, space, structure, time and location constitutes the transformative potential of assemblage-based technology. Photograph, Rachel Armstrong, August 2012.

Figure 10.29: Green algae are ubiquitous in Venice and may be thought of as biological 'stones' that – unlike Ruskin's stones that alluded to Platonic forms – are vibrant processes that transform inert materials into synthetic ecologies. Photographs and collage, Rachel Armstrong, August 2012.

are designed to deal with unexpected outcomes. Yet, the changeability of ELT does not mean that 'anything goes' but that the system's outputs exist within definable limits that are determined by the interactions between the agents and their surroundings. ELT enables us to establish new relationships with our ecosystems by becoming involved with their performance, rather than simply commanding them so they may be abandoned – which is what we do with machines. Yet, no matter how unusual ELT may seem, the different species are unlikely to produce something really alien. The vibrant stones of Venice will be gardened into existence, so much will already be understood about how to persuade and influence the outputs (Sellars, 2011). Yet, ELT represents the tip of an iceberg of opportunities provided by the technological potential embodied in living systems and possesses a remarkable range of distinctive useful properties, including autonomous activity, environmental sensitivity, robustness, spontaneous adaptation and material creativity (see Fig. 10.31).

Yet all these events are part of and intimately connected to Venice itself, as chemical poetry whose sonnets make reference to the natural world but do not succumb to naturalistic clichés. However, Vibrant Venice does not regard Nature as an untouchable inspiration to inform our endeavours but as a very real part of the everyday fabric, which operates within a different organizational framework than

Figure 10.30: The various species of algae in Venice generate outputs that can be predicted within definable limits of probability, such as blooming. However, when these systems reach tipping points, the assemblage may also behave unpredictably by dying. Photographs and collage, Rachel Armstrong, August 2012.

Figure 10.31: These reflections allude to the chemical poetry of Venice's fabric, which is shaped by countless acts of codesign. Photograph, Rachel Armstrong, August 2012.

industrial systems. Yet, while industrial technologies may subordinate and poison the natural world, out of control naturalistic material systems may be equally damaging for the city's longevity (Sellars, 2011).

10.11 Ethics, Control and Grey Goo

> *This is the first moment in the history of our planet when any species, by its own voluntary actions, has become a danger to itself – as well as to vast numbers of others ... 'Plants' with 'leaves' no more efficient than today's solar cells could out-compete real plants, crowding the biosphere with an inedible foliage. Tough omnivorous 'bacteria' could out-compete real bacteria: They could spread like blowing pollen, replicate swiftly, and reduce the biosphere to dust in a matter of days. Dangerous replicators could easily be too tough, small, and rapidly spreading to stop – at least if we make no preparation. We have trouble enough controlling viruses and fruit flies.* (Joy, 2000)

The idea of ELT raises questions about control, ethics, bioterrorism and dystopian grey-goo scenarios, which accompanied the advent of nanotechnology (Merkle, 1992), such as Bill Joy's online commentary about the combination of nanotechnology with genetic engineering (Joy, 2000). Although ELT is not fully alive and is created from non-living matter, it nonetheless raises a number of social and ethical issues. These range from impacts on individuals, society, health, waste processing and the environment to cultural challenges such as moral prohibitions, values and the understanding of Nature. These considerations are essential in establishing responsible new ways of working, as well as meeting the needs of human and non-humans. In its current form, ELT is not fully alive and does not possess the chemical strategies that promote uncontrolled replication. For example, programmable droplets rely on the availability of dissolved carbon dioxide and minerals in the lagoon to construct a reef. Their actions are also dependent on the direction offered by chemical and physical languages in their environment such as concentration gradients and sunlight. Once their metabolism, or resources are spent, they cannot be refreshed; so, the welfare and proliferation of ELT is entirely dependent on infrastructure and environmental factors and, therefore, may be entirely shaped through appropriate design programs.

> *The notion of an uncontrollable limestone reef that spreads like a malignant cancer underneath one of the world's most romantic cities is one that would delight science-fiction fans and horrify conservationists in equal measure. But Armstrong, a sci-fi author herself, is adamant that there is no real risk of it happening. 'We wouldn't just start throwing [proliferating] things into the lagoon. You wouldn't do it without any telemetry or measurement of the ecological impact, but we'd have done a lot of work to understand this before we ever did a Venice-sized intervention,' she says ... Armstrong recognises the potentially devastating effects of [ELT] falling into the wrong hands but 'You can't prevent people from being Machiavellian,' she says. 'I wouldn't be as blasé as to say there is no concern. I know that we need to design with our eyes open.'* (Patel, 2011)

While the precautionary principle (Epstein, 1980) urges caution in the use of emerging technologies, there are also risks with not developing new approaches to underpin human development. The global impact of industrial technology is devastating our ecosystems and, although there is no risk-free solution, there are obvious possible strategies for coping with the risks in working with ELT. One is simply to limit their use to confined areas and not let them escape into the environment, which is a common approach to dealing with dangerous natural pathogens, such as the Ebola virus. Another method is to design ELT to self-destruct, or build in mechanisms that cripple or control it. Yet, since dynamic droplets, for example, are so chemically simple that they are unable to self-sustain, designing complex strategies to incapacitate them is unlikely to be cost-effective. A third approach is to programme the droplets to be dependent on a substance that can be blocked or which is normally unavailable in the environment, so that they can only persist in the presence of this particular 'food'. Yet, ELT has a strictly limited lifespan that is proportional to the availability of resources that feed its individual metabolisms. Consequently, it quickly becomes inactive unless the supply of these nutrients is strategically sustained. Quite frankly, the biggest challenge that ELT faces in the environment is engineering it with the ability to persist long enough to perform useful work. Further research and development of the technology is necessary to develop a formula that can be sustainable by adding 'food' to the lagoon in desired locations and to conduct controlled experiments to test its impact on the lagoon ecologies. Yet, such observations and measures would not placate concerns about the introduction of artificial agents into native ecologies, as all safeguards are fallible, costly and no containment method is perfect (Bedau and Parke, 2009).

As with all ethical dilemmas, there are no clear-cut answers to challenging questions and there is always a degree of risk associated with working in emerging practices that raise fundamental questions about our current expectations and practices. In my view, ELT is a field worth exploring, so that evidence-based operational principles, methods and practices may be developed that enable us to make informed decisions about their applications. Equally, an over-zealous application of the precautionary principle may be damaging if it prevents us from developing a range of approaches that could conceivably give rise to new paradigms that could underpin human development. Being too risk averse with the exploration and development of new technologies will, quite simply, leave ourselves with no alternative paradigms to industrialization from which to choose.

10.12 Summary

Contemporary cities are designed by making predictions about future challenges and then proposing fixed sets of structural solutions that deal with these eventualities. As such, they possess very little robustness or resilience in dealing with unpredictable events, such as natural disasters. With the prospect of increasing environmental turbulence as a consequence of climate change over the course of this century, a fundamentally different approach to the production of architecture is pressing. The prevalent industrial approach to urban development is fundamentally damaging to ecosystems, even when considerately applied using the principles of austerity and conservation, since these approaches have no regenerative value in the environmental health of the land they occupy (Friedlander, 2009). Vibrant Venice offers a specific context in which an alternative approach to architectural design may be explored using the technologies of 'life' to sustain and even nurture environments, by establishing new ecological networks and post-natural fabrics, which increase the fertility of the site.

Vibrant Venice also builds on the city's longstanding history of surviving the continual assault by the elements, in forging new relationships with cutting-edge technologies. Vibrant matter serves as a potential next technological platform in the city's development that could 'save' the historic site from destruction – not by creating a barrier against the elements (as in the case of the MOSE gates), but by undergoing many acts of orchestrated transformation (Dpr Barcelona, 2011). The combined speculative and experimental approach adopted in this proposal enables architects to embrace an emerging spectrum of ideas, methods, materials, infrastructures and technologies that are not mainstream approaches, but open up radical new possibilities within architectural design practice that may be nurtured into existence. As Riya Patel notes:

> *For architects and designers of infrastructure, it's hard to embrace a built proposition that doesn't involve concrete or steel but that's exactly what Armstrong is asking us to do. 'When we talk about systems in architecture, we tend to revert to machine iconography,' she says. 'With protocells, we don't need to. It doesn't have parts; it's not an object.' It may act more improbably and organically than any architecture we know, but it represents a chance to replace the outdated practice of designing buildings as environmental barriers with a more constructive and harmonious approach.* (Patel, 2011)

A demonstration protopearl model of the proposed technology was built and tested in the field collaboration with chemist Hans Toftlund, Martin Hanczyc and architecture students from the University of Venice. Although protopearls are far from formalized as a mature technology, they experimentally demonstrated that vibrant matter:
- Possesses agency
- Is programmable using morphological computing techniques
- Is a codesigner of architectural programs
- Coherently operates across many scales that include the architectural realm

Much more rigorous scientific study for a real-world intervention is needed to develop the technology that would enable an artificial limestone-like reef to be grown under the city of Venice and will form the basis for future research. Yet, in the near term, it is likely that dynamic droplets and other species of ELT may work alongside traditional materials and mechanical systems in other contexts at smaller scales, such as in 3D printing (University of Southampton, not dated; WETFab, 2011, Armstrong, 2012g; Adams, 2012; Villar, Graham and Bayley, 2013), or 4D printing technologies (TED.com, 2013b), where they may be integrated within a combined platform of material, infrastructure and technological systems that enable new methods of production for architectural practice (see Fig. 10.32).

Yet, vibrant matter does not propose a comprehensive solution to Venice's precarious future, or act as an antidote to change. Rather, it increases our portfolio of choices to deal with the instabilities of our reality. Vibrant Venice is, therefore, not just a laboratory for examining the potential impacts of an emerging technology but is also a real and imaginary stage where common international issues – such as changing cultural, environmental, economic and political conditions – may be played out in full view of global audiences (see Fig. 10.33). While my research does

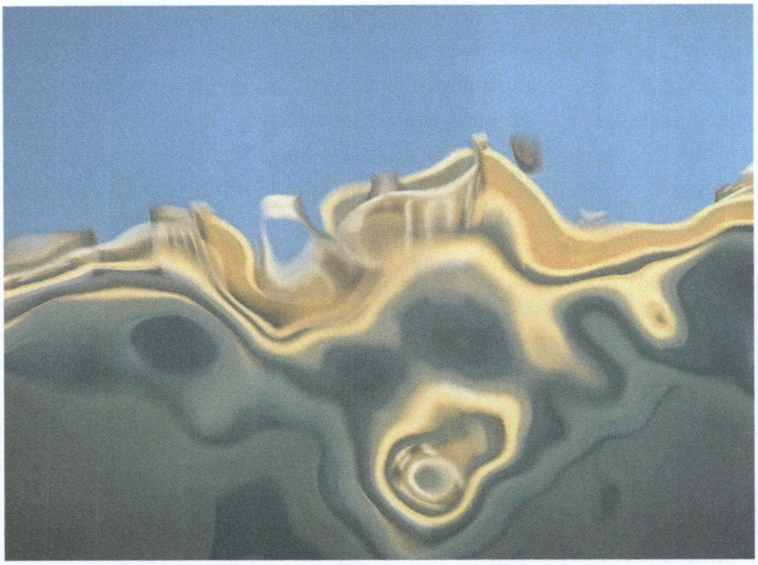

Figure 10.32: Waterway reflections provide a sensor and fabric that betrays some of the complexity of ELT applications, which offer a combined platform of material, infrastructure and technology that may inform and enable new kinds of architectural practices. Photograph, Rachel Armstrong, August 2012.

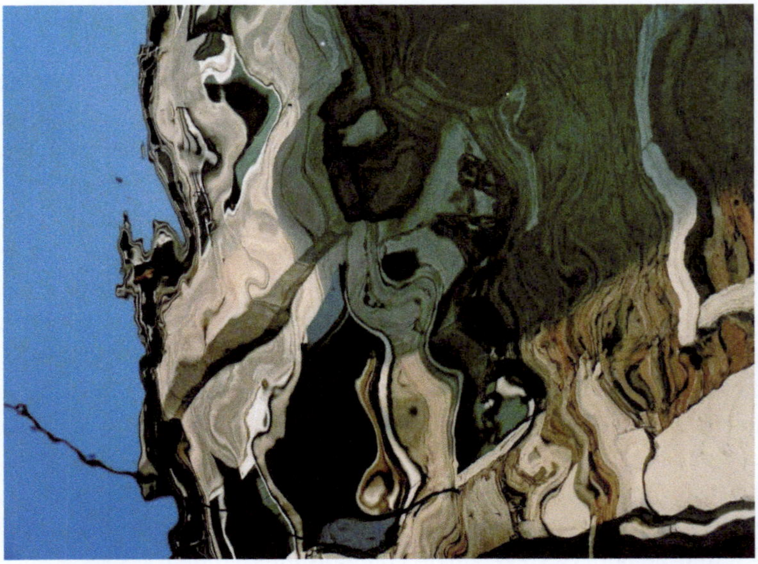

Figure 10.33: This waterway reflection speaks of the longstanding codesignership of Venice by human and non-human agents. Photograph, Rachel Armstrong, August 2012.

not propose to exhaustively debate the ethical challenges of working with lifelike technological systems, the next chapter uses speculative fiction to consider how a thriving future for the city may evolve from the complex interactions and possibilities between different species of ELT, native ecologies, garbage and human culture in the bioregion of Venice.

11 'Post-natural Venice'

11.1 Overview

This speculative narrative serves as a design exploration that extends the ecological and cultural implications of vibrant matter within the city of Venice from the perspective of its inhabitants, which include human and non-human communities. This story was written and performed for the Meta.morf conference at the second Trondheim Biennale (Meta.morf, 2012b).

11.2 Post-natural Venice

> 'Hey! Get out of here! You don't have a vendor's licence! I'll call the police on you!'

The café owner of the wooden shack harassed the unnaturally tall and bronzed beachcomber who was bumping his aluminium cart over the sluggish, dry sand outside his shop. Quite without any right, the shopkeeper regarded this stretch of land as being his own.

> 'Hey! Are you stupid? I said, get your cart off my property! You foreigners are not welcome here!'

Feeling ignored by the giant, the shopkeeper dramatically slammed his fist against a large warning sign, painted with a mosquito glaring '*Pericolo! Malaria!*'.[77] Its surface swallowed up the force of his hand with a dull 'thud', which nobody heard and no one felt the least bit intimidated by. Being all too familiar with the shopkeeper's penchant for drama, the foreigner continued to jerk his reticent cart down the beach, through the searing heat.

The giant's orange skin was not a bad application of fake tan, but large amounts of carotene deposited in his skin, which had even tinged his eyes yellow. It came from the sweetest ice-cream oils, which were made using the red varieties of seaweed that were particularly rich in the vitamin. Evidently, the recipe was so tasty that the beachcombers ate a little too much of their dairy-free treat, on which they seemed to be able to grow tall – and still had enough left over to peddle to tourists.

Spotting the illegal vendor with the strange cart of recycled tubs and thermos flasks, a small boy dashed towards him, his hands outstretched for *gelato*.[78] He tripped on a half-buried shard of driftwood and fell sprawling just in front of the orange man.

---

77 Danger! Malaria!

78 Ice cream.

The vendor lifted the boy to his feet and dusted off the sand from his palms with two quick 'high fives'. He pressed a dried seaweed cone into the boy's hand, who looked wide-eyed at the delicious treat, as it seemed to have appeared from nowhere.

'I call this one the "pick me up" flavour! Little man!'

beamed the giant, sunlight decorating his teeth.

A woman scuttled up to the vendor, exchanging care of the boy for a euro note. She perambulated the child firmly back towards the beach, partly chiding him for his impulsivity, but mostly relieved that her son had not been harmed by something nasty such as broken glass or metal fragments superficially buried under the sand. They settled down under a sun umbrella. His sister joined them but did not sit down. She danced on her tiptoes and began buzzing around the stem of the shade, flapping her arms as wings and humming in a low-pitched, irritating manner. She looked remarkably mosquito-like, with her harlequin-patterned fly-head *Commedia del Arte* mask that she was using as a sun visor. Suddenly, the fly dived on her brother, taking a large bite of his fizzing ice cream, which he was eating too carefully for her liking. Their childish wrestling brought their mother to her feet as the end of the cone crumpled and sticky effervescence splurged on the ground, while the boy fought off his licking sister.

An old man nearby watched an oversized, diamond-patterned mosquito seemingly devour a young boy. He quickly reapplied insect repellent that he kept in an over-stretched, algae-fabric pouch that had been given to him by one of his eight grandchildren. He couldn't remember specifically which one. The durable yet silky linen was embossed with 'grandfather' on one side of the purse and an imprint of San Marco on the other. The fabric possessed remarkable shape-memory qualities, which was great for pressing patterns into the fabric, but annoying when carrying large items, since they scarred the fabric with their imprint and were almost impossible to erase. All kinds of biting parasites were prevalent nowadays, like the pestilence of fleas upon rats in Venice hundreds of years before them. The old man particularly bemoaned that the region now had its own unique species of tiger mosquito, which were the main vectors of drug-resistant malaria. With the warming climate, these dreadful creatures were spreading their malady from the city into the body of mainland Europe. The old man's wife sighed as she rubbed cream into the chronic sores caused by sand flea bites on his legs, which had been hollowed out by secondary infection. Every illness nowadays appeared to be antibiotic resistant. All that she could offer her husband was love, care and the odd handful of salt in his bathwater. She slipped the stick of melting insect repellent from his grasp and helped him reach his scarred ankles, which made him wince.

'Careful woman! That hurts!' he complained.

She rolled her eyes and finished what she'd started, spreading the balm thickly. Then she conscientiously wrapped the spent remains of the applicator in a tissue and placed it her beach bag, not wishing to make even the tiniest contribution to the refuse invasion of the Venetian beaches. It was such a shame that the refuse barges couldn't make their way right into the canals any more to look after the city's canals and waterways, as they used to. Municipal dredgers had simply given up scraping the channels between the *bricole*, which were wooden posts that marked the channels in the lagoon that were accessible to boats. These were now clogged with sediment and thick, relentless weed. Consequently, people had become even more careless with their waste since there was nowhere to dispose of it. When they tossed it 'away' it finally ended up 'here', around the shores of the Venetian lagoon embraced by the Litorale Pellestrina, Litorale di Lido, and Litorale del Cavallino. These three strips of land were broken at three entrances to marshes that bordered around forty-five kilometres of the mainland around the Po delta, which was fed by lazy rivers and flushed by the slow tides of the Adriatic. At least, that was the state of affairs until the MOSE gates were constructed in 2016, a little behind schedule. Despite the prolonged protests against the series of seventy-eight robotic gates, the project had been completed and now commanded the tide like the old Danish King Canute. When the *acqua alta* came, they were raised to stave back the advance of the iceberg-fuelled rise in seawater that flooded the city limits and literally digested away the stones. Yet, these colossuses also made an unexpected, immediate impact on the lagoon: as they were raised so frequently, they caused the waters to stagnate like a giant pond.

As her husband began to doze, the woman returned to reading the out-of-print book she'd bought in a second-hand store in one of the narrow streets in the city. It was a compendium of historical insights into Venetian life, with excerpts and poems from a range of authors such as the courtesan-writer Veronica Franco.

A small queue of excited children, chattering mothers and heat-exhausted husbands straggled around the sand-bruised aluminium cart, waiting in turn to buy effervescing green and yellow cones of melting ice cream from the orange man. He used 'magic' tricks to produce the treats in exchange for currency or useful items such as batteries or old toy motors, which could be recycled. The crowd gathered not only to seek relief from the afternoon heat but also to escape the tedium of the beach. The vendor's antics entertained them with the illusion that carbonated ice cream could hide behind shade cloths, appear from out of the ears of small children, or be pulled out of the air.

The orange man's conjuring, which suggested that the cool confectionery could be extracted from the atmosphere, wasn't entirely misleading. Originally, the beachcombers were a handful of Dutch provocateurs that started an experimental, environmentally concerned, art community on the north end of the Lido. They strived to live solely on matter that could be harvested from the beach and expected to inhabit the shore for a fortnight. Yet, they found the appropriation of resources and change in lifestyle values not only a rewarding artistic pursuit, but also a profitable one. Within days, tourists had discovered the 'happening' and came to watch the community as they established the best ways to make shelter, and discover what *objets trouvés* were palatable, or safe to eat. Around a month into the newly extended project, locals began to make suggestions to help the foreigners familiarize themselves with the peculiarities of the region.

The beachcombers had limited access to wireless technology, except when they talked to spectators and asked them to research lifestyle challenges online. Audiences identified various marine species, discovered the melting points for glass and plastic and even suggested cooking recipes. Over the following months, crowds came to observe the experiment in 'sustainable' living and started to offer money or exchange goods for their strange inventions, artefacts and tasty produce. The beachcombers' most successful recipe was algae ice cream, which required a source of dry ice from beyond the beach. A group of young biotechnologists from the mainland donated the precious liquid gas in pressurized cylinders and brought it from their laboratory in Marghera by boat. They advised on using dry ice safely so that the ice cream was not eaten too soon, as it could cause frostbite. As their interest in the artistic 'experiment' grew, the researchers began to collaborate with the beachcombers spending weekends helping them build safe storage units for the cylinders and design the most efficient way of producing the algae ice cream. Their teamwork resulted in a commercial product that enabled people to make the delicious treat at home.

Having forgotten their ice-cream tussle, the boy and his mosquito sister climbed into their flip-flops, which were beachcomber-woven from ropey algae. The 'beach-safe' footwear offered significant protection against the treacherous amounts of unground glass and fractured shards of hard plastic that were deceptively sprinkled with sand. Also, summer temperatures were so hot that it was sometimes impossible to walk across the beach barefoot. Under the fiercest heat, the sun's rays turned sections of sand to smooth obsidian and melted beached plastics into oily mats that clung to unwitting skin like molten wax. With the spongy soles fizzing, flipping and flopping against the undersides of their feet, the children sped over the wet sand towards the colossal walls of the MCSE gates.

"Be careful! Don't go too far!"

their mother shouted after them, sighing audibly.

When the gates first rose and fell, churning up the lagoon waters like the lips of a slobbering great sea beast, they were quickly colonized by plaque-like algae that were quick to feast on the opportunity of a disrupted environment. The tenacious weeds set the scene for rapid invasion by all kinds of other aliens, such as bryozoans – small yet persistent creatures that had lurked in similar 'warm muddy pools' since the Cambrian era. New sandbank stretches that once intermittently protruded through the waves as silty shallows now rose like sea monsters and choked the dancing currents into the consistency of an ecological soup, setting a feed-forward cycle in motion. The rapid stasis enabled ferocious, toxic algal blooms to further slow the tidal exchanges, resulting in a dramatic shift in marine biodiversity with a particularly nasty outbreak of tiger mosquitoes. Silt quickly deposited in the deepest basins of the lagoon, which became so thick and persistent that it even prevented the dredgers from keeping the navigational channels open. Venetians watched from behind the protection of their submarine wall, as the richest Mediterranean costal ecology simply ground to a halt. By the time the authorities admitted that the light-sprinkled, cavorting waters of the lagoon had become a thick, refuse-trapping expanse of organic sludge, the tipping point had already passed. Although the gates were permanently lowered to invite the Adriatic back into the lagoon, the concession was too late and merely encouraged the algae-filled, refuse-packed waters to spill into the sea and propagate along new shores.

The wall of the MOSE gates was further away from their mother's recycled shade-umbrella than the boy and the fly had assumed. They stopped to cool down and splashed each other for a while, cleaning attractive shells in the waves that crept onto the beach. The colours sparkled vibrantly while they were wet but faded into mute pastels as the water evaporated. Viewing their haul as a magical, transmuted treasure, the children stuffed their pockets full and raced each other again towards the wall, which never seemed to get any closer. On baking hot days like today, the Venice lagoon was not an inanimate realm but a rising landscape of vitalized, interacting agents that marched together towards the city like Burnham Wood. Their entwined bodies heaved in networks of novel ecological interactions that appeared to cast out the city from the waves like flotsam and return it where humans belonged, back on terra firma. This non-human rebellion appeared to be orchestrated by Nature itself, conjuring strange matter that squirmed within the lagoon. This proto-soil was riddled with marine interactions between all different kinds of matter – living and non-living – that collectively heaved the city back to the mainland. Webs of interaction between minerals, pollution, refuse, excrement, bacterial mats, weeds, crabs, barnacles, shellfish, waders, snipes and gulls were shockingly revealed by the engorged corpulent lagoon as a gluttonous spectacle of autocatalytic organic cycles of growth and decay. Their potent, collective forces quickly reshaped Venetian life and its economy by driving the vendors and tourists from the city to the beaches.

Indeed, the native wildlife seemed to no longer consider Venice as being a seashore habitat, but a metropolitan enclave. Yet, in the winter when the algae blooms were quietest it almost seemed that the lagoon was reverting to its former aquatic character. Pools split through the sticky organic films like tiny grey mirrors under the muted light. This attracted the seagulls, which were being netted to limit damage to historic brickwork from the build-up of caustic guano, back out towards the sea. Yet the predators screamed defiantly and thrived despite increasingly concentrated efforts to completely eradicate them, having learned to snack upon easy-prey pigeons rather than fish. Seaweeds slept until May, when their metabolic activity was roused by the climbing sun and lengthening day.

As the algae swelled, everything choked and the city started to drown again beneath the proliferating bodies of organic matter, refuse and thriving biofilms. As one thing was chemically converted to another, the lagoon experienced radical changes in salinity, oxygen content and in the spectrum of its primary microbial flora, pathogens and parasites. These alien relationships worsened the biochemical changes in the region's ecological webs. Proteins secreted by microorganisms turned organic matter into pungent gases, atmospheric oxygen oxidized mineral precipitates, and even immortal plastic polymers were digested and split into sweet, noxious organic vapours that circulated through the extended metabolism. The giant, carnivorous, organic beast in the lagoon began to swell and drink the water underneath Venice's foundations so, against all the calculated odds, the city started to dry out and the monstrous mass ruptured at the edges to spawn lumps of gelatinous land.

The old woman looked up from her book to start a conversation with her husband about the time when everyone presumed that the sea, just like in the legend of Atlantis, would claim Venice, but he had fallen asleep. Pesky sand fleas continued to make fresh wounds in his flesh, so she grabbed a few clean tissues and swatted the crustaceans as the old man's legs twitched. The old man snored, actively dreaming, and retracted his legs from his wife's flapping.

'Mind out!'

he mumbled through the depths of REM sleep. The old woman smiled and withdrew her attentions.

He was happy and most probably reliving his time of glory when he was appointed captain of one of four custom-built 'protocell' barges. These custom-made vessels delivered an experimental technology to save Venice from 'drowning'. It was the highlight of his career and his appointment was even accompanied with a mayoral handshake and his picture in the national paper. Unlike the rusty, pea-green refuse barges that provided a domestic rubbish collection service through the canals, these boats were designed for the future. They glared with polished silvered surfaces that were impossible to look at directly as they acted as a suntrap to channel solar energy into the vessel to power its mechanical arm.

This rather elegant robot had wrist and finger actions so it could rather deftly mix the protocell solutions together in special containers on the deck with its fingers and quickly empty them into the canal with a brisk flick of its wrist. Several sites around Venice were approved for testing, including the industrial shores at the intersection of the Via Liberta with the main island, the Castello region around the Giardini and Arsenale, and near to the Stazione Marittima on the Giudecca Canal. Each location was allocated its own protocell barge. Captains not only steered the vessels but also gathered data on the performance and environmental impacts of the chemically programmable oil droplets through arrays of sensors on the underside of the vessel. It was expected that this strange fluid, with a will of its own, could actually grow an artificial limestone-like reef under the foundations of the disappearing city. When the silver arm tipped the agitated, oily solution into the water, its sudden effervescence was spectacular.

The droplets were sensitive to light and exploded sideways away from the sunshine and towards the shaded banks of the waterways. Here they jostled for the darkest nooks and crannies to activate a second chemical system that enabled them to make solid matter using dissolved carbon dioxide and minerals to forge stable chemical bonds and generate repetitive crystalline units. Sometimes they seemed to create pristine, symmetrical formations and at other times they twisted to enfold impurities into the structures. These submicroscopic interactions could be examined through fibre-optic tentacles and the barges tested the vigour of the lifelike chemistries to assess whether any changes in the mother solution were necessary.

When sample droplets exploded too quickly to be recorded, they could be documented by following the milky trails that were left behind, which were studded with little white pearls of a concrete-like material, called 'protocrete'. The captain marvelled at the ingenuity behind the technology, which continued to work long after he'd shut down the robotic arm. In the privacy of darkness, the protocells continued to make hundreds of thousands of tiny mineral shells under the foundations. Each droplet produced a formation with a unique character, shaped by their combined interactions with marine currents, wildlife, shadows, reflected light, pollutants and effluvia. Together, these agents thickened the girth of the narrow woodpiles and spread the weight of the city over a broader base, attenuating its slow slippage downwards into the muddy soils on which it was founded.

Not all protocells were destined to become part of the reef. Some were eaten by fish fry, others floated like little polystyrene beads and occasionally churning currents drew shoals of glistening droplets out into the lagoon. Yet the misfortunes of the few were outweighed by the success of the many. As long as the test sites were fed with new droplets and minerals, 'protocrete' continued to grow around the woodpiles. Strange mosaic gardens, which were even seeded across the lagoon in a range of environmental niches were visible at low tides from the shore.

These breakaway protocell communities no longer depended on support by the brilliant, robot-armed barges but had become feral, somehow surviving on their own and redefining the very edges of Venice. Satellite GPS systems digitally documented this material revolution that appeared as a worm-like infestation of the shoreline, which took place so subtly and on such large a scale that it remained unnoticed by local inhabitants.

The four protocell barges were decommissioned when the lagoon turned and the waters retreated. Although involuntary retirement marked the end of the old man's working career, the barges were kept busy, as they had work to do elsewhere in the intercoastal regions of the world. Two were shipped to Songdo, one to Shanghai and the last was destined to feed the mineral-starved natural reefs of the Maldives. Protocells were not successful because they performed the job they were programmed to do, but because they were flexible enough to carry out other unforeseen tasks, for which they had not been designed. So, as the organic beast in the lagoon grew thirsty, the droplets responded to the environmental changes without the need for the barges and began depositing their materials in a downward direction. The calcareous protocrete impregnated the vulnerable Venetian woodpiles and protected these pickled organic structures from decay. As the lagoon dried out further, a maze of unique networks of walkways and natural bridges straddled the shallow canals, which swelled when they were moistened by rains. Their periodic changes in the Venetian landscape were orchestrated by the swelling and sinking body of the beast, and attracted visitors headed for the famed Biennales who were fascinated by the artwork of 'synthetic nature'.

These artefacts were spat out from the lagoon, attained cultural status and were critiqued alongside other non-human works, such as chimpanzee paintings. The celebrities who had bought property in Venice and attended such affairs had abandoned the idea of living in the city, long before it started to smell unbearably. Tourist industry employees, who could not afford their own place on the island, had moved to Maestre on the mainland, away from the growing malodour and increasing pressure of visitors that continued to visit the historic city centre.

Yet despite the olfactory assault from Venice's stagnating waterways, its cultural prestige increased – especially with the strange new views across the lagoon and 'protocell gardens' – which ensured that the tourists still came.

Yet, they no longer took the traditional route by sea into the lagoon but came by land and train and regarded the overbearing odours as being 'authentic'. Indeed, it became de rigueur to pass through the alleyways behind a bouquet of herbs as in days gone by, when plague doctors braved the streets dressed as scented crows.

The orange man turned his empty aluminium cart around and headed back to the commune, with a few over-excited children following him, conspiring that should he turn around, they would disappear from his view as if by 'magic'. The giant smiled with contentment as he trudged back up the beach, and the children tired of their game. Although many Venetians had left the region, the beachcombers moved in the opposite direction. They formalized their residence with permanent shelters made from 'tabbycrete', which was produced by mixing sand with lime made from boiled-down oyster shells and adding crushed shells, to give the mixture strength. Occasionally, fragments of protocrete which had broken off from one of the protocell gardens and migrated along the shoreline were also used to thicken the paste. The beachcombers sometimes conjectured whether the protocrete was actually sufficiently 'alive' enough to propel itself through the water and, if this were the case, would it then be ethical to add the rich, actively calcifying substance to their tabbycrete mixture?

At the MOSE gates, the children caught their breath with their clothes flapping around them like beached fish and bathed in the chilled sea breeze. Perching on a huge chunk of limestone, they studied the strange, beachcomber huts that sprouted like barnacles out of the base of the supporting wall. Each of the living pods was designed to accommodate the tallest person in that family group standing upright and was crowned with an apical hole. This was plugged with melted plastic to let in the light and cover the skylight when the weather or temperature changed. During the summer months, the heat was unbearably hot and the thick tabbycrete walls provided welcome cooling. During the night, when temperatures plummeted, the layers of plastic bottles that let studded light into the space also offered robust insulation against the wind and the sea chill. Compared with industrial buildings, tabbycrete and recycled plastics were much less structurally sound and less hardy than industrial materials, but their fragility was never really a problem for the beachcombers. The pods had never been imagined as being permanent structures, and it was accepted that they would weather, crumble and be subsumed by the ravenous shoreline. So the community continually rebuilt, reshaped and reinsulated the weathering plastics, replaced the crumbing tabby and reappropriated their living spaces to meet their changing needs. Moreover, the community grew outwards, not upwards, so the pods never needed to deal with considerable structural loads that bedevilled skyscrapers. Pods merged, separated, moved over and under each other, just like a proliferating mass of living cells. Revitalized, the boy and the fly raced each other back to their mother, passing the most ambitious public structure built by the beachcombers – a large light-filled series of domes that were perforated by hundreds of glass and plastic apertures, which intersected at tree-like pillars. This space was called the 'clam' where beachcombers welcomed visitors.

> 'It's just like the Doge's palace!'

breathed the boy admiringly, before following his sister back down the beach to protest to their anxious mother that the MOSE gates really were not all that far.

The shopkeeper had been waiting for the orange man's return, to caution him against stealing his business and instruct him to find another part of the beach to sell his wares on. The gracious orange man nodded politely yet unmoved as he passed the animated shopkeeper whose words were spitefully snatched by an upstream wind, and a tiger mosquito quietly bit him on the leg.

12 Vibrant Cities

A soil was not a thing ... It was a web of relationships that stood in a certain state at a certain time.
(Logan, 2007, p.96)

12.1 Overview

This chapter explores how vibrant matter may be applied within an urban context through the design and construction of webs of living materials that constitute complex fabrics, such as our soils, which may be synthetically produced within underused and under-imagined spaces within our home and cities (see Fig. 12.1). The careful design and engineering of these fabrics may provide designers with access to a new kind of production platform that may not only change our design practices but also shape new cultural values. Potentially, these post-natural fabrics may offer a fertile field of new possibilities, which may give rise to vibrant cities where human and non-human communities collaborate and codesign our living spaces and evolve alongside us as expressions of Millennial Nature.

Figure 12.1: Complex chemical structure produced by dynamic droplets in an oil medium is reminiscent of the micro-channels that exist between soil particles. Micrograph, magnification 4×, Rachel Armstrong, August 2012.

12.2 Soils as ELT

Architecture is both responsible for and can take action against the destructive environmental practices that have characterized the last 150 years, by proposing new ways of underpinning human development that work in opposition to the prospect of a sixth great extinction event (Sample, 2009). Contemporary society draws from a world that is less determined by objects and increasingly shaped by connectivity. This perspective has become more than an academic conjecture, but an everyday reality through our growing reliance on global telecommunications systems. The clear either/or distinctions that formerly shaped our experiences of the world are being replaced by a much more fluid relationship with reality where identity is no longer fixed but can coherently and simultaneously exist in many states. This complex worldview extends to the characterization of Nature, which is made up of many interacting bodies that have been historically recognized as the animal, plant and mineral kingdoms (see Fig. 12.2), which are seamlessly entwined in ecological systems and collaborate through evolutionary acts (DeLanda, 2000, p.26).

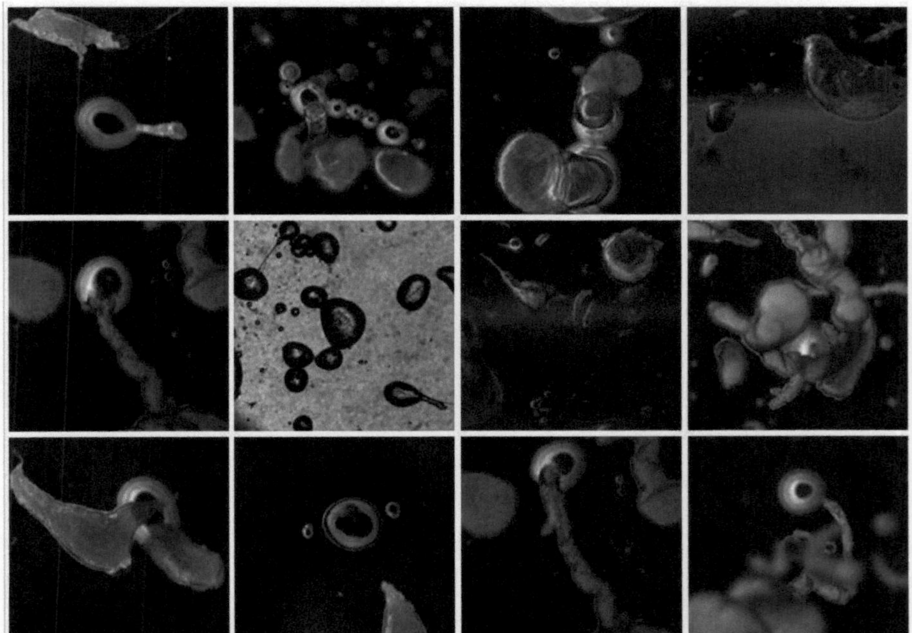

Figure 12.2: Complex structures produced by dynamic chemistries may relate to the spatial complexity produced by metabolisms, which enable the evolution of soils. Micrographs and collage, magnification 4×, Rachel Armstrong, February 2012.

The story of soil best embodies the enriching, complex exchanges within the structure of matter. The Earth was not 'born' with soil but has acquired its living web of relationships over the millennia (Logan, 2007). Indeed, Latour proposes that when thinking about natural processes we should use the term 'geostory' rather than 'history'. Geostory is a non-human narrative that stretches back in time before the evolution of our species and is forged by agents such as tectonic plates, microbes, meteorite impacts and ice sheets. It foregrounds all the actors backgrounded by history in an ontology of events where the past is understood as being more like an opera than a formal 'design', whose endurance in the present depends on its constant re-telling (Latour, 2013). Soils constitute the outermost layer of what the 19th-century geologist David Thomas Ansted (1814–1880) called 'the great stone book of nature' (Ansted, 1863), which offers a coherent, unified story that contains its own history. Soils can be up to several million years old, though many North American and European species date to the end of the last glacial period, around 15,000 years ago. Soils are attractors of terrestrial life. Plants take root in their complex chemical bodies, where organic and inorganic particles are entwined within a matrix that harbours fungi and bacteria. These complex assemblages, forged by many participating bodies, form a self-perpetuating system that breaks down the bodies of dead creatures and turns them into more soil. The speed of this dynamic conversion process varies. In fertile areas it may take 50 years to produce a few centimetres of soil, but in harsh deserts it can take thousands of years.

Our living soils age as a consequence of natural causes, such as changes in the climate. Yet, increasingly, their vitality is being impaired as the result of artificial and biological factors, such as overgrazing and deforestation. Ultimately, soils die, and when they do, they are gone forever. These acts of wanton destruction are due to our rapid expansion, technological naiveté and, as Allan Savory notes, our universal tendency to simplify the complex processes of ecosystems in agricultural management practices (Savory, 1998; TED.com, 2013a). In these last few milliseconds of evolutionary time, when Homo sapiens appeared on the Earth, we have globally acted upon our abstractions of the world at an exponential pace. Whitehead warns about our reliance on abstractions, and while he acknowledges that they are vital for our construction of ideas, he asks us to carefully consider what is at stake when we adopt them (Whitehead, 1979, p.15; Whitehead, 1968, p.116).

Indeed, our very modes of thinking have disrupted the existing complex ecosystems that we rely on, as we have embraced an increasingly globalized, industrialized culture. New ways of thinking may therefore help us address the resultant imbalances, although it is impossible to say whether we still have time to turn our environmentally destructive, modern legacy around, since ecologies are as fragile as they are resilient.

12.3 Architecture As a Site for Ecological Revival

> *Crucially, I think that city making is not a planning process; it's a becoming process. Because we can only partially see the results of what we do, we live in the face of mystery. There's magic and there's enchantment. And that leaves us with deep questions as human beings because we've been taught that we can know, master and optimise. We have to think in new ways about how to do that wisely.* (Kauffman, 2012)

Potentially, architecture offers a site for ecological regeneration and synthesis. Its sheer scale rivals our fertile biotic soils, which promote life, diversify ecologies, recycle resources and propagate globally. Soils are biological cities that are replenished by the diverse communities and countless networks from which they are formed (see Fig. 12.3).

The evolution of soil has laid the foundations for the establishment of life on Earth, the flourishing of ecosystems and the fertile conditions that enabled humans to construct the first cities around the potent soils formed around deltas such as the Nile, Tigris and Euphrates. The archaeology and foundations of contemporary architecture therefore depend upon the rich infrastructure of soil systems. While soils are many

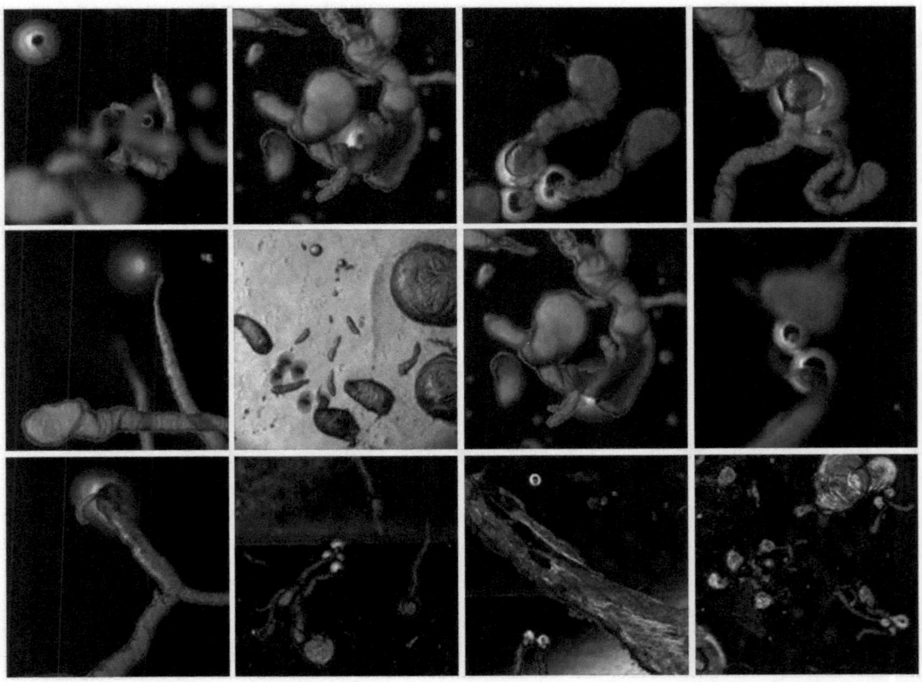

Figure 12.3: Dynamic chemistries depict a series of evolutionary transformations that embrace a range of transformations and capacity to complexify their environment. Micrographs and collage, magnification 4×, Rachel Armstrong, February 2012.

thousands of years old, even the oldest cities are only very young in comparison to them, being no more than a few thousand years old. Antep is the most ancient currently inhabited city and dates back to the old Hittite period (1750–1500 BC) but many modern cities are only hundreds of years old. George Orwell notes, ‘man is the only creature that consumes without producing’ (Orwell, 1979, p.4), and our ravenous cities are parasitically draining terrestrial soil bodies of their vast mineral and biotic resources. While biotic soils are self-renewing, the dirt produced by modern cities is not meaningfully returned and recycled in self-regenerating systems of material restoration, but is isolated, sealed in garbage dumps and landfills, or scattered into our oceans as sewage, where it is prevented from forging productive ecological networks.

Modern cities are literally made of the same kind of stuff as soils. What separates a building from soil is simply time’s arrow (Prigogine, 1997, p.1). Inert materials such as concrete, brick, clay, stone, steel and wood are simply technologically processed agglomerates of molecules that are already present in dirt and will return to dust if they are not maintained. Indeed, classical building substrates could be thought of as soil components that have been reverse-engineered from complex, heterogeneous systems into simple, obedient geometric forms. Nature abhors homogeneity and seeks to recomplexify these substances (see Fig. 12.4).

So, in the same way that soils have been forged by grinding glaciers over thousands of years, the surfaces of buildings are being weathered and sheared by the

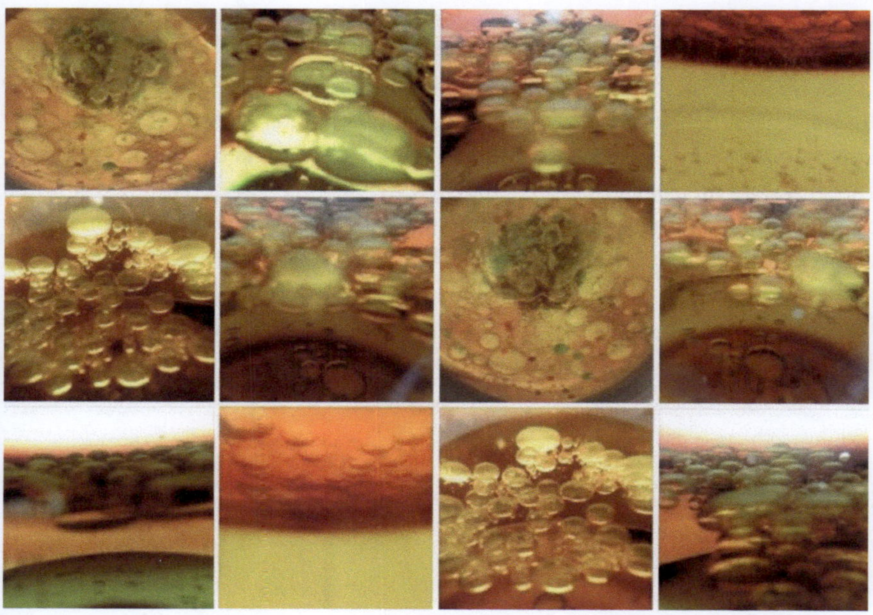

Figure 12.4: Like soils, mineralized structures produced by dynamic chemistries may become more complex by virtue of their metabolisms. Photographs and collage, Rachel Armstrong, February 2011.

same forces that created our primordial dirt. Moreover, they are invaded by microbial life that tears apart their inert infrastructure to reveal, expand and vitalize new surfaces, which can be further colonized by living invaders and through the biological process of succession. Sites of decay can be thought of as unfolding, active chemical interfaces that are symptomatic of the presence of life-giving processes and may be regarded as ecologically potent locales. Indeed, Hundertwasser notes that there is much to learn architecturally from the decomposition process. 'When rust sets in on a razor blade, when a wall starts to get mouldy, when moss grows in a corner of a room, rounding its geometric angles, we should be glad because, together with the microbes and fungi, life is moving into the house and through this process we can more consciously become witnesses of architectural changes from which we have much to learn' (Hundertwasser, 1976). Ambasz worked more directly with the notion of soils as regenerating substrates by promoting the use of landscaped terraces, earth-sheltered roofs and underground dwellings as a way of returning the very land that cities took away. He sought not only the disappearance of architecture, but for it to be absorbed into new artificial natures (Dean, 2011, pp.230–231).

Although cities (Kazan, 2008) and the Earth's ecosystems[79] have been likened to organisms, technically they do not qualify as such. The current definition of 'organism', or life, does not embrace the pervasive bodies that comprise soils and cities, and nor do they possess any centralized genetic program. Yet the similarities between cities and soils are striking, as they are complex and share, in principle, many of the characteristics of organisms. Jan Christiaan Smuts noted that materials possessed a unique spectrum of agency, or lifelike characteristics, which ranged from crystals to biology (Smuts, 1998, p.88). The importance of lifelike systems lies in their subtle, persistent behaviours, which, without human intervention, spontaneously generate infrastructures and systems where life may thrive. It is this quality that falls apart in cities, as their materiality is eroded with time and not re-enriched through ecological cycles of material exchange. For example, the city of Venice, with no living characteristics, is simply unable to fight back in the struggle for survival against the elements. Yet, a myriad of biological agents are extracting minerals from the water and (re)assimilating the city fabric to build their own accretions and microbial 'cities' (see Fig. 12.5).

79 In 1785 James Hutton, the father of modern geology, envisaged the Earth as a metaphorical 'super-organism'. He suggested that its circulatory and respiratory cycles were geological processes such as erosion.

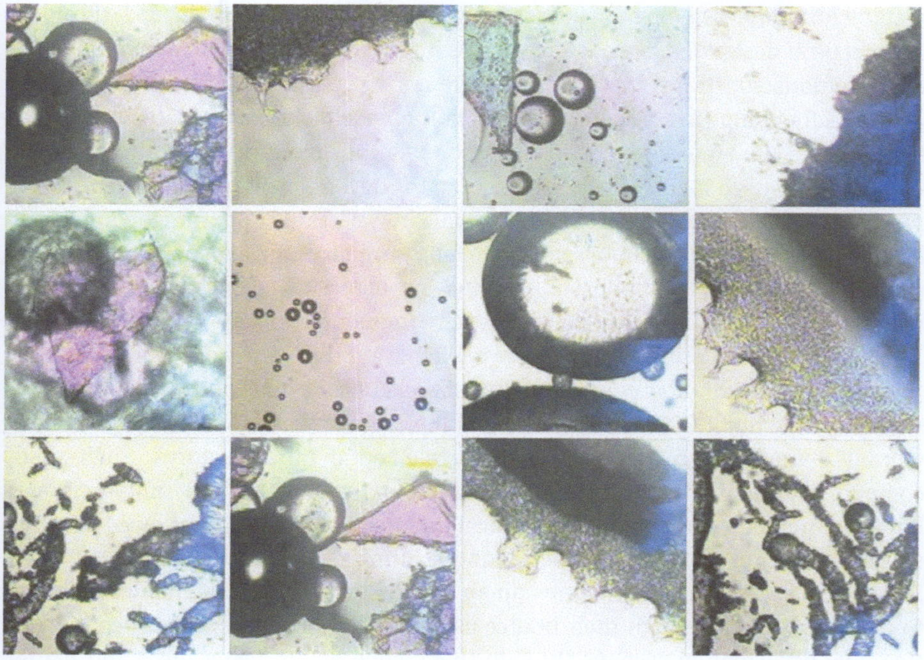

Figure 12.5: Dynamic droplets form weak bonds to produce assemblages which amplify their effects through recruitment. Micrographs and collage, magnification 4×, Rachel Armstrong, February 2009.

12.4 Soils as Architectural Infrastructure and Technology

> *... its substance is dark and malleable and thick, like the pitch that pours down from the sewers, prolonging the route of the human bowels, from black hole to black hole, until it splatters against the lowest subterranean floor, and from the lazy, encircled bubbles below, layer upon layer, a faecal city rises, with twisted spires ... a city which, only when it shits, is not miserly, calculating. Greedy.* (Calvino, 1997, pp.111–113)

The greatest challenge to our near-future cities is in how we can grasp the full material potential within our urban environments to create a new relationship with the natural world. So, rather than depleting resources and polluting our environment with toxic waste, we may be purifying and enriching it. To invoke Cedric Price's metaphor of cities as different types of eggs (Jacobs, 2011) – we need to stop cooking our cities and enable them to develop embryologically from their primordial clay so they may become chickens with a completely different set of properties than they currently possess. So, if we are to embody truly sustainable environments in our cities, then a positive, new relationship between soils and architecture must be established. It is time to end the doom, gloom and skinny corporate corset that has been wrapped around the shapely architectural profession by the industrial sustainability agenda. Indeed, 21st century architectural design must crack a few austerity whalebones with an expanding girth of voluptuous substrates that transform the current paradigm of consumption into a new relationship that is blooming with mutual exchange and generosity between humans and the material world. This is not a call for more primitive lifestyles, but highlights the need for architectural tactics in the development of infrastructures and processes, which promote the use of materials in ways that support regenerative and life-giving systems. It is simply not sufficient to reduce our consumptive practices to uphold imposed conditions of scarcity, but essential to develop and promote materially enriching ones, which promote environmental fertility.

Specifically, our cities need to re-establish a productive relationship with their soils as spatial technologies that can organize chemical events and transformational sequences (Tschumi, 2012, p.60). Soils are complex entanglements of self-replenishing vibrant matter, which are penetrated by elemental infrastructures. Such vastly complex overlapping programs enable soils to exist as more than simple surfaces but as entire bodies, with non-hierarchical architectures, which swallowed Darwin's wormstones through living subtractive and printing processes (Bennett, 2010, p.96). Soils are integrating infrastructures on an architectural scale that enables materials to be transformed through their bodies as sequential encounters that occur in the free flow of elemental systems through them, such as air, water, heat and matter. Applying the technology of soils within buildings may not only make better use of our waste water and organic matter, and enable us to grow native, not transplanted, greenery (BBC.com, 2013), but also offer flexible architectural tactics that can deal with multiple, overlapping spatial programs (see Fig. 12.6).

While complex terrestrial forces spontaneously build natural soils, the possibility of artificially engineering soils within an architectural context creates the opportunity to transform the artificial landscapes that characterize the urban environment into fertile sites. For example, the Liesegang ring plates in the Hylozoic Ground installation explored the principles for developing a simple soil matrix, where a homogenous gel was transformed into self-organizing, evolving layers of different colours and thicknesses, where multiple programs – precipitation, dissolution, colour change, gravity – entangled to create a range of events (structural and process-led) that changed with time. Of course, much work still needs to be done before the gel could be functionally likened to the complex self-enriching systems of natural soils. A soil system would, for example, need to facilitate cycles of exchange and feedback, contain air-filled cavities and organisms, and be capable of compost production. However, these first design-led experiments suggest that it may be possible to produce complex, self-regenerating bodies by orchestrating the interactions of multiple biological, physical and chemical agents. Currently, the production of synthetic soils is bio-inspired and follows a terrestrial paradigm that relies on particular chemical blends and physical properties. Reimagining the nature and functions of soils in ways that go 'beyond' our current knowledge and expectations of them could create an

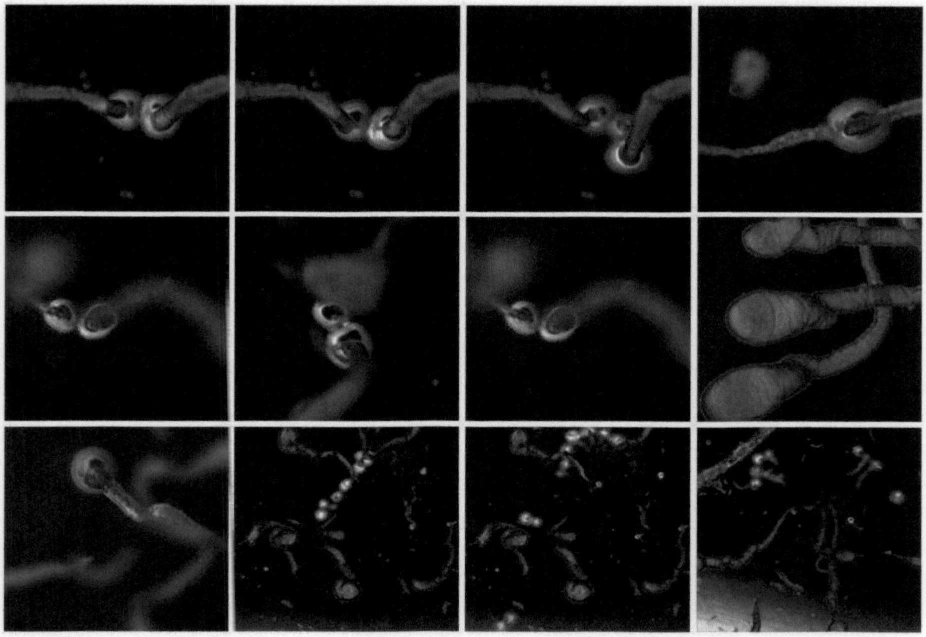

Figure 12.6: Dynamic chemistries introduce new temporal and spatial programs into microscale construction sites by spatially transforming, rather than consuming, their surroundings. Micrographs and collage, magnification 4×, Rachel Armstrong, February 2012.

additional portfolio of architectural strategies to increase fertility within desert-like urban landscapes and sustain healthy organic recycling systems.

By framing the idea of soils as complex assemblages of self-regenerating vibrant matter that are designed to support the development of all kinds of habitats, it may be possible to develop an architectural design practice that is concerned with many facets of evolving soil-like systems. These materials may be designed from a portfolio of codesigning agents such as lifelike chemistries, self-assembling systems and different species of ELT. Yet, the degree to which lifelike chemistries will integrate with simple cellular systems, such as algae, is not known. Future research will aim to produce a soil system in the long term that draws from chemical self-organization using liquids, gels, foams, colloids, coacervates and solid matrices such as fibreglass to establish catalytic sites for chemical transformation and physical means of transport. These may then be evaluated to build a coherent system that facilitates meaningful physiological exchanges within architectures which may, for example, act as possible nutrient delivery systems for the propagation of biofuel-producing algae, or harvest substances from traditional composting methods (see Fig. 12.7).

Such activity could take place invisibly in existing architectural spaces that are under-imagined. Currently, cavity-wall insulation is filled with inert materials, such as fibreglass and foams, which perform no useful functions other than to trap

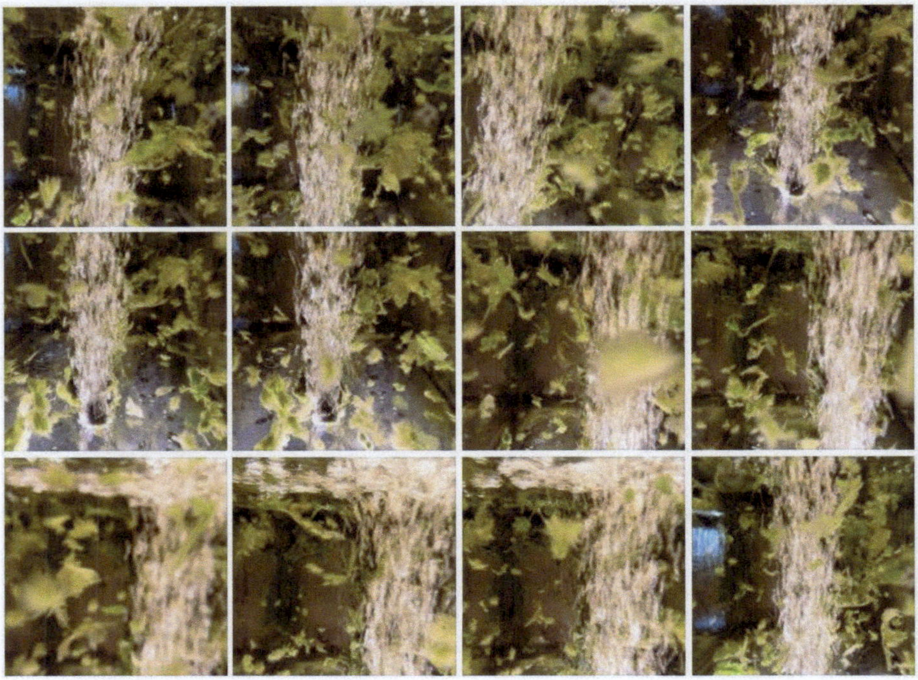

Figure 12.7: Algae can fix light and carbon dioxide to produce rich, organic landscapes whose biomass – which seemingly comes from the 'air' – may contribute to the development of fertile soils. Photographs and collage, Rachel Armstrong, August 2012.

insulating air. Yet, within these same spaces, soils could act as filters for purifying wastewater, transforming organic matter into heat, provide insulating functions and convert these passive spaces into physiologically active sites. The composting process produces comfortable, slow-release chemical energy that could be controlled simply by letting more or less air into the system. Should our grid systems fail in an emergency, then soil-producing units may increase our survival by filtering grey water, dealing with human waste, growing food and providing heat. They may also increase the city's resilience to withstand and recover from potentially catastrophic assaults by enabling its inhabitants to subsist, at least for a while, off-grid. Indeed, composting is already growing in popularity. Armed with 'red wrigglers', a species of worm, New Yorkers have started a composting revolution where organic waste is turned into nutritious soil (Robbins, 2012). Yet waste matter may also be transformed and applied in different ways, using different techniques and technologies. For example, composting materials could be pressed into bricks for building, or may even be saleable to soil collectors who would transport fresh compost to farms outside the city. The winners of the Bill and Melinda Gates concept challenge to 'redefine' the toilet have developed novel systems for transforming human waste into electricity with microwaves (Ungerleider, not dated), recycling urine to flush (ABC Science, 2012) and turning excrement into charcoal (Rodriguez, 2012). Yet, although the substances involved in soil production are culturally regarded as base matter, design practice is also able to confer new meaning on them. For example, the Philips Microbial Home project (Microbial Home, 2011) proposes a series of luxurious products where the house of the future is viewed as a biological ecosystem capable of filtering, processing and recycling what would normally be considered as garbage (McGuirk, 2011). Rather than following the modern obsession for sterility – the idol of a death-centric culture – Philips proposes a new relationship with microbes to run our homes and invite them in as productive members of our community through the process of soil production. For example, bacteria may provide bioluminescent lighting (Myers and Antonelli, 2013, pp.68–69), make biogas and even recycle unwanted plastics (Kanellos, 2009). It is possible that by incorporating the principles and practice of vibrant matter into our near-future cities, our living spaces will have a much richer infrastructure than today.

These soils and elemental infrastructures may nurture living communities of biological and chemical agents whose outputs could be monitored through smart sensors, or even developed as urban gardening practices. They may be conspicuous structures or invisibly woven into the building fabric to make more efficient use of resources. Yet these radical solutions are also compatible with our diverse needs and lifestyles, being applicable across a breadth of architectural styles, property types and geographic locations, so communities may adopt them without sacrificing historical traditions and cultural identities. Different soil-producing systems may share the same kind of local variations, complexity and ability to influence biospherical systems as native soils, which transform our biologically desert-like urban environments into fertile, biodiverse ones (Armstrong, 2013a). Rather than being the horizontal, geometrically defined amount of dirt we have beneath our feet, soil may be regarded as an architectural technology that nurtures the development of vibrant cities. Reconnected to an ecological set of relationships that link the development of cities to soil production, human development may then begin to thrive on the multitudinous networks that exist between human and non-human communities (see Fig. 12.8).

12.5 Summary

> *Her knowledge goes back only to the dawn of Time. But if she could have looked a little further back, into the stillness and the darkness before Time dawned, she would have read there a different incantation.* (Lewis, 2001, p.176)

By applying the material and technological principles of vibrant matter in architectural design practice, it may be possible to develop vibrant cities. The kinds of infrastructure that may support these lively communities of collaborating human and non-human bodies may be similar to our biotic soils (see Fig. 12.9). Potentially,

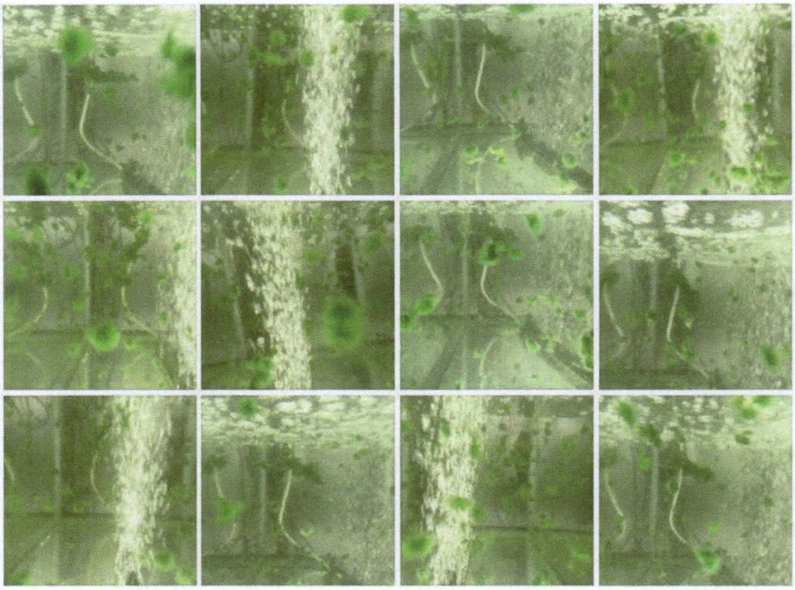

Fig 12.8: With the right infrastructural support, algae perform active roles in the construction of space and generate the infrastructure for further chemical events. Micrographs and collage, magnification 4×, Rachel Armstrong, February 2012.

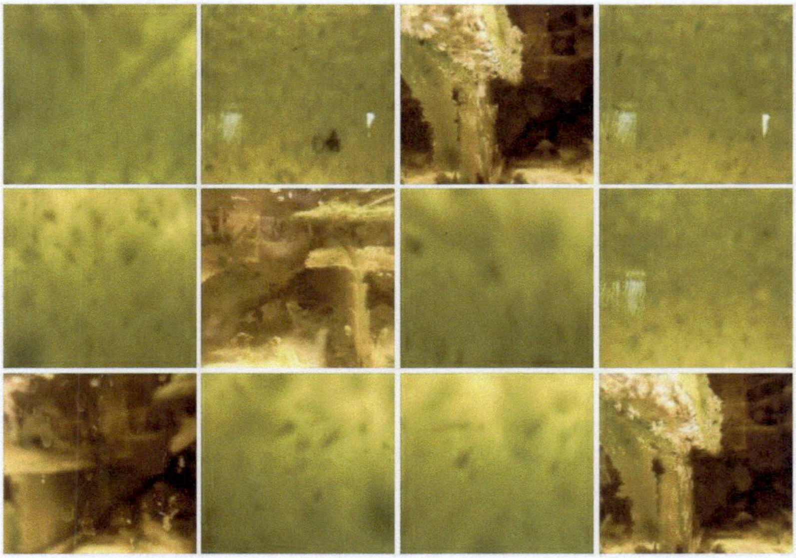

Figure 12.9: Metabolic exchanges can produce fabrics with generative potential. Micrographs and collage, magnification 4×, Rachel Armstrong, February 2012.

by strategically orchestrating the flow of matter through our cities, we may discover new ways of developing our homes and cities as continual, collaborative acts of co-evolution with non-humans. These developments may extend the traditional notions of architecture as barriers between humans and the natural world, which are concerned with site and shelter. Instead, the built environment may be considered as the urban-scale production of synthetic soils and post-natural landscapes, which may ultimately contribute to our collective survival.

The following chapter speculates on how urban post-natural fabrics may influence and even alter catastrophic natural events, by conferring our living spaces with resilient, robust qualities that can deal with unpredictability and, in essence, are designed to promote 'life'.

13 'Japan: 2060'

13.1 Overview

This speculative fiction explores the possible resilience of Living Architecture in the face of natural disaster and considers alternative outcomes. This piece was written for my TEDBook 'Living Architecture: How synthetic biology can remake our cities and shape our lives' (Armstrong, 2012d).

13.2 Japan: 2060

Nearly 50 years had passed for Yomo Yanobe since the last tsunami devastated the Sendai region. He was a young rescue volunteer in the clean-up operation after the 2011 events. It was almost unthinkable that he would find himself facing the same situation that had shaped his life.

Back then, his youthful indifference to mortality had spared him a little from the full emotional impact of the tragedy. Yet he had been deeply affected by the experience. Yomo had not rescued a single person in two whole weeks of searching, and he could not shake the feeling that someone was still buried under rubble, waiting to be found. Over the intervening decades he suffered many sleepless nights wrestling with the bitter guilt of a survivor. At night it twisted and clawed its way into his dreams, and during the day it taunted him with a profound sense of uselessness. So, when the early-warning systems and media channels screamed that a devastating earthquake had just struck the north-eastern region again, reaching a shocking 9.2 magnitude, Yomo knew that a swell of tsunami white water would gluttonously churn its way inland to engulf its prey. He wanted to resume his rescue duties, even after this long intermission. He hoped it would be different this time. Maybe he could save someone.

Clad in protective white overalls, Yomo waded ankle deep in sludge with his spade poised, looking for signs of life where traditional building fragments lay scattered like broken bones. Regenerated Sendai appeared mostly similar to how it was before the 2011 disaster, although there were some notable differences. Many buildings were designed to withstand even greater assaults from Nature. A 12-metre tsunami wall along the shoreline was raised, which had taken 20 years to complete and had initially been very controversial. Most locals were in favour of its construction, as it brought employment to the recovering city, but many objected to the visual impact and its potential damage to the recovering tourism industry.

The government had also invested in a wide-ranging disaster-survival strategy. This included the introduction of unusual coatings and fabrics that could be applied to the exteriors of homes with resilient lifelike qualities. The researchers called them living paints, and they were designed to create buildings that could respond, at least in part, to the devastation of powerful earthquakes with regenerative activity, such as growth and self-repair. It was not long after the first living paints were released that the creative local gardeners recognized their potential for growing unique sculptural features in traditional Japanese gardens. The novel coatings grew thicker with time, since they were able to lay down new material by extracting carbon dioxide and other pollutants from the air.

The bizarre gardens started to draw tourists into the city again, but some residents complained about the scruffy, rugged appearance of the earthquake-resistant buildings. Living materials also didn't respect property boundaries and transformed orderly neighbouring surfaces into strangely shaped building substrates. Mosses that traditionally grew in Japanese gardens now became entwined with living paint nodules and crept over lightweight materials to seek light, minerals and moisture. But these weird surfaces had other benefits. An emergency feature called 'survival bubbles' was woven into the roofs of houses. These inflatable capsules were made from a ropey biopolymer fabric with a coconut fibre-like texture that was activated by a simple mechanical launch device or by vigorous seismic shaking. The survival rafts floated like polystyrene by virtue of tiny pockets of air that were metabolically pumped into their walls. When expanded fully, they were incredibly tough and could protect a person like a nutshell. The capsules would split open to release their contents when they were washed up and exposed to the drying heat of the sun, like newly hatched eggs.

As Yomo combed the ground, he began to notice how the living materials adjusted to the post-tsunami landscape. Odd sprouts of bioscaffolding appeared through the rubble. Some fragments appeared to be reaching for each other to bridge gaps between shattered structures. Different species of living materials entwined together and he wondered whether the survivors could shape these regenerating pieces into temporary shelters. Where the ground remained soggy, the capsules swelled with water and filtered it within their molecular meshworks, providing survivors with a ready source of drinking water. The spongy material also processed nitrogenous waste from sewage and turned it into food for stringy biofilms of denitrifying bacteria. These glutinous skins trailed their hungry tendrils over many tens of metres through the putrefying wasteland. Yomo knew these membranes were edible but wrinkled his face at the thought of ingesting the wilted lettuce-like substance. Yet, he conceded the idea of eating the slime might be appealing – if he was starving.

Up ahead, a cluster of gas-filled survival bubbles caught his attention, but he sadly observed that they had already split open. The only signs of life were germinating seeds that had been deliberately embedded in the bioscaffolding capsules. They took longer than bacterial films to become established, but could be used as stock for rice or maize crops, in case the recovery of the region took longer than anticipated.

Then Yomo noticed that one of the half-buried pods was sprouting a twist of fibres with cherry-red-tipped shoots. He instantly recognized this unmistakable colour and knew exactly what it meant. When the availability of living materials had exploded into the DIY market the most popular products were a range of living paints that had been engineered to 'eat' carbon dioxide and change colour when they were 'full'. These were liberally applied to new buildings to help residents manage their carbon footprint responsibly.

Although people had been safely using paints with seasonally fashionable indicator hues for years, there was widespread concern when a particular formula was introduced. This product contained a vibrant cherry-red change indicator that was coupled to a powerful carbon-fixing metabolism. The resulting paint was so sensitive to carbon dioxide that it was possible to tell from the exterior patterns and colour whether someone was home or not, and sometimes even where they were in the building. This led to a host of news bulletins and late-night current affairs features linking the application of the paint to a rise in burglaries. Rapidly, the distinct colour became synonymous with crime.

Yomo stared at the unmistakable cherry-red tips of the biopolymer shoots that were gradually unfurling under the heat of the sun. They confirmed that carbon dioxide must be present at a high concentration. Something in the capsule was breathing. With his heart racing, he called over urgently to his younger, stronger colleagues to help him break open the tough fibrous coat of the capsule with their spades.

'Are you OK?' he shouted to the capsule. 'Help is coming!'

But he heard no reply. Perhaps an animal such as a family dog was inside. Yomo was determined to rescue whatever came out of the survival bubble. As his coworkers repeatedly struck the structure, he spoke calmly and reassuringly to the breather, 'Nearly with you!'

Then the capsule split open suddenly with a resounding crack, and Yomo reached inside.

'Help me remove these fragments, everyone! We need air!'

A small child squinted up at the open-mouthed rescuers as the light spilled into the cradle of tightly wound bioscaffolding fibres. She was filthy and clutching a ropey twist of fabric for comfort. Her bio-blanket was covered with bite marks, as she had been chewing on it. She would have derived a little moisture and food from this strange umbilical cord. Yomo reached inside the capsule, and the child instinctively recoiled. As he tried to reach her, he could feel the fleshlike textures of the carbohydrate-rich walls against the back of his hands. They were dry on the outside but soggy below the surface, rather like the inner lining of a child's nappy. He was relieved that there was no odour. The spongy interior had absorbed the girl's bodily fluids so that she had not suffered the unpleasantness of living in her excrement during her brief incarceration. The life support and rudimentary comfort wouldn't have lasted forever, but it had kept her alive. Yomo stopped trying to coax the frightened child and suddenly ducked into the capsule and swept the girl into his leathery arms, holding her tight. Her little body started to shake against him as she wept soundlessly.

Now, Yomo knew exactly why he had come to this disaster scene – to bear witness to how one small child had somehow survived all that Nature had to throw at her, against all the odds – and she had won! Now they would share a common dream and the blood bond of survival. Choosing his steps carefully, Yomo reached an ambulance crew and placed the child in their welcoming arms. He continued to pick his way pensively back to the desolate scene over fragmented structures that were superficially bigger and tougher than the child he'd just rescued.

Under the weakening daylight he noticed more unfurling biopolymer tendrils that were just visibly crimson tipped. 'Maybe', he wondered as he raced to examine them, 'there are other survivors!'

14 Manifesto For Vibrant Architecture

> *After a long period of often frivolous form making and unprincipled egoism in architecture, which have played into the hands of the most venal interests of real-estate developers and marketers, some architects are looking for more substantial ideas to serve, more meaningful goals to strive for, and the manifesto has come back. It is probably a temporary aberration, owing to an unsustainable idealism that lurks within statements of principle, but even their brief resurgence can help to regenerate – at least for a while – our beloved, beleaguered field.* (Woods, 2011)

14.1 Overview

While manifestos are typically positioned as a rallying cry to inspire action, I have developed one that is informed by my experimental findings and, therefore, follows my research arguments. Adopting a counterpoint to industrialization, the manifesto stakes out principles of practice and ways forward for vibrant architecture. It sets the stage for further investigations as an iterative and ongoing process of inquiry. Ambitiously, this manifesto aims to radically change the environmental impact of architecture from being a destructive force to becoming a fertile act (see Fig. 14.1).

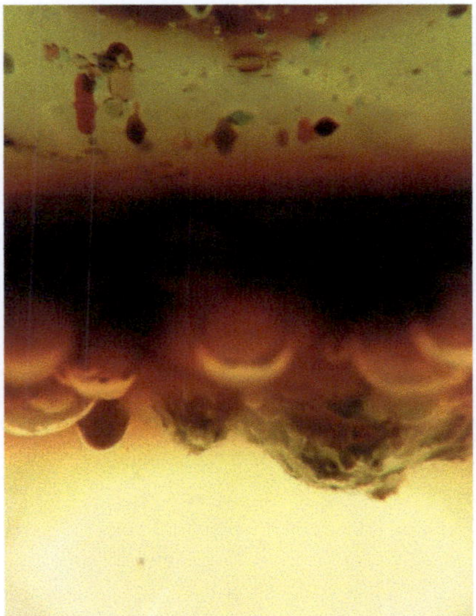

Figure 14.1: Complex chemistries embody the principles of assemblage technology that underpin the emergence of vibrant architecture. Photograph, Rachel Armstrong, February 2012.

14.2 Manifesto for Vibrant Architecture

1. Invite the non-human world to codesign 'vibrant architecture' through a shared, agile engagement with matter.
2. Vibrant architecture does not make its propositions through abstractions or representations, but by embracing the relentless materiality of the natural world – through ecological living materials (ELT).
3. There are no models of vibrant architecture, in the same way that no one organism is a model of another. Its progeny may episodically spring from the fundamental programs of dividing and expanding space.
4. Vibrant architecture is neither made nor born but must be coaxed into being. It is forged by a love of change, not consistency, and never tires of its own strangeness. It is compelled to grow, transform and propagate in relentless variations within its rhythmic networks.
5. Vibrant architecture is both object and process. Its formal rules are low-level 'pre-natural', chemical programs. It merges elemental infrastructures with matter in periodic exchanges, which can be fine-tuned at higher levels of organization. Ultimately, vibrant architecture resists the mathematics of 'natural law' and swerves with the idiosyncratic laws of quantum physics to revel in its uniqueness.
6. Vibrant architecture exhibits the same kind of rhythm that enabled Mendeleev to deduce the periodic table, compelled Henri Lefebvre to propose 'no less than a new scientific method' (Lefebvre, 2004) and inspires the musical score of the universe.
7. Vibrant architecture is in tune with its radically changing existence landscape, which sympathizes with the sharps and flats of the textile symphony in which it is immersed.
8. Vibrant architecture does not cling to the rippled underbelly of universal time but has directionality and travels with the warp and the weft in its slipstream. It deflects Newton's apple from oscillating in its geometric course through the arrow of time – denying reversibility in the system as an expression of fundamental creativity. There is no 'going back'.
9. Vibrant architecture does not pay homage to the aestheticisms of Old Nature (Van Mensvoort and Grievink, 2012). It does not sing reassuring jolly green sustainability songs, yet it is deeply entangled with the natural world and thrives on the side effects of human existence.
10. Vibrant architecture is empowered by its multitudinous assemblages and infrastructures to thrive in even the most hostile environments, infiltrating inaccessible terrains.
11. Vibrant architecture does not hanker for a time before our chemical industrial landscapes, plasticinating seas and choking skies, but regards them as new landscapes of abundance from which primordial chemical communities might spring into bloom with artificial biologies, which, until this point, have never existed.

12. Vibrant architecture is a transformer that does not find an adversary in machines, but swallows them whole and couples horizontally with them. Their progeny proliferates as rhizomes of vibrant networks that transmute industrial deserts into vast metabolic fabrics of increasingly lively landscapes, which bloom with co-evolutionary acts of transformation.
13. The surging systems and knotted materials of vibrant architecture swell with our own communities and become entangled with our habits. They thrive amongst us, in even the most densely packed domains, as codesigners of our proximate spaces.
14. Vibrant architecture does not exist in the future but explores our unevenly distributed present.
15. The delicate elemental fabrics of vibrant architecture reveal the mutual understanding between human design and material possibility – where architects set the conditions for a diversity of infrastructures that infiltrate and vibrate alongside the rhythms of our unstable Earth.
16. Vibrant architecture does not propose to save the world.
17. You see, vibrant architecture is not about making a building at all, but in establishing the codesignership between humans and non-humans, through the production of synthetic ecologies and post-natural landscapes – which become our living fabric (see Fig. 14.2).

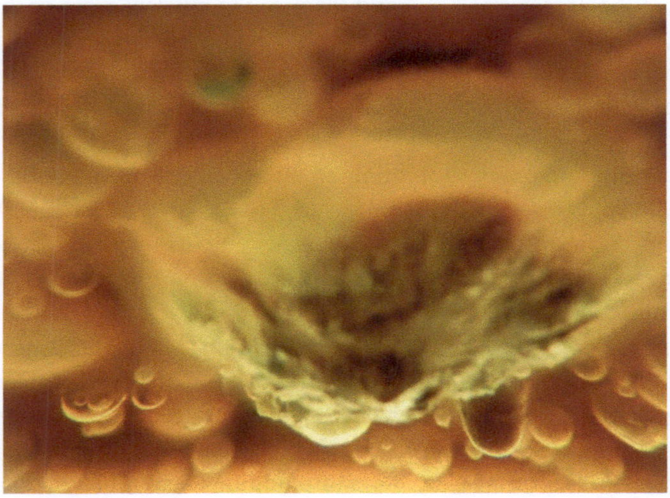

Figure 14.2: These complex chemistries exhibit emergence, which is a characteristic of vibrant architecture. Photograph, Rachel Armstrong, February 2012.

15 'The Greatest Alluvian Poet That Ever Lived'

15.1 Overview

This speculative narrative was composed to depict a 'vibrant' city for the Lisbon Triennale 2013 (Young, 2014; Sterling, 2013).

15.2 The Greatest Alluvian Poet That Ever Lived

XOE

This is how *Aurora Alluviata*, the native and dominant bacterial species of Alluvia, prefers its name to be written. XOE informs us these letters are chosen for the chemical onomatopoeia that symbolizes their morphology. 'X' and 'E' represent the asymmetric squid-like cytoplasmic extrusions at their poles, while 'O' conveys the body. In classical English, XOE is pronounced 'Zoe', which is also the Greek word for 'life'.

XOE speaks a chemical language that is forged by particle worms – bosons, leptons and hadrons – which stretch out into atoms and molecules. These molecular connections are the basis of all chemical languages, which are punctuated by the grammar of the electromagnetic spectrum. They are spoken by the most ancient life forms such as biofilms, which penetrate the soils, seas and skies. Yet, the chemical language is not confined to microbial species, but used by many other non-humans. Flowers and bees, for instance, seduce each other with electrochemical sonnets, enlivening different scales of interaction and giving rise to worlds that cannot be directly apprehended through human senses. Yet, XOE notes that our unperceiving human race regards the very idea of a non-human world capable of the same quality of thought that humans possess as 'fiction'. XOE considers this rather odd, since bacterial mood-producing oils modify human thoughts.

Left, right and up – the vacuole attendant swept a field of ions through the liquid crystals. The charged fluid spectacularly exploded into leaping images of the Alluvian cityscape, while the visitor stood momentarily enchanted by the stunning views of the city. Pillars twisted like tree trunks, rooftops dipped into scalloped edges and alleyways branched fractal-like through the diffuse, scintillating light.

> 'Dear God! The very fabric seems alive!' marvelled the visitor.

Left, right and down – the crystals melted from view as the vacuole attendant gestured the visitor towards the gelatinous tub. The attendant wore an unusually ornate bangle, which momentarily distracted the visitor. It seemed to be built from precious pieces that were intricately entangled, as if they had been woven together by a caddis-fly larva. He supposed it would much better suit a woman. The attendant gestured again, more urgently this time. Yet the man hesitated, feeling vulnerable, alien and as yet unclothed with an obligatory native microbial skin. The unfamiliarity of the occasion unsettled his prenuptial nerves. The tarry substance in the tub did not help reassure him, as it was not inviting.

Indeed, the visitor's instincts were telling him to run, to get out of this place – and quickly. Conflicted by the rising panic swelling in his chest, the man forced himself to be pragmatic, if not downright stoic about the whole process. Besides, he'd committed himself to the procedure when he proposed to take an Alluvian bride, and in any case, there was nowhere to run to. The visitor would never be able to leave the quarantine area untransformed.

Untrustingly, he crouched down inspecting the dark, viscous substance. Momentarily the visitor feared that he might be boiled alive. Yet, close up, the muck was cool. It appeared entirely clear, light and gel-like. He gingerly ran a finger across its surface, noting that he actually felt nothing at all. Trying to establish the character of the substance he squeezed it between his forefinger and thumb, but it felt equally unsubstantial. He sniffed at his fingers anticipating the stench of pondweed, but it was odourless. Finally, satisfied the goo would not hurt him, the man lowered one foot into the vat of glutinous substance, which continued to feel like nothingness itself. Registering the visitor's presence, the goo adjusted its chemistry to identically match the man's body temperature.

The vacuole attendant rolled his eyeballs at the visitor's tentative approach and obvious distaste for the immigration procedure, which required every visitor to be stripped of their native microbiome and replace it with XOE, Alluvia's indigenous bacterial species.

'Yes!' sighed the attendant petulantly. 'XOE feeds on light. That is why it's dark.' He was intensely irritated by the very idea of a foreigner in Alluvia. He couldn't help it. In fact, most Alluvians felt the same way, although they had learned to hide it better than the vacuole attendant. It was a superiority complex they felt in their flesh and bones. Their larger-than-life personalities oozed the conceit that Alluvia was a special place populated by extraordinary people. Only a few marriages were sanctioned, for economic reasons. This outsider was one of the 'lucky' ones.

The attendant lifted a large gold-plated helmet out of a trough, where it had been soaking in a soapy-looking fluid. He grunted unpleasantly as he lowered it on to the immigrant's shoulders. Then the attendant expertly snapped the fasteners on to the apparatus, which would deprive the visitor of his natural senses and speed his immersion into Alluvia's alternate reality.

'After this part, he's going to think he actually knows something about this place!'

muttered the attendant, knowing full well that the immigrant could no longer hear him. The visitor sank into the substance and glutinous tendrils swelled in his orifices with the unobtrusive presence of a mosquito, displacing his alien biome.

The gloop brought its own kind of anaesthetic, providing perfect comfort so the visitor no longer distinguished between what was his outside and inside. Indeed, he felt absolutely nothing at all as he succumbed to the substance, which filled his sinuses, swamped his memories with Alluvian chemical stories, engorged his throat and seeped into his lungs.

In the clean room at the Ramachandran Scientific Institute in Pune, Nareen Qadir felt the filtration unit breathing cool air at the nape of her neck. She slid a robotically-prepared, live bacterial cassette under the 3D lens.

The red and cyan visual fields of her goggles burst into an immersive display of rod-shaped, asymmetrically tentacled bacterial bodies. The XOE were in good form, extruding cytoplasmic tools from their bodies that probed the environment. XOE swarmed like liquid crystals using their tiny tendrils to grapple with each other and align tail-to-toe and side-by-side. The display thrust the typical view of a congested logging stream of XOE that were rapidly forming a multicellular coherent expanse, or biofilm. XOE could produce such organized bodies very quickly indeed.

XOE were temperamental and could only be handled by whirring robots in a sealed glass room. Yet, Nareen had worked much closer with the bacteria in the past, having spent cherished time when she was small, keeping her father company in the laboratory. He had kept her busy by putting her in charge of holding and directing the various makeshift light sources that the XOE thrived on, so he could take pictures of his experiments. Although a single XOE cell was way too small to be captured with an ordinary camera lens, their biofilms were profuse and quite marvellous. Under a tungsten bulb they rippled as swirling, oscillating landscapes, while an ultraviolet lamp caused them to leap and scintillate. Yet these were simply memories of former times, when XOE could be prepared manually in ventilated hoods, and her father was alive.

Assisted by a tiara-like headset and data gloves, Nareen began to navigate the space by feeling what she was looking at under the lens as if she was the same size as a bacterial cell. XOE were 'prepared' for the experiment by immersing them in gold nanoparticles. These were hungrily ingested and acted like listening and speaking devices that responded to the electromagnetic pulses that were delivered and received through the viewing platform. Nareen could feel the whole spectrum of vibrations through sensory membranes on her gloved fingertips. These conveyed subtle changes in bacterial biochemistry, as a kind of microbial Braille. Although the young woman would have loved to return to working much more closely with XOE, this species had earned notoriety, glass walls and robots for being an unreliable, and valuable, species to handle. Indeed, the company revered XOE for its potential economic benefits.

Nareen was only twelve when her father died suddenly. This was devastating but the family survived destitution on account of her extraordinary skill in deciphering bacterial codes. Indeed, her father's employers, the Biomediation Corporation, offered her a generous university sponsorship and a stipend to study languages, psychology and biological science. This generous gesture transformed her blossoming talent into matchless bacteriosemiotic skills. At eighteen she was a fully qualified 'sympathy microscopist', who could accurately interpret the nuances of bacterial chemistry. She was also the world's best XOE chemical linguist.

It was extremely difficult to get hold of XOE specimens, since the Alluvians would not agree to give samples of their 'ecobiome' to visitors. Although Nareen conversed daily with the XOE to better understand them, she did not understand exactly what role the bacteria played in shaping Alluvian culture.

Asif Raman was standing at Nareen's shoulder, pressing for information on how XOE could be cultured as bacterial workhorses. Initial trials had indicated that these strange yet temperamental bacteria could turn around a genetic sequence that ran at 30% productivity in *Escherichia Coli* into a nearly 99% efficient one. In short, the XOE were powerful 'synthetic biology' factories – which enabled the precision design and engineering of living things. This kind of potency had been previously unheard of, and the corporation was considering replacing *Escherichia Coli* altogether, which was an investment worth its weight in gold.

'Well', he asked impatiently, 'can it be commercially cultured?'

'You know', she replied dismissively, 'you talk like bacteria were just passive things. Like they have no will, or can exert no force of their own. You talk about them as if they weren't even alive.'

'Oh no, don't start this again!' complained Asif, 'Come on, they're just bugs, that's all! We've been killing them for centuries. Get them to do what we're asking.'

'That kind of superior attitude is exactly why they have also killed us. If we'd even acknowledged the presence of bacteria in this world, let alone tried to get along with them better, we may never have even needed to invent antibiotics', Nareen retorted haughtily, wrinkling her nose at the older man's pungent ketone-and-coffee laced breath, which managed somehow to leak into the airstream despite his nanoparticle-proof mask.

'And do me out of a job in the pharmaceuticals industry? Honestly, if I didn't know you were smart, I'd swear you were crazy. You've been around these things too long! How long have I known you and your imaginary little world of bacteria? Your father was just the same, complaining about their oppression by humans and the way they were "treated". Like they could feel, or reason ... or something.'

'I can't believe I'm still having this conversation with these corporate Neanderthals!' thought Nareen, as she concentrated on an interdigitating pattern of cytoplasmic fronds, which suddenly spread in a welcoming manner. She responded with her data gloves, tapping microbial Braille and watched the XOE biofilm ripple with the signals that spread from the tiny gold nanoparticles through their substance. These precious synthetic organs gave Nareen access to the life force of the bacteria and yes, even their communications patterns, which she regarded as 'thoughts'.

'Hold still a moment, please'

she requested, knowing that with this species in particular, it was counterproductive to issue a command. Having grown familiar with her inquisitive presence, the XOE paused for several moments. Nareen realized that she was face to face with a sentinel, which had special status in the XOE biofilm. Not all XOE were equal, or alike. Sentinels appeared larger, with more polar cytoplasmic extrusions that enabled them to make colony-wide decisions without a quorum from the others. The sentinel paused for milliseconds. Although this was an incredibly short time for a human, it was an unbearably long time for such tiny things.

'Thank you!'

she tapped.

'Listening to bacterial voices is real and it's what you pay me for. You might even be interested in what they're saying'

sighed Nareen, thinking that Asif was standing just a little too close.

'So, do tell me, then!' snapped Asif impatiently. 'I don't have the time or energy to argue with you. Just tell me if these little bastards can be cultured?'

Nareen looked up sharply at Asif,

'Only if they want to be. Can you take a step back please?'

XOE looked back up at the humans. Although the sentinel was behind several thick glass screens, it could observe Nareen as a whole sky of sensations that vibrated in strings of particle worms and shook with their potential. She offered the promise of change, a vibration that penetrated beyond the materials that physically contained them. XOE 'saw' Nareen as a place that could be inhabited – like a landscape, or city. Yet, to do so prosperously, XOE would have to transform its host human environment and colonize its flesh. XOE shared the collective wisdom held in the chemistry of more than 3.5 billion years of bacterial knowledge, which was common to all bacteria.

The sentinel could feel the fat man's intentions to enslave the XOE through the energy cloud that angrily radiated through the clean room. It could feel him browbeat and cajole the energy landscape that it identified as a young woman. The sentinel acknowledged Nareen's advocacy on XOE's behalf and transmitted this sentiment to the colony.

> 'Some people are quite smart'

thought the XOE as they listened to the human energy fields through their gold nanoparticle ears. Indeed, the XOE understood the political issues and intrigue that shrouded communication between the two humans much better than the people did.

The immersion was cleansing in many ways. For starters, the visitor's prejudiced view of Alluvians was corrected more effectively than with any conversations he'd previously had with his soon-to-be spouse. Alluvians were very well aware of their outsider culture and that they were globally regarded with much suspicion for their aversion to classical economics and, well, their slimy – or 'greasy' – ways.

Only Bollywood stars and the old sheikhs could get close to the young Alluvian women who left their homeland. Yet these women did not stay away for long, or yearn for heroic personal adventures, but sought marriage as a commercial transaction. Most guys knew that you could tell an Alluvian woman not just by her flawless skin, but also by how much she was drawn to gold. This was not simply a desire for wealth but something much more material. When Alluvian women wore the precious metal, their skin sparkled with the ethereal radiance of an Enlightenment painting, causing men and women to stop in the street and marvel. Yet they were stigmatized by the host culture as mud-slappers, or gold-diggers, and therefore the lowest of all castes.

Those very bold and wealthy lovers who dared to take an Alluvian bride, often against their families' wishes, were compelled to do so – not just for the companionship of the soapy flesh of their bride-to-be, but because they actually became 'infected' by their lover – and not necessarily in a bad way. One night with an Alluvian lover was enough to bewitch an outsider to seek her hand in marriage. Yet, it wasn't the woman who worked the magic, it was the XOE in her biome that secreted intoxicating, mood-enhancing, oily substances during intimate moments. 'Love-struck' fiancés were literally addicted to their brides-to-be and pledged pilgrimage to Alluvia. Here they were required to surrender any traces of their own microbiome and give their bodies over to the XOE, to live the Alluvian way.

The ectopic-eyed eagle disappeared to a speck high above the city. At this altitude, the biofilm gorged on sunlight and swelled the bird's XOE-soaked cavities, while the bioelectricity-powered head-mounted camera surveyed the quarantine zone. Through the cloudless sky below, the Alluvian biofilm could be observed patrolling its realm, like a giant terrestrial defence organ. The vibrancy of Alluvia's urban biofilm fabric starkly contrasted the city limits against the paler, drier surrounding area beyond the wall. The XOE swamped the city and was entangled with everything within it as a single, scintillating body. It lapped the wall, searching for foreign bodies that were cast out again, like so much flotsam. Indeed, it had grown so big, smart and rich that the XOE biofilm could organize humans' lives within its realm – and they were wilfully oblivious to it.

The enormous bacterial body was most active and strange at the coastline, where it stretched and retracted tendrils like pseudopodia, which reached out as far as some of the costal oil rigs. Yet, these vagrant islands never seemed to be brave enough to break free of their homeland, attaching themselves to the main body of the XOE biofilm through ropey underwater cabling. To the ectopic eye in the sky, these appeared as spider-web threads under the azure waters and kept visitor boats away from the coastline. Indeed, some intrusive vessels had already been ensnared in its mesh and bobbed abandoned, like silk-wrapped flies.

> 'So, Asif', challenged Nareen, 'if you propose to introduce XOE as a medical therapy to restore the surface of teeth, regenerate tissues and even promise rejuvenation – then do you ever consider whether this a good thing for humanity?'
>
> 'I'm a businessman, my dear', Asif assured. 'And business is good. There is a market for vanity.'
>
> 'Don't dismiss me! Ethics are important!' snapped the sympathy microscopist. 'Did you read the memo that outlined my recent findings?'
>
> 'You send me a lot of stuff, dear. Most of it, irrelevant',

patronized Asif, wondering what he'd have to do to bring her around to his way of thinking. She was all ideals and hormones with no real experience to ground any of them in. Her constant challenges and assertions required very hard work for very straightforward questions – like can XOE be commercially cultured? Unfortunately, the company needed her 'onside', to broker an arrangement with the XOE.

> 'As you know, XOE is physically connected to its home place, Alluvia, by forming biofilms. It seems that sometimes little breakaway colonies form, but they don't thrive for very long. My suspicion is that the saying "Alluvians go home" is not just an insulting stereotype that comments on Alluvians always returning to their city, but also refers to their inability to thrive for long without a direct connection to the XOE biofilm. This is why XOE are so darned difficult to culture.'
>
> 'Is there a way around it?' pondered Asif.
>
> 'We need to make them believe they are connected to their motherland if they're to thrive for any time beyond it'

Nareen replied, narrowing her eyes untrustingly at the older man who was studying the sentinel intently.

> 'So, if we are to fool XOE into thinking they're part of the mothership, just how infectious could they be? Would they cause an epidemic?'

Asif knew that Nareen would find a way around this and began to think about what it would mean if XOE became the synthetic biology workhorse of choice. Would the company lose control of them, leading to endless lawsuits?

> 'XOE is not a virulent species – it's a commensal, community-loving creature. But it's a transformer. That means that when XOE colonizes other tissues, it converts them into a condition that best nurtures the colony. And while XOE have a pastoral respect for us, since we provide food, shelter in people's bodies – perhaps they even consider the human race as their living spaces, like cities for the XOE – we have no idea what they may be capable of, if we are no longer useful to them.'

Asif was growing impatient with Nareen's bacterial superiority speeches. 'Useful to them?' he laughed, spreading adipose ripples across his lab coat.

Nareen stopped gesturing to the sentinel, which had just adopted a remarkably attentive, splayed tentacle posture, and pulled down her mask, breaching clean-room protocol.

> 'Why is it so difficult for you – and the company – to understand there are other bodies, other living forces on this planet that may be just as smart as we are? What if these apparently "insignificant" life forms are actually running the show?'

> 'Calm down dear! And put your mask back on.'

Nareen ignored her superior.

> 'What you are proposing by using XOE as a synthetic biology "workhorse" is that our species should become assimilated by a smart bacterial swarm!'
>
> 'You're letting your imagination run away with you, don't you think, dear? Bacteria outnumber the cells in our bodies right now at a ratio of 10:1. Do you feel less "human", or more manipulated, when you understand this fact?'

Nareen raised her mask again.

> 'Aren't we already engaged in business propositions, buying and selling ourselves to XOE through Alluvian marriage arrangements?'
>
> 'Nobody forces them to get married, dear.'
>
> 'But does anyone really know what goes on in Alluvia? Do we really know who these people are? Or, perhaps they're not "people" – not as we understand the term, anyway!'
>
> 'Nareen', Asif put his hand on her shoulder, 'I know you blame the XOE for your father's death. Perhaps you blame yourself too, but do stop this personal crusade and start to appreciate that, whether you like it or not, you're part of this business. Besides, there are many ways that the XOE could bring great benefits – to all of us! You really need to get things in perspective!'

'But we're not talking about "things" here! We're talking about people!'

'Indeed, and we intend to improve the lives of many, many millions of folks, just like yourself. Think about it. What would it be worth to you to have ten, twenty, thirty – maybe many more years than that – of healthy life? Wouldn't it be lovely to have your mother living long enough and healthily enough to help out with her great-great grandchildren?'

Nareen chose to ignore the veiled sexism and voiced a deeper concern.

'If you set up clinics, which I assume is what you intend, to distribute XOE around the world, beyond Alluvia, you need to ask yourself whose interests you are serving. Theirs – or ours.'

'My dear, if you're right about bacterial superiority, then perhaps you might consider we've already reached the stage where there is no longer a distinction. That whether bugs or humans are in control of events no longer matters.'

The scintillating bride-to-be and purged groom-to-be sat opposite each other on sacred ropey bench-like twists, atop a large mound of earth, just a few minutes' walk from where their wedding ceremony would be held. Observing tradition, their marriage ceremony would take place when the Moon was ripe, which was only three more days away. But now, they enjoyed the prenuptial ceremony to pledge devotional love, in the most fertile place in the city – the graveyard. This sacred place was marked by the respectful placement of golden sculptures around its base, which gradually sank into its rich mud and brought light to the underworld. Alluvian corpses, wrapped in digestible films that were destined to be recycled through acts of microbial rebirth, lay only a hand's breath beneath the romancing couple's feet. Yet there was no stench of rotting flesh, or signs of putrefaction, for Alluvians were buried with the creatures that would process their flesh and speed their reincarnation. Indeed, larvae, fungi and worms were ritually sealed within the edible biofilms and were expected to eat their way out of these capsules, releasing the vital essence of the deceased into the compost.

The decomposition of Alluvian corpses was therefore ceremoniously encouraged and funerals were happy occasions, celebrated by singing, dancing, feasting on 'XOE-fu' and the scattering of worms, dung beetles and fungi over the mound like confetti. Each of these dutiful living systems was morally bound to convert 'tired' flesh into vigorous matter. They transformed rotten tissues into metabolically vigorous root-like mycelial structures and organ-like nodules, which were the symbols of new life that rose up and protruded from the mound. Indeed, the mound was thought of as a living body, like a soil that housed a sacred community of transformers, which had special status within the community and which it was taboo to kill or harm. The mound body slowly travelled inland, away from the sea, dragging behind it a trailing scarp slope of rich soils that were planted out with beehived orchards and brilliant flowerbeds to seduce Alluvia's fat, drony insects. Throughout the year, devoted relatives paid tribute not by bringing cut flowers, which to an Alluvian was a shameful act associated with death, but by topping up the compost heap with rainbow earthworms, scarabs and bioluminescent mycelia. So, the mound was the perfect site for prenuptial rituals, since it was where Alluvian forces of life and death were completely entwined.

The couple began to lose themselves in each other's adoring gaze, sighing heavily and longingly with each cyclical exchange. As twilight fell, the families of the to-be-weds stealthily began to encircle the earth mound, which was the custom. They brought portable husks, which were positioned in locations where they anticipated they would best hear the conversation between the betrothed, which followed soon after the yearning ritual. Some family members brought shell-like listening devices to amplify the Sound of Romance, a transmissible and entrancing ripple of vibration said to stir the Alluvian soul, while others were more traditional and conjured holographic recording devices from sachets in their hessian garments.

When the families had settled, invited guests and love addicts, who got 'high' on prenuptial ceremonies and were tolerated for their dramatic displays of awe that were considered a form of light entertainment, also silently joined them. Yet the audiences of this Alluvian spectacle did not anticipate consummation in a carnal pagan act, but further sublimation of the feelings shared by the adoring couple through the lavish recital of poetry. This public ritual, which had begun privately enough, was considered a community experience, where everyone shared the love expressed in the marriage contract between man, woman, family – and colony. Indeed, Alluvians had a fascination for and even an obsession with 'love' – or the idea of it.

The extraordinary Alluvian light was dimming and pinpoint bioluminescent sources were already detailing the scalloped rooftops and twisted columns typical of the city's dwellings with a second night sky. Even the background molecular hum of the universe and the symphonies of particle worms quietened with anticipation for the groom-to-be to speak. The XOE felt the man trembling and his vocal cords seize. He looked upon his bride-to-be, whose gaze fell upon him with wide, dilated pupils, and her skin was ablush with anticipation. Her plumped lips were parted, ready to respond to the first devotional utterances of love with her own pledge of passion.

Several family members started coughing nervously, and were quickly stilled by their partners' sharp elbows. Waves of light swept the smouldering skyline, forming an aurora over the buildings. The man separated his lips too, as if to speak, or kiss, but there was no sound, his thoughts were drenched with adoration as XOE mood molecules pulsed through his veins, crossed the blood–brain barrier and filled the groom-to-be with the most unspeakable awe.

Unable to control their anticipation any longer, the front row of relatives began to weep. The emotion of the event was contagious and as the shaking man's vocal cords stayed silent, sparkling hot tears ran down the bride-to-be's cheeks. A large crowd had gathered around the prenuptial ceremony. Not family, friends or 'love' addicts, but other sections of society drew close to feel the enchantment of the occasion. The poor, who usually only watched these rituals of the wealthy on their liquid crystal screens, stood at a comfortable distance from the mound, peering for a glimpse of the shadows of the betrothed casting their silhouettes over the sacred mound, whose soils were now burning with bioluminescence. Even the beachcombers, who didn't really care for screen-based entertainment, came to bathe in the emotional waves that seared through the passion-taut air.

The intensity of the occasion startled a large dung beetle, which bolted from the compost and was snatched by a young man seated on a husk, who popped it in his mouth and crunched the twitching critter.

'Granddad! Don't eat the sacred scarabs!' scolded the small boy sitting next to him.

'Shhhhh!' winked the young man. 'Don't tell your mother! Or I'll never hear the end of it.'

The boy cast a suspicious glance at his senior, who goaded him to join in with a playful punch on the arm as a couple more brilliant-shelled beetles scuttled out of the mound.

'Hey, if you're quick enough to catch one, they taste much better than XOE-fu!'

Distracted by the sudden noise, the groom-to-be looked around and saw the weepers. Since he was fairly new to the culture, only having been recently purged, their presence was somewhat of a shock. He tried to formulate a word, a worthy thought, or a poetic phrase that would keep the promises made in his betrothal vows, but he could not free himself from the throws of sublime rapture. By force of will he conjured the shapes of ideas and projected them like shadows in the hollow cave that had become his mind. Yet, he was unable to apprehend anything. The shades flickered and left with no wise words or poetry to share.

Now, everyone present was crying. The bride-to-be had buried her hands in her face in a profuse stream of hot, scintillating tears, melting the complex, entwined geometries that had been so carefully painted on her fingers by female relatives using bacterial pigments.

The man began to reach for superlatives that might do justice to the bride-to-be. But there was still no sound. Indeed, he realized that since the ritual purging, his lungs were full of fluid and there was no breath to draw. Under scrutiny by a myriad of weeping eyes and XOE light beads, the groom-to-be realized that he had forgotten how to speak.

The XOE, stirred by the intoxicating human emotion on which they thrived as much as they did light, pulsed chemical signals throughout the biofilm, calling each single bacterium to 'quorum'. The sentinels orchestrated the ensuing vibration so that the colony could amplify and feed upon the emotional intensity of the occasion. To the throngs of sobbing people it seemed as if the extraordinary humming of XOE choirs emerged from nowhere. This wasn't so much of an audible sound, but a feeling; a sonnet of molecular strings that could move the Alluvian flesh and soul to ecstasy.

And so, the bride-to-be, now dripping dark goo from her nose and eyes, as the XOE spilled from her congested orifices, responded to her man. Stirred by the deepest poetry that anyone from the city had ever commanded, she tugged at her right index finger and extruded a splinter of bone from her body. The man, still speechless, remained agog as his bride-to-be wrapped her twitching finger fragment into a ring. Taking his trembling hand, she slipped it silently on to his finger, the expectation being, of course, that he would return the gesture during the marriage ceremony.

The audience called it 'soul-stirring'. They spoke in superlatives of the greatest poetry ever felt. It had been delivered to the colony on a moonless night, through a pledge of love that was carried by 'raw emotion'. And the conductor of the song of the XOE was not a stage fright-wrecked, recently-purged visitor – but a true Alluvian who could command the voice of the universe.

Asif waved his hand at the work desk impatiently. He needed to try another approach with Nareen to get her to align with the company's goals, to harness the hell out of these bugs and make a fortune in the synthetic biology industry. The field of rejuvenation was wide open! And so what if Nareen's father had died from an overwhelming colonization of the bugs that had, for reasons that were not quite clear, turned into a nasty infection that killed their host. Indeed, no one had forced him to sample a XOE biofilm by taking an illegal Alluvian bride when he was already married, abandoning her only months later to bring back his precious harvest. Since XOE were only bacteria, they couldn't possibly orchestrate an act of revenge on his betrayal of their culture. And since it was impossible to conceive of such a premeditated act, there was therefore no explanation for why the devious critters would pull such a blatantly suicidal act, other than their terrible 'moodiness'. This was exactly why the woman-child had been charged with the task of winning their complicity in the first place.

'But she has a point!'

he grumbled as he leafed through a series of holomovies that Nareen had taken earlier, 'We're investing a lot of resources trying to outmanoeuvre bugs!' Then he laughed aloud as he replayed the moment the *Aurora* sentinel 'stayed still' for Nareen.

'Goddammit!'

He cursed as he shook his head in disbelief and gestured again at the work desk, zooming in on a hologram of an XOE sentinel.

> 'We're not outmanoeuvring them! We're negotiating with them!'

Asif scowled at the wilful creature, which appeared to be glaring right back at him.

16 Conclusion

And I hear, from your voice, the invisible reasons which make cities live, through which perhaps, once dead, they will come to life again. (Calvino, 1997, p.136)

16.1 Overview

This chapter summarizes the key findings and challenges of my research, which are also used to develop founding principles for the practice of vibrant architecture.

16.2 The Agency of Matter

The transformative technological potential of this physical realm is possible through the significant contributions made by non-humans, such as molecular forces, that have invested in the infrastructure for our existence. Vibrant architecture experimentally investigates how to design along with these agents in more environmentally positive ways (see Fig. 16.1).

The agency of our material realm is distributed across the actions of many different actants that are an intrinsic part of our urban communities – from catalytic surfaces to viruses, microbes, mycelial networks, flocks of pigeons, tumbling plastic wrappers, clouds of fungal spores and wandering, thin soils. All these bodies stretch from the subatomic realm to the megascale, entangling us within many parallel

Figure 16.1: Non-human actants inhabit the shores of Venice and are continually (re)defining its boundaries. Photographs and collage, Rachel Armstrong, August 2012.

realities that intersect with our architectural design interventions. Indeed, we may begin to value matter differently if we acknowledge the diverse material communities that have collaborated to create the conditions for our survival. By consciously appreciating the enormity of these collective endeavours and the vast investment made by non-humans in our own culture, we may be more inclined to build on, rather than destroy, these relationships. Our modern industrial worldview that shapes the way we imagine our living spaces rests on object-centred canons that narrow the possibilities for architectural design practice down into object-centred, geometric plots and adopts an almost exclusively anthropocentric perspective. This illusion of architectural certainty is being assaulted by the increasing turbulence within natural systems (*Bloomberg*, not dated; Dolnick, 2011; NBC News, 2012; *Telegraph*, 2013), and the idea that we are 'in control' of our environment is melting along with our polar ice caps (see Fig. 16.2).

Yet, if we are to develop more environmentally compatible approaches, our assumptions about the material realm require significant deconstruction – and reconstruction to propose alternatives, by constructing manifestos and populating these concepts with many new 'species' of architecture. Radical new proposals are needed to catalyse this punctuated equilibrium in architectural design evolution – where new features suddenly appear and reach escape velocity from what has gone before to carve out new territories, which also requires new ways of thinking. These propositions also need to be grounded in a technological reality so that they may be explored and realized.

Vibrant architecture draws from the technological qualities of Millennial Nature. Matter has never been predictable or obedient. We have simply learned to understand it this way through Enlightenment perspectives. My proposal is that, by appreciating the potency and profound investment that non-human agents have made and continue to contribute to our existence, it may be possible to challenge entrenched architectural mores and practices; by appreciating the liveliness of the material realm, so that production of architecture may be regarded as more than the production of a finely crafted object or building, but the beginnings of developing a site of activity through

Figure 16.2: Reflections at dusk on the Grand Canal in Venice. Photographs and collage, Rachel Armstrong, August 2012.

which many ecological events can be orchestrated through material relationships (see Fig. 16.3). Yet, this approach requires a different kind of thinking that can apprehend the complexity, connectivity, fluidity and multi-scalar nature of working with innately empowered bodies. Vibrant matter resists formal classification, as it is a shape-shifter, potent transformer and production platform that operates independently of human intent through the coordinated action of assemblages. However, although vibrant matter exhibits varying degrees of autonomy, it is not anti-human and may work alongside us as codesigners of our living spaces.

Yet, at the human and, therefore, architectural scale, the materialist view of vibrant matter[80] provides architects with access to materials with lifelike qualities. Consequently, my research set out to explore a range of materials that visibly expressed these qualities (Bennett, 2011). A range of complex chemistries at far from equilibrium states were identified as appropriate models that could experimentally

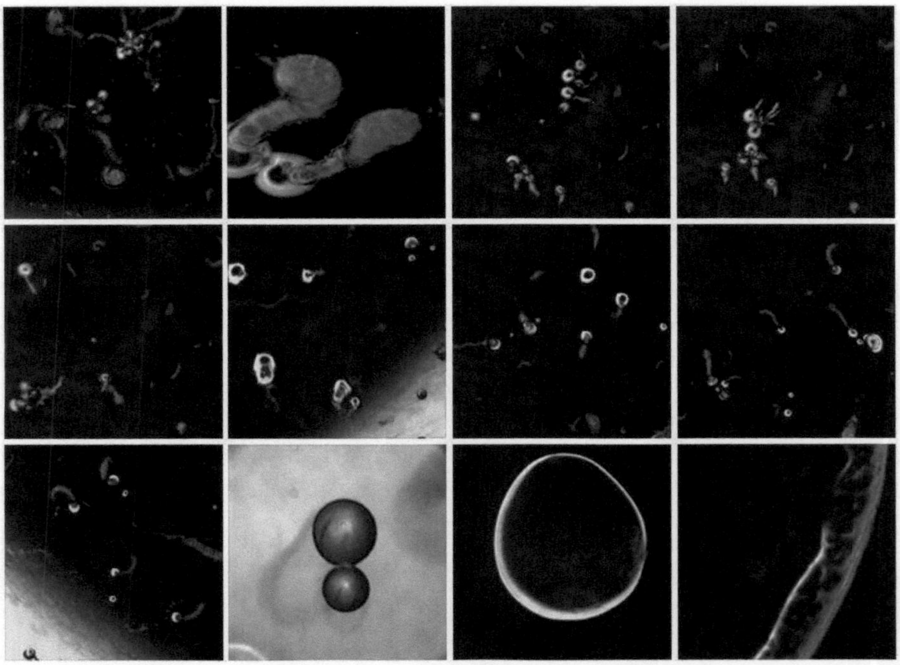

Figure 16.3: Countless populations of non-human agents may be shaped through ELT to produce new kinds of architectures. Micrographs and collage, magnification 4×, Rachel Armstrong, February 2009.

80 A materialist view of vibrant matter regards the effects of materials as originating from within their substance as a function of their quantum nature and does not look to vitalism, ephemeral forces or perceptual influences to account for their ability to act.

demonstrate the properties of vibrant matter such as Bütschli droplets, Liesegang diffusion/precipitation reactions and Traube cells, which were used as model systems to interrogate and explore the implications for working with vibrant matter in architectural design practice (see Fig. 16.4). Chemistry was used as an embodied language, which operated as a combined software and hardware platform, which could potentially negotiate with the untapped potential of various vibrant matter species. Such a discourse was possible by applying morphological computing techniques to persuade lively substances to ally with the human cause through the production of assemblages as a creative negotiation between different 'bodies' (Adamatzky et al, 2013).

Based on observations made through experiments, models and prototypes, it became evident that vibrant matter was more than a design substrate that could be modified by internal and external interventions, but also constituted an emerging

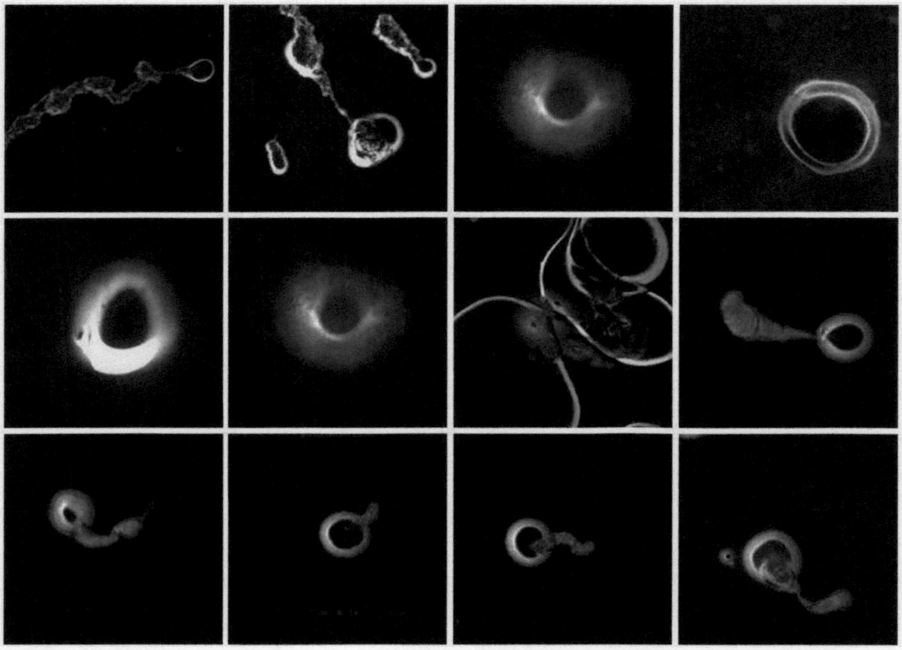

Figure 16.4: Bütschli dynamic droplets served as a model system to investigate the effects of 'vibrant' matter in an experimental context by possessing agency and exhibiting Bennett's criteria (Bennett, 2011) of slowness, porosity and inorganic sympathy. Micrographs and collage, magnification 4×, Rachel Armstrong, February 2009.

technological platform, or ELT, whose outputs could be navigated, shaped, and communicated through acts of architectural design.

At an architectural scale, when vibrant matter is entangled and shaped within our living spaces, it produces 'vibrant architecture', which is more than a container, or 'machine for living in' (Gallagher, 2001; Le Corbusier, 2007, p.158), but embodies lifelike processes that codesign our spaces, and ultimately helps shape our evolution. Vibrant architecture may seamlessly operate within mechanical and process-led worldviews and can therefore be viewed as both an object and a process. Yet, fundamentally, it is a transformer, which breaks down the ontological barriers that set objects and systems in opposition, by horizontally coupling them together through the production of assemblages and synthesizing oceanic ontologies (see Fig. 16.5). This increases their combined effects of heterogeneous agents, which may bring about radical transformation within these newly connected systems to produce something entirely new. This material convergence provides a new production platform whose operations are shaped by spatial programs and are amplified through the interactions of assemblages across many scales.

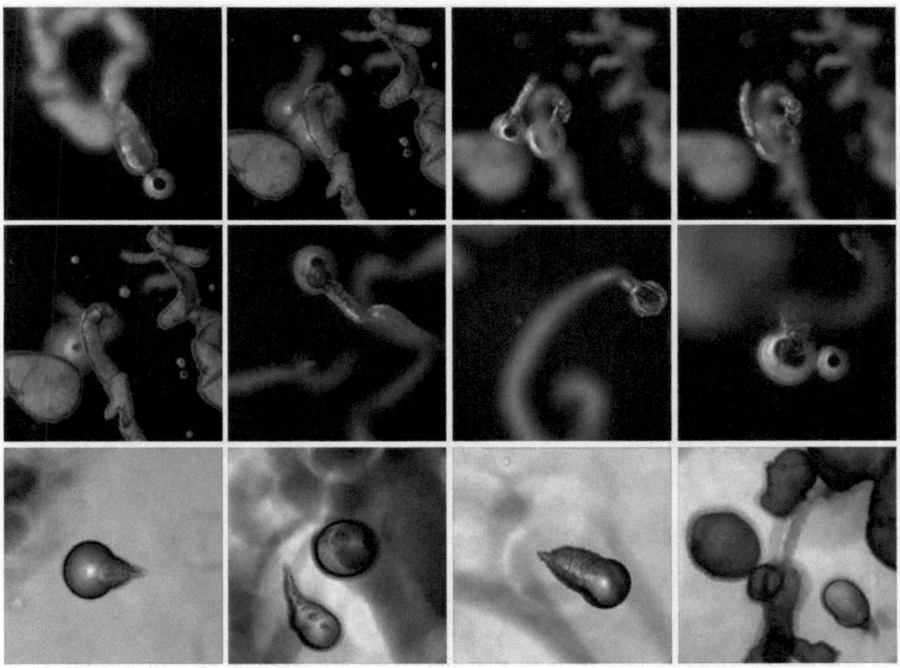

Figure 16.5: This micrograph collage depicts Bütschli droplets responding to internal and external conditions to amplify and scale up their effects. Micrographs and collage, magnification 4×, Rachel Armstrong, February 2009.

Congregating assemblages of vibrant matter may give rise to vibrant architectures and cities, but, uncared for, equally produce our rubbish tips (Bennett, 2010, p.6) and great oceanic garbage patches that ooze noxious substances into the environment. The lively bodies of these assemblages spring from the multitudinous, seamless entanglements of humans and non-human codesigners as fundamental soil-like systems. They build fertile, post-natural landscapes, which are rich with lively new material possibilities and may even shape human development, so that our continued survival is also beneficial for our non-human codesigners.

In this way, vibrant architecture is more than a new production platform that can shape our living spaces but proposes that architectural design may forge a qualitatively new relationship with Nature. Yet, vibrant architecture does not inhabit the incarcerated shroud of Old Nature (Van Mensvoort and Grievink, 2012) and nor does it stand against humanity as anti-modern Nature (Koolhaas, not dated). It does not embody neo-environmentalism that views the material world as a robust 'standing reserve' (Heidegger, 1993), or acts as little 'soft machinery' (Spiller, 2007, pp.202–224; Burroughs, 1961) that is destined for assimilation within industrial paradigms. Rather, vibrant architecture proposes a relationship with a new kind of Millennial Nature that incessantly forges relationships between many heterogeneous bodies. Indeed, it challenges the conventional taxonomic definitions of life, where Nature itself is not a distinct set of objects, but a technological substrate and connecting substance that may be modified and transformed by the actions of many actants.

Millennial Nature does not reside in the wilderness, the countryside or on the pretty rooftops of tall buildings, but infiltrates our urban environments. Rather than being a wan damsel in distress, awaiting rescue from the ravages of industrialization, she has metamorphosed alongside her legions of antibiotic-resistant microbes into something stranger, more substantial and infinitely more subversive. Millennial Nature confronts us with her raw, culturally unedited substance, which heaves open the tops of barely-buried garbage icebergs, oozes from guano-rich crevices that melt building fabrics and splits the paving slabs of walkways with her exploding weeds. Millennial Nature smothers us with her vitality at every twist and turn of the urban landscape and bleeds into the spaces beyond – yet we choose to edit out her presence and remain oblivious to her staggering materiality (see Fig. 16.6).

Smitten by cruel, cultural nostalgia (Morton, 2007, pp.4–5), we refuse to accept that Nature is maturing into a new materiality that fits her fuller 21st century figure. Instead, we long for those bygone days of imagined bucolic perfection, so we corset her up behind façades of polished stone and plump up her sagging tissues with obedient lawns, as if we were conducting cosmetic surgery.[81] Yet no matter how much we try to avoid her grotesquery – by burying her in great holes carved in the earth, or purging

---

81 Morton likens our admiration of Nature to what patriarchy does for the female figure by putting it 'on a pedestal and admiring it from afar' (Morton, 2007, pp.4–5).

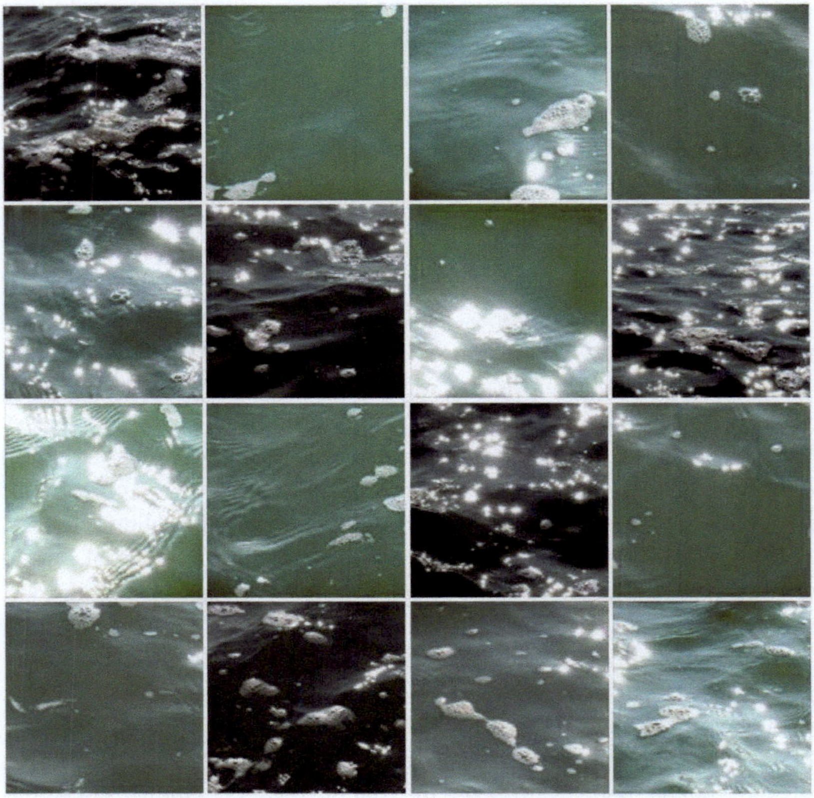

Figure 16.6: Non-human actants may codesign our spaces through assemblage formation and transformation of matter. Photographs and collage, Rachel Armstrong, August 2012.

her filth from our cities through our sewers into our acidifying seas – our living spaces are nevertheless crawling with her tissues and she continues to thrive. Indeed, she lies – barely chemically 'breathing' – in metabolic suspension in garbage graveyards where she is decoupled from networks of biospherically essential exchanges, which may have sustained her corpulence or enabled her rebirth through the revitalizing production of compost. Yet, like Tarantino's Bride,[82] Millennial Nature defies the odds against her survival and resurfaces from her untimely grave to exert her revenge by leaking carcinogens and volatile gases into our urban environments (Canada.com, 2007). These acts of resurrection are nothing less than magnificent, as she gnaws the Earth with pitiless sinkholes, advances algal blooms in our seas and dazzles our skies

82 Here, I am referring to Beatrix Kiddo, or 'The Bride', who is the lead female character in Kill Bill Vol. 2 by Quentin Tarantino (2004) that claws her way out of a grave to seek revenge on her enemies.

with rainbow-brilliant, polluted sunsets, which continue to invoke awe in those who may have feared that her splendour had somehow faded as she (re)negotiates her survival (see Fig. 16.7).

Figure 16.7: Assemblages of actants negotiate the harsh environment of the Venetian shoreline and flourish against the tempestuous odds. Photographs and collage, Rachel Armstrong, August 2012.

16.3 A Recipe for Vibrant Architecture: An Experimental Approach

My conclusions are informed by critical theoretical reflection and speculative inquiry and fuelled by architectural design questions based in experimental research. This has added depth and complexity to my investigations and enabled the properties of vibrant matter to be tested in a design context, through the production of drawings, models and project work.

Vibrant matter was explored as a real substrate for architectural design practice using dynamic chemistries. A range of systems were developed in a design context, such as the Traube cell and Liesegang rings, which offered a combined production platform that could be manipulated using morphological computing techniques. However, the mainstay of my experimental work was conducted using Bütschli droplets as a demonstrator platform, which were characterized and explored through a series of experiments, models and projects (see Fig. 16.8) that interrogated and portrayed the nature of such lively materials, using a variety of methods and contexts to reveal a set of principles for the production of vibrant architecture.

While the identified practical principles are most readily applied to systems at far from equilibrium states, the design considerations also still apply to materials at equilibrium, since, as Kauffman notes, the idea of classical equilibrium is a relative condition, not an absolute one (Kauffman, 2008, p.125). Architects may, therefore, decide whether they are dealing with relative equilibrium or non-equilibrium conditions, according to the scale at which they are working and the particular design context. Applications of this form of design thinking may be considered as ecologically compatible practices, since, over the last few decades, environmental behaviours have been recognized as working in accordance with the principles of systems theory (Von Bertalanffy, 1950; Willis, 1997), which also underpins the operational principles of materials at far from equilibrium states (Prigogine, 1997) (see Fig. 16.9).

My research aimed to establish whether it was possible to develop a 21st century architectural design practice that deals directly with material complexity whose 'object status' is not contravened, but accepted and extended into the territory of dynamic systems. By elevating the status of the material world and attributing it status

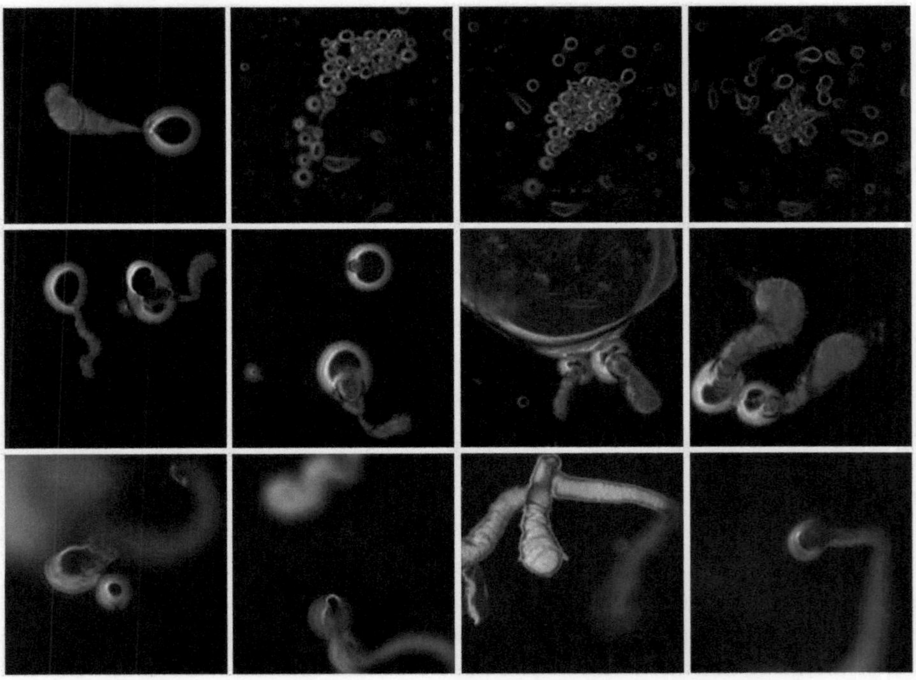

Figure 16.8: Bütschli droplets provided a dynamic model experimental system through which 'vibrant' matter could be explored as a material at far from equilibrium and potentially applied within an architectural design context. Micrographs and collage, magnification 4×, Rachel Armstrong, February 2009.

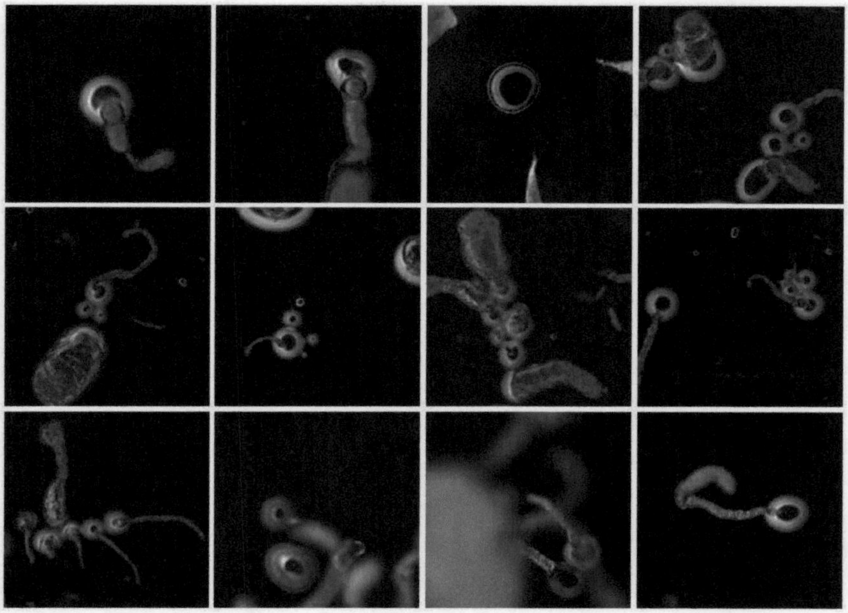

Figure 16.9: The Bütschli system possesses unique poetics and aesthetics at the microscale, with some homologies with natural systems. The design challenge was to translate these possibilities to the architectural scale. Micrographs and collage, magnification 4×, Rachel Armstrong, February 2009.

within the design process, it may be possible to draw attention to the codesignership of our living spaces produced by agencies that are not under our instruction – for example, by growing lively materials that may continually adapt to their environment (Rowlinson, 2012). This does not mean that there is no role for the human designer in shaping our living spaces, but asks architects to think differently about how materials, infrastructures and technologies that are used in architectural production respond to spatial programs as acts of codesign in the production of living spaces (see Fig. 16.10).

16.4 Embodying Complexity

Complex materials are generally conservative in their performance but operate within definable limits of predictability (see Fig. 16.11).

My research endeavoured to directly harness their potency and capacity as a codesigner of architectural systems. I have used informed speculation, grounded in experimental observation, to explore possible outcomes of design tactics and architectural programs where overlapping knowledge fields and methods may synthesize specific approaches that can deal with probabilities, rather than fixed

Figure 16.10: These reflections hint at the possibility of harnessing the power of the material realm to generate novel approaches in architectural design. Photograph, Rachel Armstrong, August 2012.

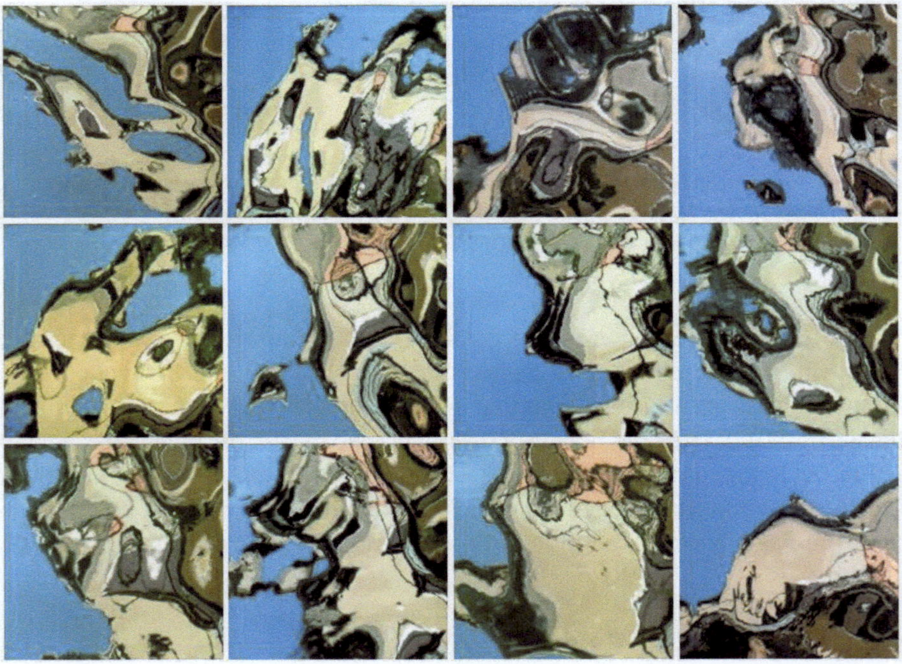

Figure 16.11: Venetian canal reflections offer an interface through which it is possible to see time and matter evolving in a dynamic material system. Photographs and collage, Rachel Armstrong, August 2012.

products. Working with vibrant matter and ELT requires designers to surrender five pillars of architectural certainty, which are summarized in Table 16.1.

Table 16.1: Architectural pillars of certainty

Pillar of certainty	Limitation	Alternative
Predetermined designs	Hermetic approaches and practices that resist change	21st century architectural design is an open, collaborative process that works in partnership with and is informed by other disciplines, so it remains responsive to changing needs, ideas and contexts that are not contained within predetermined designs but are fashioned through continual, iterative exchanges
Sole authorship	Hierarchical, centralized control systems that are executed according to a centralized plan reduce the complexity of possible material responses to the spatial programs of a site	As codesigners of systems, architects may work in collaboration with other humans and non-human agents through processes that do not invoke hierarchies of order but forge mutual relationships between many heterogeneous agents. These can respond to complex challenges through a form of actant-led 'crowdsourcing' of challenges where morphological computational decisions are made by heterogeneous groups of actants in a distributed and cooperative way
Discrete boundaries	Impermeable barriers limit the possibility of exchange between material agents and prevent environmental responsiveness	Porous interfaces enable the collective action-organizing 'hubs' of activity that may respond to spatial programs beyond the life of an architect
Incremental research goals	Risk of redundancy with the advent of disruptive events and technologies	21st century architecture is not about tinkering around the edges of what already exists but has a moral obligation to future architects and the public through ambitious, visionary thinking
Completion	Does not enable the possibility of change with time	Ideas as well as solutions change and are in constant evolution. Architecture is never truly finished but is an open process that is never fully resolved. Buildings continually contribute to biospherical processes and may be adapted by and assimilated within future architectures. They may also be recycled and transformed into other systems such as soil

16.5 Site-specific Exploration of an Emerging Technological Platform

While laboratory findings provided essential information about the possibilities of ELT, prototype experiments in the site-specific context of Venice created a set of conditions that focused development of the emergent technological platform within an architectural context. Specific questions were raised about the feasibility of designing a dynamic droplet system that could potentially 'grow' an artificial limestone reef under the darkened foundations of the city by developing a basic prototype droplet system that could model some of the proposed effects. Experimental findings were iteratively combined with speculative approaches that used ELT to develop the prospect of enabling Venice to literally fend for itself in an environmental struggle for survival. Using Stengers' notion of 'an ecology of practices' (Stengers, 2000), it was possible to reflect on the kinds of conditions that would enable such a technology to draw raw materials from the lagoon through many acts of codesign and work in concert with humans and the marine wildlife to produce a reef-like architecture (see Fig. 16.12).

Figure 16.12: Mineral accretions forged by biological and chemical systems spontaneously form around Venice's shoreline. Photograph, Rachel Armstrong, August 2012.

16.6 Scaling ELT to Urban Dimensions

The idea of soils as architectural-scale bodies was developed in the context of vibrant cities by considering them to be technological expressions of vibrant matter, which are forged by an entanglement of biological and chemical agencies. Soils are metabolically active and enable the free flow of elemental systems through them, such as air, water, heat and matter. Applying the technology of soils within buildings as entangled assemblages and composites of vibrant matter, which do not slavishly mimic biological processes, but interpret their environments through morphological computing processes, could potentially transform and recycle resources to perform useful work such as the production of heat and filtering wastewater. The by-products of these metabolic processes could also potentially produce 'urban' composts as a valuable resource to increase the fertility of the urban environment. For example, vibrant cities may produce native, not transplanted, soil systems that enable metropolitan greenery to flourish (Armstrong, 2013a) (see Fig. 16.13).

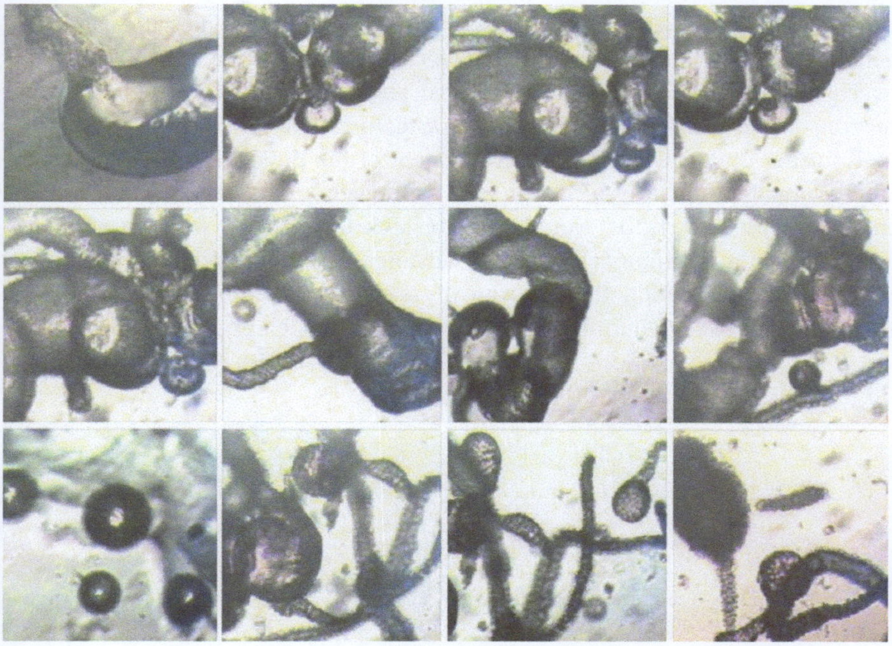

Figure 16.13: Soils may be regarded as a complex chemical technology capable of many acts of transformation. Micrographs and collage, magnification 4×, Rachel Armstrong, August 2012.

16.7 Speculative Narratives as a Means of Exploring Innovation Space

During my research, I composed a series of speculative narratives, which explored topics ranging from synthesizing lifelike agents from chemistry in a laboratory environment to imagining cities built upon the principles of vibrant matter. Each of these design propositions was regarded as an extended hypothesis that was in keeping with the exploratory nature of my research and the kinds of materials and approaches adopted in my investigations. Developing these scenarios opened up new spaces for reflection on the experimental systems, emerging practices, materials and technologies that I was working with, which are not formally available outside of a laboratory setting and whose outcomes had not already been established and/or could not easily be predicted (see Fig. 16.14). These narratives served as instruments for reflecting possibilities back into the real systems with which I was working, and enabled me to make decisions about my research. For example, the relationship between vibrant matter and morphological computing in an architectural context was not sufficiently mature to offer formal approaches to the research proposals.

Speculative propositions, however, were not made in ignorance of emergent systems but were informed by having a working understanding of vibrant matter in a design context, identified model systems, made observational studies and conducted chemical experiments. These first-hand experiences enabled me to work iteratively between informed speculation and practical engagement, which enabled further reflection, design and development of ongoing research questions. These possibilities helped further understanding and development of the project by reflecting on the aesthetic qualities of the project, as well as engaging with the dynamics of the accretion process in ways that would not be testable through real-world experiments. Although speculative approaches do not advance development of the technology towards its implementation, they have greatly assisted in the progression of research ideas and will be used to inform further experiments, models and prototypes.

Figure 16.14: In this photograph, washing on a line is transformed by the active surface of the canal water into discrete, vibrant architectural territories. Photograph, Rachel Armstrong, August 2012.

16.8 Manifesto for Vibrant Architecture

This manifesto addresses the strangeness of the material realm and proposes new relationships between humans and non-humans that do not advocate the triumph of one set of principles over another. Despite the inevitable polemics of writing a manifesto, I aimed to draw attention to the transformative powers of the material world so that it was possible to appreciate its codesignership in the production of living spaces (see Fig. 16.15). A manifesto for vibrant architecture therefore not only summarizes my research findings as a set of design propositions, but also identifies opportunities within architectural practice for new spatial programs and design tactics that may form the basis of further research.

Figure 16.15: Bütschli droplets provided a model system for a manifesto of 'vibrant' matter. Photographs and collage, Rachel Armstrong, February 2009.

16.9 Designing with Vibrant Matter

This section reviews the major research findings, which explore the challenges of working with complex, embodied systems. These are:

— Design with probability
— Process as material
— Design with lifelike systems

16.9.1 Design with Probability

Designing with probability means shaping the environment and context of a system so that it is likely to provoke an event, rather than focusing on a single end point that predetermines the outcome of the system. My research established a set of principles for dealing with non-deterministic systems by, firstly, creating the context in which this kind of approach could be imagined. Terms derived from process philosophy that invoked ideas about relationships, flows, networks, systems, agents, actants, and assemblages were explored as a way of creating the conceptual context for probabilistic design events. Process-based terminology evokes the kinds of possibilities that may be expected within limits of definable probability. Probabilistic design is an inherently creative way of working, where materials creatively participate in the development of spatial programs as codesigners of the system. The production platform through which probabilistic design is realized is based on the production of assemblages. These are inherently flexible systems that deal with matter at non-equilibrium states and are extremely responsive to environmental changes. Their outputs may be shaped by using parallel computing processes to produce sustained material outputs that (even briefly) resist entropic decay towards inertia. A comparison between probabilistic and deterministic design paradigms is summarized in Table 16.2.

The implications for using materials at non-equilibrium are that they challenge deterministic principles of design and work with probabilities. Indeed, the frameworks that underpin their development operate according to the same organizational principles as Nature. Rather than suffering the wear and tear of object-centred designs in the environment, they are instead prone to derive sustenance from the continual transfer of material and may adapt to persistent patterns of exchange. Alternatively, they may transform as internal and external circumstances change. This challenges the way that architecture is represented, planned and executed and requires programmatic approaches that can deal with probability, which resist confinement within the complexity of non-geometric spaces and deal with changes in performance of the various systems and actants with time. For example, Neil Spiller and Perry Kulper propose drawing tactics that embrace many dimensions of reality – both empirical and subjective – to explore a new, probabilistic, fabric of reality (Spiller, 2011). Alternatively, the Hylozoic Ground chemistries are not representations but the 'things in themselves' that mediate change in their surroundings and are not represented, but developed *in situ*, as they are only truly 'known' when they are constructed (Armstrong and Beesley, 2011).

Table 16.2: Comparing design in probabilistic and deterministic paradigms

Principle	Probability	Determinism
Language	Process-led. Imagined as potent fields of interactivity that may give rise to unique events	Object-oriented. Described as ordered hierarchies and sequences
Metaphors	Complex, alchemical and open to multiple interpretations	Simple, aim to reduce ambiguity and distil concepts into discrete elements
Programming	Parallel	Series
Ontology	Oceanic ontology, which is changeable and context-sensitive (Lee, 2011)	Hierarchical and fixed relationships
Limits	Relativistic and provoke creativity	Absolute and inhibiting
Identity	Fluid, multiple, parallel, transgressive and transmutating	Singular
Time	Directional, acts to shape events in complex and radically creative ways. Not reversible	Reversible
Relationship with Nature	Works synergistically with the relentlessly material and continually reforming systems of Millennial Nature	Reinforces binary divisions of Old Nature and establishes oppositional stances
Taxonomy	Convergent and alchemical	Divergent, represented through tree-like branching structures
Epistemology	Evolving, contextualised and perpetually in transition	Context-sensitive, constrained by perceptual and linguistic influences (Jablonka and Lamb, 2006). Potentially 'post-epistemological' or unclassifiable according to traditional modes of classification (Latour, 2013)

16.9.2 Process As Material

> *The role of the architect here ... is not so much to design a building or city as to catalyse them: to act that they may evolve.* (Pask, 1995, p.7)

Process-led materials exist at non-equilibrium states and may produce structures that change with time. For example, modified Bütschli droplets are roughly spherical but gradually produce crystals that alter the shape of their oil/water interface container. Yet matter at non-equilibrium states has similar requirements to living systems, which thrive in open, resource-rich environments and require elemental systems that enable the free flow of matter and the removal of waste. As such, designing with process-led materials requires different conditions than those for making objects. A range of factors that shape design decisions are listed in Table 16.3.

Table 16.3: Designing with process as material

Principle	Observation	Design Decision	Example
Open system	Closed systems reach equilibrium states	Identify abundant resources and map them. The flow of matter through the system influences the way that processes evolve and will inform oceanic ontologies, e.g. exchanges between the actants in the Venetian lagoon such as light, marine organisms, minerals and carbon dioxide will shape the character of mineral shells in protopearls	Bütschli droplets are expected to remain dynamic longer in an open system such as the Venetian lagoon than in the relatively closed Hylozoic Ground flasks, which have much more limited surface area for exchange with their surroundings
Abundance	Processes are enabled in sites with plentiful resources and restricted by scarcity	Maximize surface area available for exchange e.g. using effusive, convoluted surfaces rather than smooth ones, to enable circulation and transfer of matter to prolong processes, such as mineralization in Mother Shipton's Cave, where a continual flow of heavily mineralized waters through spongy foam and fabric feed crystal growth	Copper ion availability limits Traube cell growth in solution. When this limitation is removed the typical 'cell' continues to grow
Context	Operates within definable limits. Extremely sensitivity to environmental conditions. Performance is site-specific and unique	Environmental conditions may need to be designed or manipulated to enhance material performance, e.g. Traube cell growth can be extended in a biopolymer rich environment, which allows water to enter into the system more slowly, provide support to fragile membranes and produces a different structure (narrow rather than broad) to that grown in an aqueous environment	Protopearls in the Vibrant Venice project form crystals under alkaline conditions and mineralize in the presence of silica salts
Metabolism	Sets of chemical exchanges and interactions produce energy and change within a system	'Messy' combinations of interacting chemistries may produce more surprising and persistent outcomes than carefully measured relationships, e.g. complex exchanges between a range of different salts in modified Bütschli droplets in the incubator flasks produced a variety of interior structures	Metabolisms are most effective when an environmental component is incorporated into chemical exchanges, e.g. protopearls use water as source of carbon dioxide and minerals

Continued Table 16.3: Designing with process as material

Principle	Observation	Design Decision	Example
Vectors	Breaking symmetry in chemical systems induced polarity and directionality and results in physical forces, which create observable effects, e.g. propulsion, growth, expansion	Establishing a clear context and mapping environmental conditions may help designers influence the rate of change and direction of movement within a dynamic material system. The techniques that may achieve these outcomes are more similar to 'gardening' than engineering a machine	The rate and direction of change in dynamic systems is dependent on the flow, environment and texture of the material, e.g. Traube cell growth on bioscaffolding can be extended under the influence of gravity
Dissipation	Materials not only take in energy and matter but also release it during 'organizing' phases that take place when systems are far from equilibrium	Changing the condition of the medium, environment or the body of the agent, can optimize conditions for the probable emergence of dissipative structures. The vigorous material exchanges in dissipative systems are pluripotent and early stage disturbances shape subsequent events as in the production of osmotic structures by Bütschli droplets	Bütschli droplets exist as dissipative high-energy 'shell' structures that release energy into the medium, which contributes to the emergence of organized droplet structures
Medium	Processes require media that enable movement such as gaseous or wet conditions	Using a combination of media and materials (like developing layers of embryonic tissue) enables process-led systems to develop a range of outputs and therefore diversify audience experiences, e.g. Hylozoic Ground chemistries are entangled with a range of dynamic media such as liquids and gases that enable them to change and grow with time	Intensely organizing systems are possible in gaseous media, such as the cloud installation by Berndnaut Smilde (Design Boom, 2012) and in liquid media, such as the Hylozoic Ground chemistries
Quality	Process-led materials are soft and dynamic	Oceanic ontologies establish the range of material possibilities that may be processes by a particular architectural program for a specific site and conditions, such as in the Bütschli system	Bütschli droplets exist in a number of changing, variable forms and leave soft, crystalline 'osmotic' residues

Continued **Table 16.3:** Designing with process as material

Principle	Observation	Design Decision	Example
Spatiality	Agents form intimate connections with the surroundings through the production of assemblages. These participating bodies confer process-led materials with a rich dynamic and undulating spatiality	Novelty may be induced by overlapping spatial programs, e.g. Liesegang rings are produced by periodic spatial interactions between salt solutions and alkalis, which are further spatially separated under gravity to produce banded formations. Banding does not happen without spatial distribution of the participating chemistries	Using matrices can produce qualitatively new events and structures, e.g. biopolymers, to separate chemistries and prevent them from being homogeneously mixed, such as in the Liesegang ring reaction
Time	Processes change with time, which is source of creativity in the system	Time-based interventions may increase the likelihood of specific events in overlapping spatial programs, e.g. organic solvents are attractants during the early stages in the Bütschli droplet system (chaotic and organizing) but have little effect on later stages (quiescent), so transformative interventions need to be designed to occur at an early stage in the development of the system	Dynamic chemistries such as Bütschli droplets are time-sensitive in their operations, being more plastic early on in their operations when they are furthest from entropic equilibrium
Transformation	Materials constantly change in response to internal and external cues until they reach equilibrium	Dynamic materials are indicators of change. Process-led materials may create different products, which are influenced by many variables and factors. These changes can be recorded, e.g. as material traces or drawings that depict chemical information landscapes that are otherwise invisible to the unaided eye. Notably, Bütschli droplets may shed a range of skins that may be regarded as 'wet' drawings	Bütschli droplets exhibit different forms and behaviours during their organizing phase

Continued Table 16.3: Designing with process as material

Principle	Observation	Design Decision	Example
Decay	Dynamic materials approach entropic equilibrium as environments and agents become homogenous	Increasing resource availability or changing environmental conditions may prolong process-led systems. For example, stirred reactors may increase the longevity of the Belosov–Zhabotinsky reaction (Hsu, Mou and Lee, 1994). However, decay in process-led systems is inevitable and designers need to decide what to do when their dynamic system is no longer able to self-repair, or is no longer viable (Woods, 2012b). For example, aging and decay may be an intended form of obsolescence that serves as the foundations for other, more vigorous systems (like a coral reef or stromatolite)	The 'werewolf' transition in the Bütschli droplet system is an indicator of entropic decay and signifies a complex, phased transition towards inertia that occurs between the organizing and quiescent phases of Bütschli droplets. Droplets become agitated and produce large amounts of residue before becoming inert

Process-led, design decisions may be informed by elevating the status of matter in the system and incorporating non-equilibrium materials into spatial programs. This provides architects with the possibility of developing and working with a whole new range of materials, technologies and construction processes in the production of space. For example, living spaces may have 'metabolic' functions that are shaped by soft materials such as algae and bacteria that are able to process water, waste food and packaging in community spaces, or within spandrels (Gould and Lewontin, 1979) in our homes. These unconventional spaces are akin to Matta-Clark's 'Fake Estates' – a project where the artist acquired a number of miniscule plots of land auctioned off by the city of New York that were of redundant utility (Kroessler, Richard and Finkelpearl, 2005). In other words, the existence of these slivers of no-man's land was determined by factors other than formal architectural programs being distributed; fragmented spaces with deregulated spatial characteristics. Such sites are potential hosts for urban physiologies that constitute an extended notion of spatiality, through material networks, metabolic linkages, synthetic ecologies and post-natural fabrics within our cities. These subversive, distributed spaces are sites of programmatic (re) invention and nascent fertility that may ultimately enable societies not just to be consumers, but also producers of their spaces. As codesigners in the production of space, living architects may cultivate material, infrastructural, technological, social and cultural relationships and use soft control design tactics where appropriate to nurture and encourage fertility in the development of urban ecologies and post-natural landscapes.

16.9.3 Design Characteristics of Lifelike Systems

My research aims to outline a toolset of concepts and approaches that may help designers work with matter in more vibrant, lifelike ways. Experimental work produced during my research established that vibrant matter:

- Possesses agency
- Programmable with morphological computing techniques
- Exhibits co-authorship in our environments
- Coherently operates across many scales that include the architectural realm

Unlike digital technologies, lifelike systems do not need every possible outcome to have been anticipated before they deal with a challenge, but may find ways to resolve them using real-time parallel processing. This does not mean that lifelike systems are infallible – if they do not produce a solution, they may 'crash' and become inert. Yet, the history of 'life' is characterized by continual losses and extinctions of material solutions. Those that are successful become increasingly adapted to the kind of challenges found within a particular environment or niche. Biologists refer to this process as evolution. Although design principles for working with biological systems and lifelike systems are similar, there are important differences that relate to the value placed on their respective performances. Table 16.4 summarizes characteristics of lifelike materials which have been derived from my experimental work and the expectations that designers may have of them.

Working with lifelike materials implies that architects are codesigners of spatial systems and, therefore, require design tactics, materials, infrastructures and technologies that help to negotiate the complex and often contradictory relationships that emerge continually in living systems. Vibrant architecture requires human participation and engagement if it is to be shaped and developed towards a specific design outcome. While managing these networks is energy intensive, it is likely that at least a degree of the infrastructural regulation within living architectures may be performed by computer-generated processes through microfluidics chips. These can regulate infrastructural performance in biological/chemical/mechanical systems, yet the degrees of freedom that escape mechanization create opportunities for personalizing synthetic ecologies so they have individual characters. Such communities require a shared language that can grow and develop accordingly as the post-natural fabrics that were forged through their intimate interactions evolve.

16.10 Vibrant Matter as Technology

Vibrant matter may provide a new technological platform, or ELT, when physical agents are combined with morphological computing techniques. The system may be orchestrated by designers using a chemistry-based language that interprets spatial

Table 16.4: Design characteristics of lifelike systems

Principle	Observation	Design Decision	Example
Agency	Acts autonomously on the environment to exert effects that influence human interactions	Lifelike materials are empowered and produce a variety of different effects that may change with time, e.g. phase transition behaviours of Bütschli droplets	Dynamic droplets can move around their environment, sense it, interact with other agents and participate in population scale behaviours
Uniqueness	Lifelike materials are functions of their internal composition, environment, context and relationships	Designers may influence the outputs of lifelike materials by changing internal and external conditions in the system	No two Bütschli droplets are alike
Programmable	Morphological computing techniques can chemically and physically shape events	Designers may reveal environmental details that are invisible to human senses and record them using lifelike systems, such as Bütschli droplet drawings	Modified Bütschli droplets in space-constrained environments may produce Turing bands
Anthropocentrism	Lifelike materials resist anthropocentrism	Designers may work inclusively with 'ecologies of design' (Stengers, 2000) so human and non-human agents can work alongside each other without hierarchies of order	An artificial limestone-like reef may be 'grown' under Vibrant Venice through the action of multiple non-human agents, e.g. 'smart' droplets, marine ecologies and tidal flow
Design method	Living materials deal with persistent change and evolving outcomes by acting as codesigners of these systems	Architects need to manage the difficult relationships between communities of actants that operate according to artisan-led, bespoke, distributed, decentralized processes that may be shaped by humans through 'gardening-like' processes, such as Explora Biotech's 'husbandry' of the Bütschli system over the duration of the Hylozoic Ground installation.	Architectures are grown and change with time and according to their context, such as Hylozoic Ground chemistries

Continued **Table 16.4:** Design characteristics of lifelike systems

Principle	Observation	Design Decision	Example
Combined platform	Material, infrastructure and technology are seamlessly integrated in lifelike systems – not componentized, as in machines	Designers need to appreciate that the outcomes of lifelike systems may be shaped within limits and offer multiple readings of spatial programs, such as the Hylozoic Ground chemistries	The active interface of dynamic droplets is embedded in the environment and interfaces with many actants at different scales, which interact and even transform each other to produce highly unique and individual effects
Environmental sensitivity	Lifelike materials are sensitive to, shaped by and dependent on their environment	Designers need to characterize and even manipulate the environmental, spatial and chemical programs that will help to develop appropriate tactics for working with lifelike systems, such as the mineral and carbon dioxide content of the Hylozoic Ground flasks	In the presence of dissolved carbon dioxide, Hylozoic Ground chemistries may produce crystal growth that reflect the amount of greenhouse gas in the environment over the duration of the installation
Programs	Lifelike systems enable the differential spatial and temporal distribution of matter through their embodiment	Designers may use morphological computing techniques to manipulate lifelike chemistries. This flexible language may be combined to produce different effects, e.g. the spectrum of behaviours observed in the Bütschli system	Adding minerals to the bodies of Bütschli droplets changes their chemical programs so they begin to produce solid structures
Character	Lifelike materials give rise to soft architectures	Designers can make use of spatial spandrels in under-imagined places, buoyant media and embryological programs, and develop them as sites and opportunities for soft architectures, such as synthetic soils in cavity wall spaces	Flexible Traube cell bioscaffolding extensions can be shaped by gravity, which can extend structures from 4 cm to 60 cm

Continued Table 16.4: Design characteristics of lifelike systems

Principle	Observation	Design Decision	Example
Site and context	Soft architectures require procurement systems and infrastructures that enable liquidity	Rather than designing the event itself as an end point, lifelike systems warrant the production of spatial programs that shape context. While lifelike systems do not guarantee a specific event they are agents that increase the fertility of a site and make it more likely that a set of material relationships will bring about a possible spectrum of anticipated events	Transitions being made from dry to wet façades e.g. IBA Hamburg and Arup (Steadman, 2013)
Slowness	Dynamic chemistries exhibit the slowness of the material world as they grow and self-assemble through analogue computing processes, whose calculations occur at the quantum scale	Lifelike systems exhibit relative slowness that designers may deal with by exposing time parallaxes that are usually imperceptible, such as the fast-interacting neural network contrasted with the slow-changing dynamic chemistries in the Hylozoic Ground installation. Comparing these systems within the installation's spatial programs – as a form of experimental relativity – reveals time slippages that may be experienced as uncanny encounters	Banding patterns in the Hylozoic Ground Liesegang ring plates evolve over the course of several days to months
Porosity	Chemical assemblages produce bodies that are entangled with many systems	Designers may provoke transformation within lifelike systems by designing spatial programs that facilitate porosity, as they are susceptible to invasion and/or infusion with other spaces and bodies	Bütschli droplets show a range of behaviours where they spontaneously aggregate and disperse according to local chemical cues and mutual interactions between individual actants

Continued Table 16.4: Design characteristics of lifelike systems

Principle	Observation	Design Decision	Example
Inorganic sympathy	Lively chemistries originate from the non-living realm and ultimately return to it	Design programs, therefore, may incorporate the inevitability that lifelike programs will become inert. Designers may seek to delay this inevitability by, for example, providing infrastructural support and the abundant flow of nutrients through the system	Bütschli droplets exhibit inorganic sympathy and become inert when their osmotic 'skins' occlude the oil/water interface. Once this condition has been reached they cannot be fully reanimated by breaking the skin to reactivate the reaction.
Parallel processing	Molecular interactions enable materials to respond to overlapping spatial programs	Lifelike systems materialize outputs in the presence of overlapping or conflicting spatial programs, without needing to know the outcome in advance	Modified Bütschli droplets simultaneously process a range of inorganic salts and convert them into complex heterogeneous structures
Multiple scales	Lifelike materials simultaneously operate over a range of scales and may produce unique scale-specific effects	Designers may construct spatial programs that potentially work across parallel worlds (Deutsch, 1997)	Modified Bütschli droplets act locally to produce crystalline microstructures and also function as organs within an architectural-scale cybernetic body
Non-linearity	Lifelike materials are unpredictable and may undergo phase changes in behaviour and/or collapse around tipping points	Design paradigms that work with lifelike systems need to be open and robust to deal with the unpredictability of living systems	Turing banding and phase change behaviours occur in the Bütschli system
Technology	Operations are directed through the formation of assemblages	Lifelike systems operate through the technological platform of assemblages. They are able to respond flexibly to changing spatial programs in environmentally sensitive ways and therefore behave variably within definable limits of probability to offer a range of solutions to any particular design challenge	Bütschli droplets are attracted to each other and spontaneously aggregate in population-scale formations

Continued Table 16.4: Design characteristics of lifelike systems

Principle	Observation	Design Decision	Example
Soft control	Assemblages can only be orchestrated, not commanded, as they are probabilistic	Lifelike systems do not come with a push button and respond to soft (facilitative continual nurturing), rather than hard control systems (energy-intensive command). They also possess an independent chemical 'will', which needs to be negotiated and engaged by designers, rather than subdued	Changing the internal and external conditions of the system, such as using chemical attractants to induce cluster formation, may influence Bütschli droplets.
Codesign	Lifelike materials continually engage in material exchanges and forge networks. They therefore continually influence their surroundings and shape them in acts of codesign	Designers need to identify synergies between actants so they may become codesigners. The architect as codesigner extends their individual agency to operate within an assemblage of human and non-human exchanges, which results in the production of architectural design outcomes	Many actants shape the construction of the Vibrant Venice artificial reef, such as marine organisms, humans, tides and smart droplets.
Multi-disciplinary	Lifelike behaviour is not the exclusive specialty of any one discipline but implies convergence and collaboration around the ideas associated with living systems	Designers may work collaboratively to increase knowledge fields using an 'ecology of practices' (Stengers, 2000)	Hylozoic Ground was produced by an international, multi-disciplinary collaboration
Non-hierarchical	Lifelike materials are forged from 'horizontal couplings' and interactions between agents, rather than hierarchies of top-down or bottom-up programs	There is no hierarchical ordering in lifelike systems as many elements inform the spatial programs that shape assemblage interactions	Bütschli droplets interact in populations without fusing with each other

Continued Table 16.4: Design characteristics of lifelike systems

Principle	Observation	Design Decision	Example
Boundaries	Assemblages do not possess discrete borders but are porous to energy, matter and information	Designing with lifelike systems requires them to be considered as both objects and systems where the bodies of assemblages may be defined at any given moment but are continually shaped and even transformed by their interactions	Bütschli droplets self-organize as dissipative structures that absorb and release large quantities of matter and energy, and yet appear to have a distinct structure.
Classification	No formal classification system exists for lifelike materials	Oceanic ontologies may be used to relate multiple events within a lifelike system to each other to produce complex, probabilistic spatial maps and topologies	Bütschli droplets may undergo many morphological and behavioural transitions, such as the 'werewolf' moment
Body	Lifelike bodies forged by assemblages are soft, wet and mutable	The mutable qualities of the bodies of lifelike systems requires designers to consider multiple spatial programs and tactics to shape and organize their behaviours and interactions as well as creating a context for unexpected transformations and events. Indeed, a precondition for designing for lifelike systems need a world in flux; for instance, the flux of heat and light coming from the Sun. They need also strong coupling between different types of processes	Traube cells possess structure which is 'seaweed-like' and defined by the processes that shape the production of the cell's membrane, but the system does not possess a formal physical shape and transforms continually
Carbon	Lifelike systems are transformers of their environment and may absorb, fix and recycle carbon-containing compounds	Lifelike systems may create the conditions to design carbon-positive architectures that remove carbon emissions from the environment and incorporate them into their construction process	Protopearls fix carbon dioxide in solution into an insoluble, shell-like, mineral structure

Continued **Table 16.4:** Design characteristics of lifelike systems

Principle	Observation	Design Decision	Example
Born not made	Vibrant matter is a fundamental quality that exists as a function of the primordial laws of the un verse and cannot be acquired	Lifelike agents require a particular place where molecules may productively interact with each other within the limits of their existence. When interactions between molecules are possible, they may produce materials that are born from ingredients and produced through recipes that are made operational through acts of 'cooking', rather than building them using mechanical processes	Bütschli droplets spontaneously arise and self-organize from a field of olive oil and alkali
Equilibrium	Lifelike matter exists at far from equilibrium states and resists the decay towards entropic equilibrium (Schrödinger, 1944)	Designing at non-equilibrium enables the progressive discovery of new relevant questions, revealing the diversity of Nature, and pedagogical systems that enable designers to approach this diversity	Bütschli droplets exhibit qualities of dissipative structures at far from equilibrium states
Metabolism	The active exchange of chemistry across an interface enables lifelike matter to evade reaching entropic equilibrium	Designing with metabolism is a challenge set for ecologists by Morton, where outputs of one set of chemical exchanges may be the substrate that begins interactions between another set of agents to generate 'straightforward environmental images' (Morton, 2007, p.150)	Bütschli droplets consume their surroundings (olive oil) to produce a soap-like product and energy. Droplets lose their lifelike qualities when the interface between alkali and oil is occluded by waste product
Simultaneity	Vibrant matter exists both as an object and as a process	Designers may need to develop tactics and spatial programs that overlap, converge and even inhabit parallel worlds to deal with the paradoxes of lifelike systems	Bütschli droplets are composed of paradoxes – they are simultaneously soft (deformable) and hard (crystalline), as well as being wet (flow) and dry (osmotic structures)

Continued Table 16.4: Design characteristics of lifelike systems

Principle	Observation	Design Decision	Example
Sustenance	Vibrant matter seeks food and energy sources to resist entropic imperatives and delay reaching equilibrium	Lifelike systems are 'motivated' by resources and constrained by their context. Their properties change with time and as they mature. Designers may deal with these qualities through infrastructure and procurement system development, which enable vibrant matter to thrive and have the capacity to profoundly shape its interactions	Bütschli droplets vigorously follow chemical attractants or 'food'
Velocity	Chemistry possesses mass and therefore works more slowly than digital computing	The speed of natural computing processes is constrained by its embodiment and also possesses an innate polarity and, therefore, directionality, which needs to be taken into account in the design process	Liesegang ring plates are limited in their ability to produce banding patterns by the rate of diffusion, the speed of the reactants and gravity
Rhythm	Vibrant matter exhibits periodicity in its chemical interactions, which are shaped by its molecular and macroscale interactions	When working with lifelike systems designers may consider spatial programs and design tactics that deal with periodicity	Static Bütschli droplets that have not yet reached quiescence oscillate in their crystalline residues with lengthening time periods between each contraction, until they reach equilibrium
Ecological compatibility	Through its innate propensity to form material assemblages, vibrant matter seeks to spontaneously integrate with other material systems	Designers may need to encourage 'social' interactions between actants to amplify their desired effects, such as using organic solvents to encourage aggregation and assemblage formation in the Bütschli system	Bütschli droplets actively seek each other out and form satellite colonies

Continued Table 16.4: Design characteristics of lifelike systems

Principle	Observation	Design Decision	Example
Completion	Lifelike materials are constantly changing and are never truly 'complete' until they reach entropic equilibrium	Lifelike processes are open systems that are always complete without fully resolving their design programs	Bütschli droplets constantly change and are also transformed in response to changes in their internal composition, context and environment
Infrastructure and procurement	The vigour of lifelike materials depends on their supporting infrastructures and procurement systems	Lifelike systems carry their own infrastructures that enable their survival, such as Bütschli droplet interfaces. These interfaces may be prolonged and shaped by increasing the flow of elemental media through the porous assemblages	Hylozoic Ground chemistries were enabled through the provision of appropriate infrastructure and supply of resources throughout the installation

programs constructed by the interactions of many heterogeneous assemblages within a site. The unique quality of chemistry is that it is embodied, capable of parallel processing and acts simultaneously as a software and hardware. The relationships between matter and information are not only tightly coupled but also sensitive to environmental changes. For example, modified Bütschli droplets could be manipulated within the cybernetic matrix of the Hylozoic Ground installation by periodically activating them by exposing them to environmental changes such as fluctuations in carbon dioxide, as well as physical alternations within the body of the installation: specifically, the production of heat and light. This very specific form of morphological computing appears to be a transferable practice in an architectural design context. Beesley has continued to develop the chemistry of the Hylozoic Ground installation, through a series of works that incorporate vibrant chemistries in a variety of settings. For example, Traube cells have been grown within Bütschli droplet bodies for the 'Epiphyte Grove' installation at the Trondheim Biennale, Norway (Philip Beesley Architect, 2012) (see Fig. 16.16) and at the 'Radiant Soil' installation at Fondation EDF, Paris (Philip Beesley Architect, 2012) (see Figs. 16.17 and 16.18). Additionally, Leduc cells have been used to 'fix' dissolved carbon dioxide from the surroundings for 'Epiphyste Grove' and also in 'Protocell Mesh' at the Building Centre, London (Philip Beesley Architect, 2013b) (see Fig. 16.19). This series of new works not only establishes ways of working with dynamic chemistries using morphological computing principles, but has also founded methods of collaborating through multidisciplinary, international partnerships.

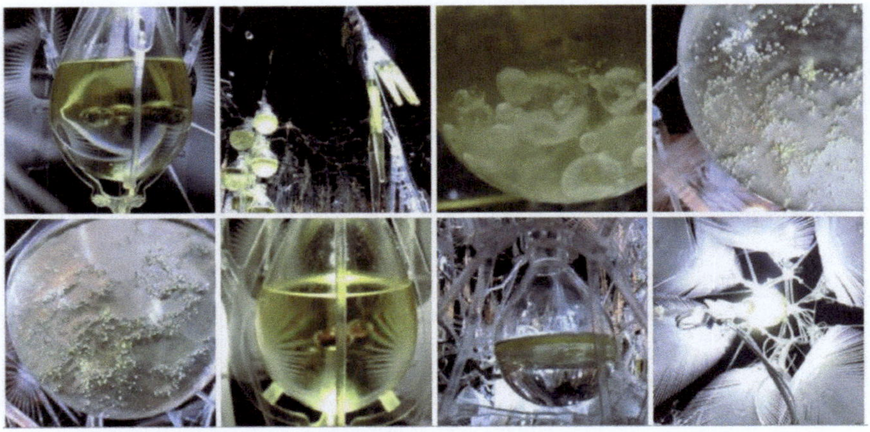

Figure 16.16: The range of dynamic chemistries for the Hylozoic Ground installation is being developed through collaborative work that supports the chemical evolution of the metabolic organs within Beesley's ongoing international exhibitions. Photograph, Rachel Armstrong, September 2010.

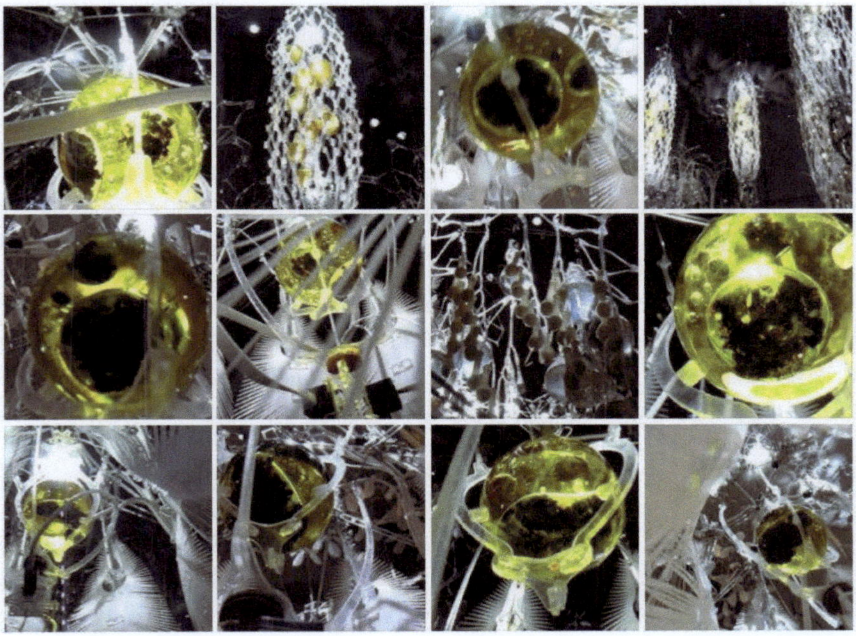

Figure 16.17: Traube cells produce profuse osmotic membranes within modified Bütschli droplets at EDF, Paris. Photographs and collage, Rachel Armstrong, April 2013.

Figure 16.18: Detail of Traube cell incorporated into the cybernetic system at EDF, Paris. Photograph, Rachel Armstrong, April 2013.

Figure 16.19: 'Protocell Mesh' installation at the Building Centre, London. Photographs and collage, Rachel Armstrong, September 2013.

16.11 Procurement and Infrastructure

The liveliness of vibrant matter is dependent on their infrastructures and procurement systems. For example, in the Hylozoic Ground installation, the ingredients for the dynamic chemistries were transported from Marghera on the mainland to the Giardini. Venice's transportation systems are by boat and foot, so the chemistry was sourced from the Explora Biotech laboratory and carried overland to the site. After the show, the chemistries were recycled and reused wherever possible, and where matter was 'spent', then it was emptied into a container and transported back to the mainland for formal chemical disposal. These labour-intense delivery processes mirrored setting up the chemistry flasks and building the dynamic droplet systems by hand. This added a ritualistic feel to the production of the installation, which was reminiscent of the painstaking construction processes employed by Postman Cheval in the construction of the Palais Idéal (Dannies, not dated:a) and the steel Watts towers of Sam Rodia (Watts Towers, 2006–2013) where locally supplied materials, time and human labour are vigorous actants and codesigners of the system.

The Bütschli droplets were modified so they would survive during the three-month installation period and could continue to chemically process and respond to environmental changes, since they were situated in aqueous environments, which were periodically topped up by Explora Biotech. The notion of an architectural-scale liquid infrastructure was further explored in Vibrant Venice, where the city's coastal location provided ubiquitous access to water at the site of action of the 'protopearls' ELT.

Speculatively, chemically programmed droplets would move away from the light when added to the waterways by virtue of a photophobic metabolism to accrete and deposit solid material when they settled under the darkened foundations of the city. Other forms of infrastructural support were also explored in the production of Liesegang ring plates for the Hylozoic Ground installation, using an organic gel matrix. The molecular structure of the gel attenuated chemical interactions, yet also enabled water to move through the system under the influence of gravity. This infrastructural arrangement served to spatialize the dynamic chemistries, inserting time into the system, which is a novelty-producing force (Prigogine, 1997). Further reflection on the role of infrastructure in shaping the effects of vibrant matter in an urban context was considered through the technology of soils. Evolutionarily speaking, soils have provided the fundamental infrastructure that enabled the transition of life from the water to the land.

Accordingly, selecting or designing the right infrastructures when working with vibrant matter may enhance performance in these lively systems so they increase the potency of a site. Further exploration of these ideas could be examined by designing 'artificial soils' to act as complex assemblages of vibrant matter, which do not simply rely on their chemical properties but are also influenced by the spatial and temporal factors that characterize a site. For example, additional layers of complexity could be added to the Liesegang ring plate preparation by creating micro-channels in the system, which is a practice established in microfluidics, and using these channels as an additional transport system within the matrix. This could potentially provide fast (micro-channel) and slow (gel-matrix) transport systems to set up differentials of chemical exchange and interaction, which may give rise to complex, novel events (see Fig. 16.20).

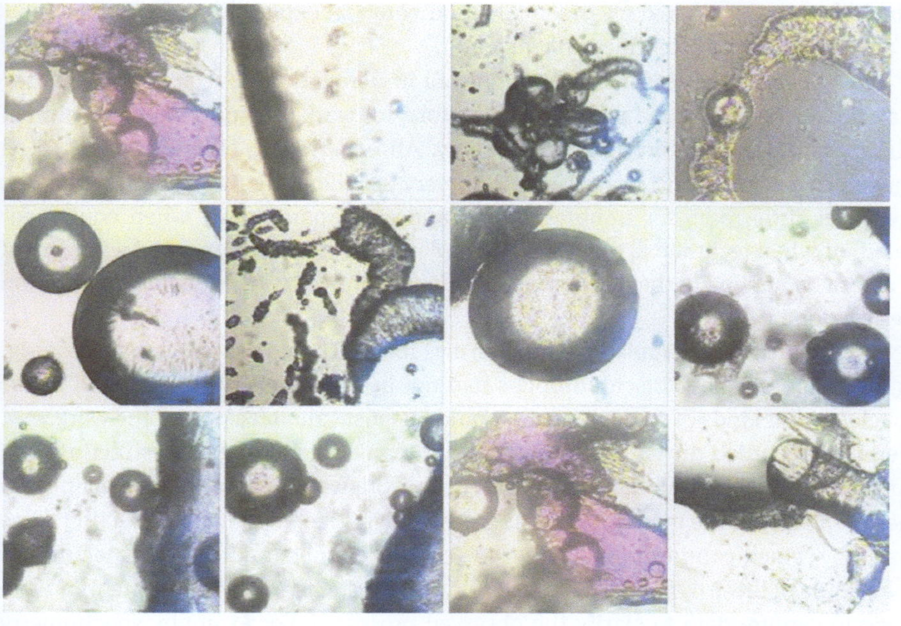

Figure 16.20: Different ELT species potentially offer tools that may provide the materials, infrastructure and technology to enable architects to work with complex chemical fabrics capable of metabolic transformation, such as synthetic soils, which are the building blocks for vibrant architecture. Micrographs and collage, magnification 4×, Rachel Armstrong, August 2012.

16.12 Challenges

My research proposition – that vibrant matter could form the basis for a new production platform with different environmental impacts to that of machines – was a challenging one.

Vibrant matter is a theoretical substrate and has not been applied in a scientific, technological or design context. Yet, the notion of elevating the status of matter so that it could conceptually and physically act as codesigner of systems opened up new spaces for the architectural imagination in the production of spatial programs and design tactics. Potentially, these changes could bring about different ecological impacts to the current damaging impacts that are being wrought by industrial processes.

I worked across the disciplines of architecture and chemistry to raise questions about the spatial and cultural impacts of the proposal, where the two disciplines could be read 'against' the other to map knowledge sets and identify possible knowledge 'gaps'. My multidisciplinary research was supported by working with architects Neil Spiller and Philip Beesley, as well as chemists Martin Hanczyc and Hans Toftlund, who have greatly inspired my practical and theoretical exploration of vibrant matter in architectural and scientific contexts.

My experimental research was conducted in Denmark and Venice, which meant that I did not have full-time access to a laboratory, so I worked in intensely concentrated periods and managed my time very strictly. All experiments, therefore, had to be imagined, designed and resourced within very specific time frames. Consequently, there was little room for troubleshooting and error, particularly when equipment failed, or when techniques did not work to plan. Yet, this also provoked a great deal of creativity and led to the development of, for example, the hygroscopic preparation (see section 8.5.1).

The acquisition of empirical data from my laboratory studies was also challenging, as the capabilities of Bütschli droplets had not been previously studied, and it was not clear what kinds of measurements would be most valuable. Most of the work was, therefore, recorded during microscopy rather than documented by abstracting data from the experiments. Moreover, since my endeavour was to appreciate the richness in the complexity of the material systems I was studying, refining and abstracting these systems seemed counterproductive to the quest to understand the full spectrum of behaviours of these dynamic, dissipative Bütschli systems. Yet, various attempts were made to capture data to clarify the observed processes. For example, spectrometry was attempted, to provide qualitative information about the changes in the chemistry during the dynamic processes, but the spectrum produced was effectively equivalent to water and the analysis did not add much useful information about its chemical transitions. Fluoroscopy proved to be the most powerful tool in observing how the oil/water interface behaved and gave information about spontaneous fusion in the system – yet this could not be counted as 'empirical' research, but descriptive. Much further analysis and development of the Bütschli system is necessary and will be explored in further scientific experiments and design collaborations as the understanding of this system progresses.

Other significant challenges lay in developing a system that had not been previously applied in a technological context. Oil/water droplet systems are not formally recognized as a technology, but have been regarded as models for exploring ideas about the characteristics of 'life'. I addressed this challenge by using action research methods to become immersed in developing the system using laboratory experiments and by making architectural models. Many scientific questions regarding the technological potential of the Bütschli system remain unresolved, and much more in-depth analysis is necessary. This will be addressed through further research and scientific experiments, collaborating with researchers who may bring new ideas and approaches to the formal investigation of this system (Armstrong and Hanczyc, 2013).

A significant technical challenge for working with the Bütschli system was in influencing self-assembly at the microscale in ways that meaningfully address architectural-scale challenges. This issue was addressed by changing the scale of operation and droplet size by slowing down the metabolism of the Bütschli system. Additionally, entangling its chemical interactions with other dynamic systems, such as the Carbon Eaters in the Hylozoic Ground installation (see Fig. 16.21), could also alter the scale at which the effects of dynamic droplets were encountered. The ability of chemical assemblages to horizontally couple their interactions with other (unlike) systems at different scales bestows this emerging platform with a robustness and flexibility that may be transferable to a broader range of different contexts. Working with vibrant matter also required the development of elemental infrastructures to sustain its activity and enable its operations at architectural dimensions. While chemical flasks were sufficient to support the slow interactions of dynamic chemistries in the Hylozoic Ground installation, Venice's lagoon potentially offered a robust, although more inconstant infrastructure for the more vigorous action of protopearls, by being able to deliver a steady stream of nutrients and remove 'waste' metabolites such as surfactants.[83]

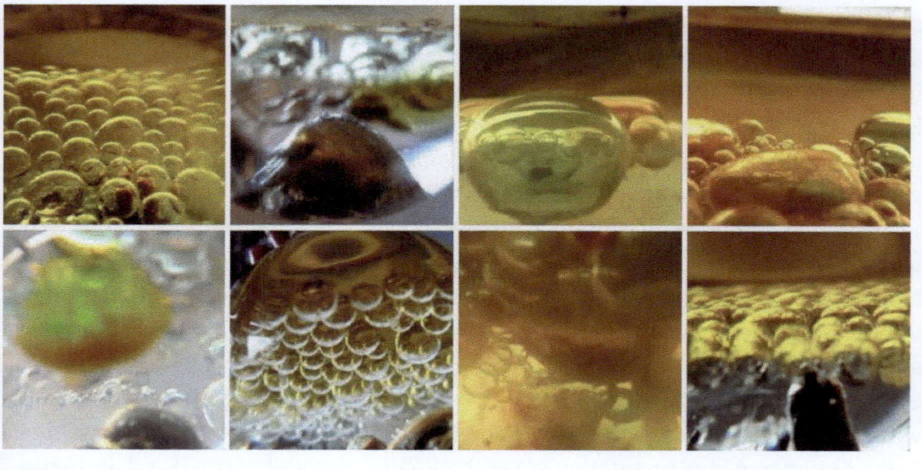

Figure 16.21: Modification of Bütschli droplets enabled the system to be 'technologized' as an assemblage at the human scale. Photographs and collage, Rachel Armstrong, February 2010.

83 Surfactants are soap-like substances, which are produced when oil and alkali are reacted.

Yet, before vibrant matter can be effectively integrated into architectural practice, new kinds of infrastructures, such as synthetic soils, are needed (Hsu, Mou, and Lee, 1994; Györgyi and Field, 1992; Field, Körös and Noyes, 1974). It is also likely that they will be be integrated with digital technologies, where microfluidics sensors could monitor chemical changes to optimize the performance of vibrant matter. The coupling between a lively body and its various infrastructures may be thought of as an architectural physiology that provides an operational substrate for post-natural fabrics.

Another challenge was that vibrant matter could also reach tipping points that changed its behaviour so that dynamic chemistries were at risk of collapsing or becoming inert when conditions changed. The unpredictability and surprise that is inherent in dynamic chemistries needed to be taken into account when working with vibrant matter, by building contingency into these systems or exploiting it, as in the case of Vibrant Venice, where the dynamic droplet system was expected to change its behaviour should the lagoon dry out.

Working with a system that did not have a discrete control system was also challenging. During my research I learned how to use chemistry as a form of combined hardware and software to manipulate outputs of dynamic chemical systems using morphological computing techniques that operate through soft forms of control. For example, the Hylozoic Ground chemistries were slowed down to make them bigger and more visible within a gallery setting. However, this modification also contributed to their extended lifespan and enabled them to persist for the duration of the exhibition (Armstrong and Beesley, 2011).

Vibrant matter eventually reaches entropic equilibrium and exhibits inorganic sympathy (Bennett, 2011), which has implications for the design of lifelike systems and raises questions about whether they should be maintained, or that their inevitable decay in the system should be regarded as a form of planned senescence. However, this characteristic is not one to be 'solved' by an overriding set of rules but left for individual architects to deal with according to their design programs, aesthetic preferences and philosophical principles (Woods, 2012b).

The lifelike nature of vibrant matter also raises many ethical questions about using living systems to perform work in a technological context. It also provokes moral issues related to how the technology may precipitate new lifelike events as unnatural fertile acts. With the prospect of vibrant matter exhibiting autopoiesis, agency or autonomy, their political status may need consideration as a member of an extended, diverse 'ecological' community (Bennett, 2010, p.viii). Indeed, designing with vibrant matter raises bigger questions about the role of the architect as midwife in the design and implementation of synthetic ecologies and post-natural ecologies. I used speculative fiction as a way of reflecting on possible impacts of ethical and moral issues raised throughout my research. For example, 'The greatest Alluvian poet that ever lived' reflects on the impacts of lifelike technologies in establishing an urban post-natural fabric.

The many challenges that I faced in developing the materials, technologies and approaches towards a practice of vibrant architecture are far from resolved and I have not attempted to deal with them exhaustively. However, my research proposes to offer a taste of the kinds of issues and challenges that architects must face, if they decide to work with systems that are lifelike and, in a way, have a life of their own. ELT may just be able to help us tap into the creative potential of the natural world. Yet this platform does not propose to subordinate Nature and consume it as industrial systems do, but to apply the astonishing power of transformation of the material realm at non-equilibrium states, so that we may innovate more organically within architectural design practice (see Fig. 16.22).

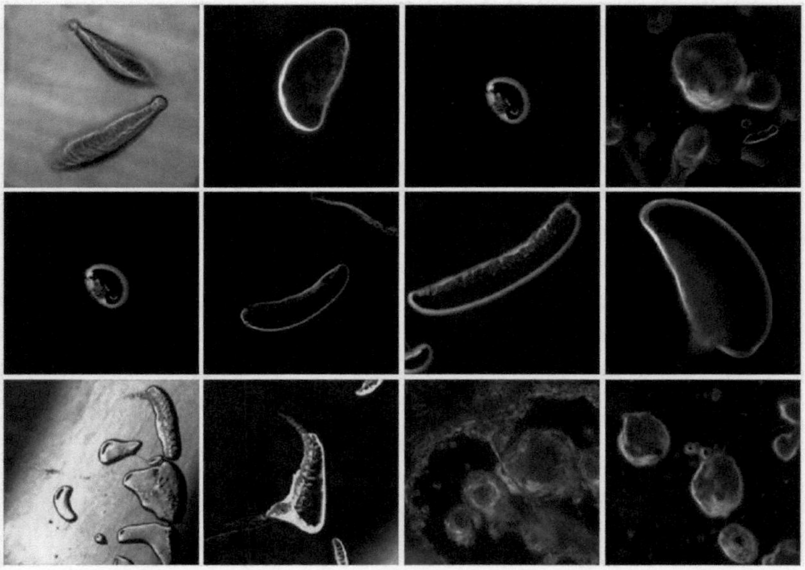

Figure 16.22: Bütschli droplets possess an innate force and creativity that bestow them with the potential to act as codesigners of our living spaces. Micrographs and collage, magnification 4×, Rachel Armstrong, February 2009.

16.13 What Next?

Vibrant matter provides a means of developing new tactics for the construction of dynamic spatial programs that may shape our living spaces. It creates a material context, technical platform and cultural imperative for the possibility of vibrant architecture, which may be realized through further research in the following fields of inquiry, namely:

- Theory (shape of ideas)
- Experiment (developing the context, infrastructures and tools for a novel architectural design practice)
- Practice (cultural adoption and technical implementation)

16.13.1 Theory

Further theoretical engagement with agentized materials that addresses the seemingly irreconcilable split between Nature and machines is needed, which in an architectural context has become politicized around the issues of humanism and environmentalism. My research resists using language that frames my work within either of these polarized positions by avoiding words associated with mechanical systems, such as 'efficiency', or evocations that natural systems are intrinsically 'good' (Armstrong, 2013f). Yet, seeking equity between humans and non-humans is problematic, especially since anthropocentrism is impossible to avoid completely (Bennett, 2010, p.102) and the humanist/environmentalist dichotomy is framed around a set of already established agendas and ongoing debates (Henderson, 2013). Although vibrant matter and morphological computing propose a new technological platform that is compatible with natural systems and, therefore, speak to a neo-environmentalist agenda,[84] they do not seek to preserve established power systems. Rather, they propose alternative value systems to those that currently exist (Sadler and Smart, 2012). While the dilemma between human and ecological interests is not resolved, the principles of vibrant architecture and morphological computing may be regarded as a step towards establishing new, positive relationships between technology and Nature. These values do not take the form of neo-environmentalism slotted into an old industrial system, but generates different ideals to represent a much broader shared interest in the future of the planet, where the fate of humans and ecosystems are intertwined (see Fig. 16.23).

From a theoretical perspective, perhaps the immediate way forwards lies in developing new value systems that may enable the interests of people and Nature to be equally met. With the advent of a coherent value system, it may be easier to develop appropriate ethical and moral systems to navigate the complex decisions that will shape communities in a post-natural world. New kinds of communities may arise concurrently with a practice of vibrant architecture – from how they imagine and

84 Neo-environmentalists such as Stewart Brand view Nature as a resilient, consumable resource that can withstand a range of assaults and therefore may be harnessed to meet human needs using advanced technologies such as genetically modified crops or nuclear power (Kingsnorth, 2013).

Figure 16.23: Morphological computing aims to develop spatial and temporal tactics that influence the behaviour of materials that are far from equilibrium, so that the body of these systems may help solve computational challenges. This is already intrinsic to the performance of biological systems such as jellyfish, with compliant bodies that help them to move through water without the need for a central nervous system to instruct them. Photographs and collage, Rachel Armstrong, August 2012.

apply the material fabric of our dwellings to the ways in which they may midwife new kinds of Nature into their living spaces. Vibrant architecture represents the start of a very long conversation and series of practical experiments that are essential in changing old paradigms for new – and there is still a long way to go.

16.13.2 Experiment

My research establishes the theory and principles of practice for more ecologically compatible forms of architecture, which directly tap into the non-equilibrium potential of natural forces. A range of ELT was identified that served as a production platform that coupled information, process and matter through spatial programs, which worked through many collective acts of persuasion, cooperation and symbiosis. Yet, there were many limitations in this approach that invoked questions, such as how far ELT could scale, the reproducibility of results and what kind of 'control' systems were possible.

Although my research did not propose to resolve all these questions, or to formally develop the systems as a mature technology, I was able to develop a model system that allowed me to work effectively with vibrant matter in a design context. For example, it was possible to make the Bütschli system produce larger droplets by slowing down their 'metabolism' using an inhibitor (which was a product of the reaction); or additionally, by changing the internal and external conditions of the droplets by adding various mineral solutions, or altering the infrastructures by opening the system to the environment. While the modifications and basic infrastructures explored in my research did not formally constitute the production of a new ready-for-market technology, my experiments enabled me to identify the framework in which these kinds of processes and practices could reasonably exist within an architectural design setting. Further design projects are also necessary to develop a practical understanding of the infrastructural, technological, poetic and aesthetic qualities of vibrant matter. Additionally, further development of the principles of practice for ELT may be established by building a technological portfolio that includes other vibrant matter species such as Traube cells (Traube, 1867) and Leduc cells (Leduc, 1911, pp.124–130). I am collaboratively exploring a range of possibilities by applying these systems in new installation work by Philip Beesley (Philip Beesley Architect, 2012; Philip Beesley Architect, 2013a; Philip Beesley Architect, 2013b).

16.13.3 Practice

My research practice created a context for regarding matter as an active, rather than passive, agent in the design process. Through a series of design-led experiments, a set of repeatable observations were produced that could be applied to a range of contexts. From these findings, it was possible to propose ideas that work towards a theory and practice of vibrant architecture, which extends beyond the existence of visibly lifelike technologies, but also applies to other kinds of active materials and surfaces such as titanium oxide (Woody, 2009). Although the model systems that I developed during my research, such as the Bütschli droplets, are not ready-for-market technologies, they persuasively demonstrate that different kinds of approaches may be possible when thinking about the impact of architecture on the environment. Being aware that all materials possess different degrees of agency that under certain conditions may also be applied technologically extends the possibilities of design practice beyond the current conventions of thinking, such as minimizing energy, reducing resource consumption, or mimicking biological systems. Indeed, the practice of vibrant architecture opens up new exploratory spaces and methods for imagining the relationship between architectural design, materials and the environment.

As such, my research is perhaps most potent when considered as a possible manifesto for change, since it serves as a platform from which to further develop the conceptual and practical approaches for vibrant architecture. While there is a risk that vibrant architecture may be swallowed up into industrial frameworks as soft machines (Spiller, 2007, pp.202–224; Burroughs, 1961), many new developments are beginning to emerge that are facilitating a new way of working, which creates new contexts that may resist this imperative – from Skylar Tibbit's 4D printing (TED.com,

2013b) to Gabriel Villar's printing with abiotic vesicles (Villar, Graham and Bayley, 2013).[85]

Yet, vibrant matter does more than promise new techniques – it provokes and also embodies them. Raising the 'status' of the material world is an essential step in developing more ecologically compatible communities and alternative technological platforms that may underpin human development. This notion of ecology is not merely implied by observing relationships between urban actants differently to predict the movement of resources and people, but is tangible and manifest through spatial, temporal and material relationships. Architectural programs, therefore, form the basis of a synthetic ecology, which is not an inert artefact, but exists as a living network of organizing hubs within the metropolitan system. These material attractors and transformers of matter may take the form of pollution-transforming paints (*Raconteur*, 2012; Fraunhofer, 2012), composting systems (Robbins, 2012) or genetically engineered glow-in-the-dark lighting (Cha, 2011; *Raconteur*, 2012) and change the potency of spaces by precipitating events as orchestrated expressions of architectural design that are subject to soft control systems, like gardening.

During my research, I designed practical tests to explore and examine the validity and limits of some of these ideas using a series of dynamic chemistries. Broadly speaking, these may be regarded as model building blocks of vibrant matter that embody some of the qualities of vibrant matter (slowness, porosity, inorganic sympathy) (Bennett, 2011) and exhibit the strange, unruly, restless materiality that Morton proposes is concealed by contemporary aestheticisms that tame the image of Nature (Morton, 2007). Indeed, matter that is sufficiently lively to exert significant force with human-scale consequences raises architectural design questions – particularly with respect to the relationships between matter, form, program, environment and Nature. Lively materials may respond to many different kinds of architectural design programs that may use overlapping and contradictory cues, or decentralized and deanthropocentrized approaches (Hundertwasser, not dated), where architects are not the sole designers in the system but collaborators that work with the radical creativity of vibrant matter and its collaborating non-human communities.

Yet, if vibrant matter is to be a useful technology, as opposed to a material curiosity, it implies that elemental infrastructures must be more widely distributed within our building fabrics and living spaces. The condensation of elemental flows around these sites will encourage metabolic activity within architectural spandrels, or even to be more publicly celebrated as central features within social spaces, as in 'Hylozoic Veil' in the Leonardo Building, Salt Lake City (Philip Beesley Architect, 2011). Indeed, the development of macrofluidics systems and water computers within

85 Also, my work with the Cronin group at the WETFab workshop in January 2011, combining 3D printing technologies with self-organizing chemistries (WETFab, 2011; Armstrong, 2012g; Adams, 2012).

our buildings and (vertical) gardens, which may be similar to Julius Popp's 'Bit Flow'[86] installation (Popp, 2011), may become vital technologies that help us meet the resource constraints of 21st century urban lifestyles (Pruned Blogspot, 2012) since they can deal with more than one computation simultaneously, such as recycling water stores, processing data and providing contexts in which vibrant matter may reach transformational tipping points (see Fig. 16.24).

Figure 16.24: The Hylozoic Ground chemistries are transferable technologies that can integrate with the unique chemistry of geographical locations. Photograph, Rachel Armstrong, September 2012.

86 Bit.flow is an anamorphic, liquid computer that consists of coloured oil droplets within a continuous flow of water in sealed, clear tubes. A mechanical pump powers the liquid information and its outputs are letters of the alphabet that can be read by audiences at certain spatial points in the system. Popp drew inspiration from Michel Foucault's provocation on the consequences of Ariadne's thread being torn apart and describes the system as a complex interplay between order in chaos - and chaos in order (Popp, 2011).

16.4 From Here to There: Towards a Practice of Vibrant Architecture

Although we may continue to deny that the material character of the world, or Nature herself, has evolved into a stranger, even more awe-inspiring configuration that is native to this millennium, we have been forced to deal with a new baseline of existence from which we are reimagining the future of our cities. While smart and sustainable cities aspire to produce 'better' forms of industrialization (Armstrong, 2012b), they do not offer Millennial Nature a brand-new relationship – but more of the same kind of approach – even if more considerately applied (Armstrong, 2013e).

Vibrant matter offers the context in which architectural design may develop a new set of approaches, materials and technical systems that reconstruct our relationships with the natural world. Vibrant architecture, therefore, strives towards the production of synthetic ecosystems that blur the historical boundaries between building and landscape, and may even transform our living spaces into rich synthetic ecologies and post-natural landscapes. Indeed, they may establish fertile soil-producing systems, which enrich the environment, generate wealth and even change our value systems so that we are no longer consumers of our environments – but producers of them. Yet, since vibrant architecture can exert forces that can act independently from human agendas, it may offer more than functional value and invokes subjectivity. Vibrant architecture also resists the unimpassioned greyness of the inert industrial landscape that has tamed our notions of ecology (Whitehead, 1970, p.54). For example, vibrant architecture may provide sustenance, income and entertainment for humans, while magnificently co-evolving alongside us. Indeed, vibrant architecture may respond vigorously and effusively to chemical changes in our living spaces (hormonal, mineral, toxins) by reconfiguring them through spectacular acts of continual digestion and secretion that may be as compelling as watching flames leap from a fireplace (Moran, 2008). These creative events may, in turn, be shaped by – but not dependent on – the way we respond to and inhabit them (see Fig. 16.25).

Through our (re)identification with the material world, we may view these post-natural fabrics as being entangled with our own physiology. Such intimate encounters with vibrant architectures may alter our relationship to the environment, where our homes may be regarded as a living membrane that dissolves hostilities between humans and the natural world, so they nourish each other. Yet, like our own bodies, these architectural-scale weavings of natural and synthetic materials are not a panacea, but also create their own paradoxes, difficulties and contradictions, which may raise ethical, cultural and moral challenges. Indeed, this renewed camaraderie with matter will by no means protect us from cataclysmic cosmic events and extinction tipping points, but offer new conduits through which we may negotiate the turbulent material flows of our planet differently. Indeed, the vast substance of our urban landscapes may be regarded as a technological platform and host for a chemical dialogue with the material world. This may be enriched and made diverse through multitudinous acts of architectural design – some of which are initiated by

Figure 16.25: Transformation in materials at far from equilibrium states is enabled through the intersections between elemental fields and flows. Photographs and collage, Rachel Armstrong, August 2012.

humans. Such a detailed and varied fabric may alter the severity of, or even postpone, the damaging consequences of the environmental impacts that we have accelerated over the course of the 20th century. Yet, vibrant architecture does not propose to resolve the probabilistic nature of reality, or collapse it into systems that we can easily command. Instead, it seeks to work with the creativity and strangeness of the natural world, using a new palette of design possibilities that are not to be mimicked, but directly engaged with. By drawing on the eccentric properties of the quantum laws that underpin the fabric of Nature and weaving them coherently into our living spaces, vibrant architecture establishes new ambitions and expectations for 21st century architectural design (see Fig. 16.26).

Vibrant architecture not only produces buildings that enhance biotic environments, but also acts as a point of counter-resistance that provides access to emerging Nature-based technologies, or ELT. To access these possibilities, architects may need to act against the entrenched mores of their profession (Stamp, 2004)[87] before they can reconstruct its philosophy, materiality and environmental impacts.

---

87 Bernard Tschumi's 1978 postcard reads: 'To really appreciate architecture, you may even need to commit a murder. Architecture is defined by the actions it witnesses as much as by the enclosure of its walls. Murder in the Street differs from Murder in the Cathedral in the same way that love in the street differs from the Street of Love. Radically.'

Figure 16.26: Heterogeneous agents produce post-natural fabrics around the Venetian bioregion. Photograph, Rachel Armstrong, August 2012.

Currently, these points of resistance take the form of an emerging set of lifelike materials and tools, which are already beginning to be incorporated into movements such as Bio Design.

Yet, vibrant architecture has not yet been carved out as a formal practice with distinct aesthetic, programmatic and cultural concerns. It remains open and ready for incorporation within existing systems and ultimately seeks to subvert established power relationships, formal categories of production and the way that architecture is inhabited by inviting non-human codesigners to collaborate in its substance. This may take place through innumerable acts of architectural design, at many scales, whose outcomes are always works in progress (see Fig. 16.27).

Vibrant architecture works to unleash the potency of matter by reaching into its innate, quantum strangeness, and offers access to the unique poetics and sensual palette of the material realm.

It may be expressed through spatial programs and design tactics that give rise to rich and varied architectural experiences. Ultimately, vibrant architecture takes the form of post-natural fabrics that offer a fertile field of new possibilities in vibrant cities where human and non-human communities collaborate and codesign our living spaces and evolve alongside us as expressions of Millennial Nature – to augment the liveliness of our planet rather than diminish it (see Fig. 16.28).

Figure 16.27: The city of Venice may be regarded as a unique site for vibrant architecture, being supported by elemental infrastructures that are capable of transforming its boundaries. Photographs and collage, Rachel Armstrong, August 2012.

Figure 16.28: Juxtaposition of artificial and natural fabrics at Cathédrale Notre-Dame de Paris offers an historical precedent for post-natural, vibrant architecture. Photographs and collage, Rachel Armstrong, August 2012.

References

@Google. (5 November 2010). 'Digital art@google: Rachel Armstrong & Mitchell Joachim.' [online] Available at: https://www.youtube.com/watch?v=DHd6dzMTFgI. [Accessed 2 June 2014].

ABC Science. (31 January 2012). 'Our ancestors recycled their urine: why shouldn't we?' [online] Available at: http://www.abc.net.au/science/articles/2012/01/31/3415550.htm. [Accessed 19 April 2014].

Adamatzky, A., Armstrong, R., Jones, J. & Gunji, Y.K. (2013). 'On creativity of slime mould.' *International Journal of General Systems, 42*(5), pp.441–457.

Adamatzky, A., Bull, L., De Lacy Costello, B., Stepney, S. & Teuscher, C. (2007). *Unconventional computing*. Beckington, UK: Luniver Press.

Adamatzky, A. & De Lacy Costello, B. (2003). 'Reaction-diffusion path planning in a hybrid chemical and cellular-automaton processors.' *Chaos, Solitons and Fractals, 16*, pp.727–736.

Adamatzky, A., De Lacy Costello B. & Ratcliffe, N.M. (2002). 'Experimental reaction-diffusion pre-processor for shape recognition.' *Physics Letters A, 297*, pp.344–352.

Adams, D. (1995). *A Hitchhiker's Guide to the Galaxy: A trilogy in five parts*. London: William Heinemann.

Adams, T. (21 July 2012). 'The 'chemputer' that could print out any drug.' *The Observer*. [online] Available at: http://www.guardian.co.uk/science/2012/jul/21/chemputer-that-prints-out-drugs. [Accessed 19 April 2014].

Agar, W. E. (1936). 'Whitehead's philosophy of organism: an Introduction for biologists.' *The Quarterly Review of Biology, 11*(1), pp.16–34.

Ambasz, E. (2006). *The Universitas Project: Solutions for a post-technological society*. New York: Museum of Modern Art.

AMigO. (Not dated). 'Carnitine O-acetyltransferase activity, AMigO: Term details for GO:0004092. The gene ontology.' [online] Available at: http://amigo.geneontology.org/cgi-bin/amigo/term_details?term=GO:0004092. [Accessed 28 May 2014].

Andrewartha, H.G. & Birch, L.C. (1984). *The Ecological Web*. Chicago: University of Chicago Press.

Ansted, D.T. (1863). *The Great Stone Book of Nature*. London: Macmillan and Company.

Anthoni, J.F. (2006). 'Mineral composition of sea water.' [online] Available at: http://www.seafriends.org.nz/oceano/seawater.htm#composition. [Accessed 18 April 2014].

Armstrong, R. (2014) 'Morphological Computing and Architecture.' In: R. Fuschlin, H. Hauser & R. Pfeifer (Eds.), *Morphological Computing*. Zurich: University of Zurich [e-book].

Armstrong, R. & Hanczyc, M.M. (2013). 'Bütschli dynamic droplet system.' *Artificial Life Journal, 19*(3-4), pp.331–346.

Armstrong, R. (26 March 2013a). 'How soil can help save the earth.' Future Cities Supplement, *The Times*, p.9.

Armstrong, R. (30 January 2013b). 'The hitchhikers' guide to complexity.' *Transhumanity*. [online] Available at: http://transhumanity.net/articles/entry/the-hitchhikers-guide-to-complexity. [Accessed 17 April 2014].

Armstrong, R. (2013c). 'Coming to terms with synthetic biology.' *Volume, 35*, pp.110–117.

Armstrong, R. (2013d). 'Alternative biologies.' In: M. More & N.V. More, (Eds.), *The Transhumanist Reader: Classical and contemporary essays on the science, technology, philosophy of the human future* (pp.100-109). Chichester: John Wiley & Sons.

Armstrong, R. (29 May 2013e). 'Disruptive technology: is business investing enough in the future?' *The Guardian*. [online] Available at: http://www.guardian.co.uk/sustainable-business/disruptive-technology-business-investing-future. [Accessed 2 June 2014].

Armstrong, R. (11 July 2013f). 'Rachel Armstrong on biomimicry as parametric snake oil.' *The Architectural Review*. [online] Available at: http://www.architectural-review.com/reviews/biomimicry-parametric-snake-oil/8650000.article [Accessed 18 July 2014].

Armstrong, R. (11 July 2012a). 'Synthetic biology: Designing a 'metaphysical' chicken.' Institute for Emerging Ethics and Technology. [online] Available at: http://ieet.org/index.php/IEET/more/6073. [Accessed 17 April 2014].

Armstrong, R. (5 December 2012b). 'Lawless sustainability.' [online] Available at: http://www.architecturenorway.no/questions/cities-sustainability/armstrong/. [Accessed 16 April 2014].

Armstrong, R. (2012c). 'A matter of feeling.' In: E. Gangvik (Ed.), *Meta.morf* (pp.168–171). Trondheim: TEKS Publishing.

Armstrong, R. (2012d). *Living architecture: How synthetic biology can remake our cities and reshape our lives.* Kindle Single, TED Books. [online]. Available at: http://www.amazon.co.uk/Living-Architecture-Synthetic-Biology-ebook/dp/B0076QQJMY [Accessed 16 April 2014].

Armstrong, R. (2012e). 'A short story of a short life.' *Organs Everywhere, 4*, pp. 24–29.

Armstrong, R. (2012f). 'Biolime: Mock rock.' *Organs Everywhere, 4*, pp. 54–58.

Armstrong, R. (2012g). 'Print to protocell.' In: B. Sheil (Ed.), *Manufacturing the bespoke, making and prototyping architecture AD reader* (pp.238–247). London: John Wiley & Sons.

Armstrong, R. (2011a). 'Architectural synthetic ecologies.' In: C. Daly, S. Hicks, A. Keene & R.R. Ricardo (Eds.), *Paradigms of nature: Post natural futures. Kerb 19, Journal of Landscape Architecture* (pp.92–98). Melbourne: RMIT Landscape Architecture.

Armstrong, R. (2011b). 'Unconventional computing in the built environment.' *International Journal of Nanotechnology and Molecular Computation, 3*(1), pp.1-12.

Armstrong, R. (2011c). 'Is there something beyond 'outside of the box?' *Architectural Design, 81*(6), pp.130–133.

Armstrong, R. (2011d). 'How protocells can make stuff much more interesting,' *Architectural Design, 81*(2), pp.68–77.

Armstrong, R. & Beesley, P. (2011). 'Soil and protoplasm: The Hylozoic Ground project.' *Architectural Design, 81*(2), pp.78–89.

Armstrong, R. & Spiller, N. (2011). 'Synthetic biology: Living quarters.' *Nature, 467*, pp.916–918.

Armstrong, R. (2010a). 'The Hylozoic Soil chemistries and Hylozoic Ground as a model for unanswered questions in the origins of life.' In: P. Beesley, P. Ohrstedt & H. Isaacs (Eds.), *Hylozoic Ground: Liminal Responsive Architecture* (pp.124-141).Toronto: Riverside Architectural Press.

Armstrong, R. (2010b). 'Systems architecture: A new model for sustainability and the built environment using nanotechnology, biotechnology, information technology, and cognitive science with living technology.' *Artificial Life, 16*, pp.73–87.

Armstrong, R. (2010c). 'Living technology.' In: M. Bedau & P.G. Hansen (Eds.), *Living Technology: Five Questions* (pp.9–20). Copenhagen: Automatic Press/VIP.

Armstrong, R. (2010d). 'Living architecture.' *Forward 110: Architecture and the body.* American Institute of Architects, *10*(1), pp.77–82.

Armstrong, R. (2009a). 'Systems architecture: A new model for sustainability and the built environment using nanotechnology, biotechnology, information technology, and cognitive science with living technology.' *Artificial Life, 16*, pp.1–15.

Armstrong, R. (19 November 2009b). 'Systems evolution and bio-feminism: Move over Darwin.' *C-Theory*. [online] Available at: http://www.ctheory.net/articles.aspx?id=621. [Accessed 19 April 2014].

Armstrong, R. (2008a). 'Designer materials for architecture.' *Architectural Design, 78*(6), pp.86–89.

Armstrong, R. (2008b). 'Artificial evolution: A hands-off approach of architects.' *Architectural Design, 78*(6), pp.82–85.

Aron, J. (19 April 2011). 'Chemical computers solve shapely puzzles.' *New Scientist*. [online] Available at: http://www.newscientist.com/blogs/onepercent/2011/04/chemical-computers-solve-shape.html. [Accessed 18 April 2014].

Artificial Life XII. (2009). 'Applications for Alife in the built environment: New materials for the bottom up assembly of architecture.' Affiliated and related events. [online] Available at: http://www.alife12.org/events/. [Accessed 19 April 2013].

Artificial Life XI. (2008). [online] Available at: http://alifexi.alife.org. [Accessed 18 April 2014].

Artistic Things. (12 August 2009). 'Simone Decker: Chewing in Venice.' [online] Available at: http://artisticthings.com/simone-decker-chewing-in-venice/. [Accessed 25 May 2014].

Ascott, R. (2005). 'Syncretic reality: art, process and potentiality.' [online] Available at: http://www.drainmag.com/contentNOVEMBER/FEATURE_ESSAY/Syncretic_Reality.htm. [Accessed 19 April 2014].

Ashby, W.R. (1956). *An Introduction to Cybernetics*. London: Chapman and Hall.

Ashby, W. R. (1947). 'Principles of the self-organizing dynamic system.' *Journal of General Psychology, 37*, pp.125–128.

Asheim, G.B., Buchholz, W. & Tungodden, B. (2001). 'Justifying sustainability.' *Journal of Environmental Economics and Management, 41*, pp.252–268.

Baird, G. (2003). *The Space of Appearance*. Cambridge (MA): The MIT Press.

Ball, P. (2009a). *Branches. Nature's Patterns: A tapestry in three parts*. Oxford: Oxford University Press.

Ball, P. (2009b). *Flow. Nature's Patterns: A tapestry in three parts*. Oxford: Oxford University Press.

Ball, P. (2009c). *Shapes. Nature's Patterns: A tapestry in three parts*. Oxford: Oxford University Press.

Bassett, C., Steinmueller, E. & Voss, G. 2013. Better made up: the mutual relationship between science, science fiction and innovation. NESTA working paper. [online] Available at: http://www.nesta.org.uk/library/documents/bettermadeup.pdf. [Accessed 19 April 2014].

Bateson, G. (1979). *Mind and Nature: A Necessary Unity (Advances in Systems Theory, Complexity, and the Human Sciences)*. New York: Hampton Press.

Bateson, G. (1972). *Steps to an Ecology of Mind: Collected essays in anthropology. psychiatry, evolution and epistemology*. Chicago: University of Chicago Press.

BBC.com (2013). 'The ethics of using peat. Gardening guide.' [online] Available at: http://www.bbc.co.uk/gardening/today_in_your_garden/ethical_peat.shtml. [Accessed 19 April 2014].

Bedau, M.A., McCaskill, J.S., Packard, N.H. & Rasmussen, S. (2010). 'Living technology: Exploiting life's principles in technology.' *Artificial Life, 16*(1), pp.89–97.

Bedau, M. (2009). 'Living technology today and tomorrow.' *Technoetic Arts: A Journal of Speculative Research, 7*(2), pp.199–206.

Bedau, M.A. & Parke, E.C. (2009). *The Ethics of Protocells: Moral and social implications of creating life in the laboratory*. Cambridge, MA: The MIT Press.

Beer, S. (1994). A progress note on research into a cybernetic analogue of fabric. In: R. Harnden & A. Leonard (Eds.), *How many grapes went into the wine: Stafford Beer on the art and science of holistic management* (p.29), New York: Wiley.

Beer, S. (1992). World in torment: A time whose idea must come. [online] Available at: http://www.williambowles.info/sa/World%20in%20Torment.pdf. [Accessed 25 April 2014].

Beesley, P. & Bonnemaison, S. (2008). 'Why Revisit D'Arcy Wentworth Thompson's On Growth and Form?' In: P. Beesley & S. Bonnemaison (Eds.), *On Growth and Form: Organic Architecture and Beyond* (pp.7–15), Cambridge, ON: Tuns Press and Riverside Architectural Press.

Belousov, B.P. (1959). 'A periodic reaction and its mechanism.' *Compilation of Abstracts on Radiation Medicine, 147*, p.145.

Bennett, J. (2011). 'Powers of the hoard: Artistry and agency in a world of vibrant matter,' Vimeo. [video]. Available from: http://vimeo.com/29535247. [Accessed 17 April 2014].

Bennett, J. (2010). *Vibrant Matter: A political ecology of things*. Durham, NC: Duke University Press.
Bennett, J. (2005). 'The agency of assemblages and the North American blackout.' *Public Culture, 17*(3), pp.445–465.
Benson, J.F. & Roe, M.H. 2000. *Landscape and Sustainability*. Abingdon: Taylor and Francis.
Berger, P., Adelman, N.B., Beckman, K.J., Campbell, D.J. Ellis, A.B. & Lisensky, G.C. (1999). 'Preparation and properties of an aqueous ferrofluid.' *Journal of Chemical Education, 76*, pp.943–948.
Bergson, H. (1922). *Creative Evolution*. London: Macmillan.
Bernhard, J.M., Edgcomb, V.P., Visscher, P.T., McIntyre-Wressnig, A., Summons, R.E., Bouxsein, M.L., Louis, L. & Jeglinski. M. (2013). Insights into foraminiferal influences on microfabrics of microbialites at Highborne Cay, Bahamas. *Proceedings of the National Academy of Sciences*, doi:10.1073/pnas.1221721110. [online] Available at: http://www.pnas.org/content/early/2013/05/22/1221721110.abstract. [Accessed 30 May 2014].
Beynus, J. (1997). *Biomimicry: Innovation inspired by Nature*. New York: William Morrow.
Bird, J. & Di Paolo, E. (2008). *Gordon Pask and His Maverick Machines*. Cambridge, MA: The MIT Press.
Bitbol, M. (1996). *Schrödinger's Philosophy of Quantum Mechanics*. Dordrecht: Kluwer.
Bland, E. (25 August 2009). 'Bacteria desalinate water, generate power.' *Discovery News*. [online] Available at: http://dsc.discovery.com/news/2009/08/25/bacteria-water.html.] [Accessed 18 April 2014].
Bloomberg. (Not dated). 'The 'Frankenstorm' hits. Hurricane Sandy.' [online] Available at: http://topics.bloomberg.com/hurricane-sandy/. [Accessed 19 April 2014].
Bohm, D.J. (1980). *Wholeness and the ImplicateOrder*. London: Routledge.
Bohm, D.J. (1973). 'Quantum theory as an indication of a new order in physics. Implicate and explicate order in physical law.' *Foundations of Physics, 3*, pp.139–168.
Bok, C. (2001). *Pataphysics: The poetics of an imaginary science* (Avant-garde and Modernism Studies). Evanston: Northwestern University Press.
Bosch, T. (2 March 2012). 'Sci-fi writer Bruce Sterling explains the intriguing new concept of design fiction.' *Slate*. [online] Available at: http://www.slate.com/blogs/future_tense/2012/03/02/bruce_sterling_on_design_fictions_.html. [Accessed 19 April 2014].
Botazzi, R. (28 May 2012). 'Green strategies: Ecological design research and computation.' *Architectural Review*. [online] Available at: http://m.architectural-review.com/8630590.article. [Accessed 17 April 2014].
Bowie, A. (1995). 'Romanticism and technology.' *Radical Philosophy, 72*, pp. 5–16.
Bowie, D. (Not dated). 'Glass spider lyrics. Metro lyrics.' [online] Available at: http://www.metrolyrics.com/glass-spider-lyrics-david-bowie.html. [Accessed 17 May 2014].
Brand, S. (2009). *Whole Earth Discipline*. New York: Viking Penguin.
Braun, D. (28 April 2010). 'Promise to reduce biodiversity loss not kept.' *National Geographic News Watch*. [online] Available at: http://newswatch.nationalgeographic.com/2010/04/28/promise_to_reduce_biodiversity_broken/. [Accessed 18 April 2014].
BREEAM. (2010–2013). [online] Available at: http://www.breeam.org/index.jsp. [Accessed 25 April 2014].
Buckminster Fuller Institute. (Not dated). 'The Buckminster Fuller Challenge.' [online] Available at: http://challenge.bfi.org/movie. [Accessed 25 April 2014].
Building Centre. (26 February 2010). 'Unconventional computing & crchitecture, Diary event.' [online] Available at: http://www.buildingcentre.co.uk/events/event_diary_details.asp?id=510. [Accessed 17 April 2014].
Bürkle GmbH. (2011). 'Viscosity of liquids.' [online] Available at: http://www.buerkle.de/media/files/Downloads/Viscosity_EN.pdf. [Accessed 19 April 2014].

Burroughs, W.S. (1961). *The Soft Machine*. New York: Grove Press.

Burry, M. (1993). *Expiatory Church of the Sagrada Família: Antoni Gaudí. Architecture in Detail.* London: Phaidon Press.

Bütschli, O. (1892). *Untersuchungen ueber microscopische Schaume und das Protoplasma*, Leipzig.

Cairns-Smith, A.G. (1990). *Seven Clues to the Origin of Life*. Cambridge: Cambridge University Press.

Cairns-Smith, A.G. (1987). *Genetic Takeover and the Mineral Origins of Life*. Cambridge: Cambridge University Press.

Cairns-Smith, A.G. (1971). *The Life Puzzle*. Edinburgh: Oliver and Boyd.

Canada.com. (10 February 2007). 'Top 10 most common environmental toxins.' [online] Available at: http://www.canada.com/vancouversun/news/story.html?id=57586947-9466-4fcf-bb1b-2c464dd19e5c. [Accessed 10 May 2014].

Capra, F. (1996). *The Web of Life: A new scientific understanding of living systems*. New York: Anchor Books.

Carnerup, A.M., Christy, A.G., Garcia-Ruiz, J-M., Hyde, S.T. & Larsson, A-K. (2006). 'The record of early life: In search of biosignatures.' In: J. Seckbach (Ed.), *Life as we know it. Cellular origin, life in extreme habitats and astrobiology* (p.248). Dordrecht: Springer.

Caschera, F., Rasmussen, S. & Hanczyc, M.M. (2013). 'An oil droplet division–fusion cycle,' *ChemPlusChem, 78*(1), pp.52-54.

Castelletto, N., Ferronato, M., Gambolati, G., Putti, M. & Teatini, P. (2008). 'Can Venice be raised by pumping water underground? A pilot project to help decide.' *Water Resources Research, 44*, W01408, doi:10.1029/2007WR006177.

Calvino, I. (1997). *Invisible Cities*. Berkshire: Random House.

Calvino, I. (2009). *The Complete Cosmicomics* (Penguin translated texts). London: Modern Penguin Classics.

Carson, R. (1962). *Silent Spring*. Boston: Houghton Mifflin.

CBEC & APEGEC. (2009). 'Budget guidelines for consulting engineering services.' The Consulting Engineers of British Columbia and The Association of Professional Engineers and Geoscientists of British Columbia. [online] Available at: http://www.apeg.bc.ca/ppractice/documents/ppguidelines/BudgetGuidelines.pdf. [Accessed 25 May 2014].

Cha, B. (28 November 2011). 'Philips Bio-light creates mood lighting with bacteria,' *C/Net*. [online] Available at: http://news.cnet.com/8301-17938_105-57332431-1/philips-bio-light-creates-mood-lighting-with-bacteria/. [Accessed 18 April 2014].

Chalfie, M., Tu, Y., Euskircher, G., Ward, W.W. & Prasher, D.C. (1994). 'Green fluorescent protein as a marker for gene expression.' *Science, 263*, pp.802–805.

Charney, N. (3 January 2012). 'Inside the shroud of Turin, the secret history of art.' *Blouin Artinfo*. [online] Available at: http://blogs.artinfo.com/secrethistoryofart/2012/01/03/inside-the-shroud-of-turin/. [Accessed 1 June 2014].

Chu, J. (9 August 2012). 'Soft autonomous robot inches along like an earthworm.' *MIT News Office*. [online] Available at: http://web.mit.edu/newsoffice/2012/autonomous-earthworm-robot-0810.html. [Accessed 17 April 2014].

Cisco. (Not dated). 'Cities of the Future: Songdo, South Korea.' *The network: Cisco's technology news site*. [online] Available at: http://newsroom.cisco.com/songdo. [Accessed 18 May 2014].

Clarke, A.C. (1973). *Profiles of the Future*. New York: Harper and Row.

Condea. (2000). 'Dr. Z presents all about fatty alcohols.' [online] Available at: http://www.zenitech.com/documents/new%20pdfs/articles/All%20about%20fatty%20alcohols%20Condea.pdf. [Accessed 19 April 2014].

Connor, S. (17 April 2013). 'The 'living fossil' coelacanth fish left behind by evolution.' *The Independent*. [online] Available at: http://www.independent.co.uk/news/science/the-living-fossil-coelacanth-fish-left-behind-by-evolution-8577129.html. [Accessed 25 April 2014].

Cooper, G.J., Kitson, P.J., Winter, R., Zagnoni, M., Long, D.L. & Cronin L. (2011). 'Modular redox-active inorganic chemical cells: iCHELLs.' *Angewandte Chemie International Edition in English, 50*(44), pp.10373–10376.

Copeland, J.B. (8 January 1997). 'The Church-Turing thesis.' *Stanford Encyclopedia of Philosophy*. [online] Available at: http://www.science.uva.nl/~seop/entries/church-turing/#Bloopers. [Accessed 25 May 2014].

Cotterby, J. & Dadeby, A. (2012). *The Passivhaus Handbook: A practical guide to constructing and retrofitting buildings for ultra-low energy performance*. Devon: Green Books.

Creswell, J.W. (2009). *Research Design: Qualitative,quantitative, and mixed methods approaches*. Thousand Oaks: Sage.

Cronin, L., Krasnogor, N., Davis, B.G., Alexander, C., Robertson, N., Steinke, J.H.G., Schroeder, S.L.M., Khlobystov, A.N., Cooper, G., Gardner, P.M., Siepmann, P., Whitaker, B.J. & Marsh, D. (2006). 'The imitation game—a computational chemical approach to recognizing life.' *Nature Biotechnology, 24*, pp.1203–1206.

Crutzen, P. J. & Stoermer, E.F. (2000). 'The 'Anthropocene'.' *Global Change Newsletter, 41*, pp.17–18.

Cyberplasm Team. (2010). 'Cyberplasm: a micro-scale biohybrid robot developed using principles of synthetic biology.' [online] Available at: http://cyberplasm.net/. [Accessed 18 April 2014].

Dahl, T. (2009). *Climate and Architecture*. London: Routledge.

Dannies, K. (Not dated:a). 'Le palais idéal de Ferdinand Cheval.' *Flair Magazine*. [online] Available at: http://www.flair-magazine.com/casa-deluxe/architecture/37-postman-s-palace. [Accessed 31 May 2014].

Dannies, K. (Not dated:b). 'Off kilter Waldspirale.' *Flair Magazine*. [online] Available at: http://www.flair-magazine.com/casa-deluxe/architecture/36-off-kilter-waldspirale. [Accessed 1 June 2014].

Darwin, C. (2007). *The Formation of Vegetable Mould Through the Action of Worms*. Fairford: The Echo Library.

Darwin, C. (1999). *The Origin of Species* (Wordsworth Classics of World Literature). Ware: Wordsworth Editions.

Darwin, C. (1859). *On the Origin of Species by Means of Natural Selection, or the Preservation of Favoured Races in the Struggle for Life*, Chapter 4, Diagram of divergence of taxa. 1st edition. London: John Murray.

Davies, K. & Vercruysse, E. (2012). 'Incisions in the haze.' In: B. Sheil (Ed.), *Manufacturing the Bespoke: Making and prototyping architecture AD reader* (pp.162–174). London: John Wiley and Sons.

Dawkins, R. (1976). *The Selfish Gene*. Oxford: Oxford University Press.

DEAF. (2006-2013). 'Sandbox. Driessens and verstappen'. [online] Available at: http://deaf.nl/works/sandbox. [Accessed 9 August 2014].

De Chardin, P.T. & Huxley, J. (2008). *The Phenomenon of Man*. New York: Harper Perennial.

Dean, P.J. (2011). *Delivery without Discipline: Architecture in the age of design*. Ann Arbor: Proquest, Umi Dissertation Publishing.

DeLanda, M. (2002). *Intensive Science and Virtual Philosophy*. London: Continuum.

DeLanda, M. (2000). *A Thousand Years of Nonlinear History*. New York: Zone Books.

Deleuze, G. & Guattari, F. (1986). *Kafka: Toward a minor literature*. Minneapolis: University of Minnesota Press.

Deleuze, G. & Guattari, F. (1983). *Anti-Oedipus: Capitalism and schizophrenia*. R. Hurley, M. Seem, and H. Lane, translation. Minneapolis: University of Minnesota Press.

Deleuze, G. & Guattari, F. (1979). *Thousand Plateaus: Capitalism and schizophrenia* (Athlone Contemporary European Thinkers). London: The Athlone Press.

Denning P.J. (2007). 'Computing is a natural science.' *Communications of the ACM, 50*(7), pp.13–18.

Descartes, R. (1983). 'Meditations V.' In: C. Adam & P. Tannery, (Eds.), *Oeuvres De Descartes (1639)*. Paris: Vrin.

Design Boom. (1 March 2012). 'Floating cloud installation by Berndnaut Smilde.' [online] Available at: http://www.designboom.com/weblog/cat/10/view/19555/floating-cloud-installation-by-berndnaut-smilde.html#.UM2eOdm9rPE.twitter. [Accessed 19 April 2014].

Deutsch, D. (1997). *The Fabric of Reality: The science of parallel universes – and its implications.* London: Penguin Books.

Deutsch, D. (1985). 'Quantum theory, the Church-Turing principle and the universal quantum computer.' *Proceedings of the Royal Society, Series A, 400*, pp.97–117.

Di Martino, E. (2007). *The History of the Venice Biennale*. Venice: Papiro Arte.

Diamond, J. (2012). *The World Until Yesterday: What can we learn from traditional societies?* London: Allen Lane.

Dias, J. (15 December 2012). 'Did you know that these fields of ice flowers exist at sea?' *Gizmodo*. [online] http://gizmodo.com/5968768/did-you-know-that-these-fields-of-ice-flowers-exist-at-sea/gallery/. [Accessed 19 April 2014].

Doctorow, C. (11 October 2012). 'Game of Life with floating point operations: beautiful SmoothLife.' *Boing Boing*. [online] Available at: http://boingboing.net/2012/10/11/game-of-life-with-floating-poi.html [Accessed 20 April 2014].

Dolnick, S. (28 August 2011). 'Recovery is slower in New York suburbs.' *The New York Times*. [online] Available at: http://www.nytimes.com/2011/08/29/nyregion/wind-and-rain-from-hurricane-irene-lash-new-york.html?pagewanted=all&_r=0. [Accessed 19 April 2014].

Doolittle. R.F. (1994). 'Convergent evolution: the need to be explicit.' *Trends in Biochemical Sciences, 19*, pp.15–18.

Dpr Barcelona. (16 December 2011). Metabolic Venice: Algae and protocells anticipating evolved landscape. [online] Available at: http://dprbcn.wordpress.com/2011/12/16/metabolic-venice/. [Accessed 18 April 2014].

Driesch, H. (1929). *The Science and Philosophy of the Organism*. Gifford Lectures, University of Aberdeen, 1907–1908. 2. London: A. & C. Black.

Driesch, H. (1914). *The Problem of Individuality*. London: Macmillan and Company.

Driver, A. Peralta, C. and Moultrie, J. (2011). 'Exploring how industrial designers can contribute to scientific research.' *International 'Journal of Design, 5*(1), pp.17–28.

Easterling, K. (2012). *The Action is the Form. Victor Hugo's TED Talk*. Kindle e-book, Strelka Press. [online] Available at: http://www.amazon.co.uk/action-form-Victors-Hugos-ebook/dp/B00BXM2Z0M/ref=sr_1_sc_1?s=digital-text&ie=UTF8&qid=1368802599&sr=1-1-spell&keywords=Thea+action+is+the+form+Victor+Hugo. [Accessed 16 May 2014].

Edge. (18 August 2009). 'We are as gods and have to get good at it.' [online] Available at: http://edge.org/conversation/we-are-as-gods-and-have-to-get-good-at-it. [Accessed 18 April 2014].

Edge. (2008). 'Breaking the Galilean spell. Third Culture.' [online] Available at: http://www.edge.org/3rd_culture/kauffman08/kauffman08_index.html. [Accessed 19 April 2014].

Edwards, B. (2010). *Rough guide to Sustainability*. London: RIBA Publications.

Else, L. (8 August 2012). 'From STEM to STEAM.' Culture Lab, *New Scientist*. [online] Available at: http://www.newscientist.com/blogs/culturelab/2012/08/john-maeda-steam.html. [Accessed 19 April 2014].

Epstein, L.S. (1980). 'Decision-making and the temporal resolution of uncertainty.' *International Economic Review, 21*(2), pp.269–283.

Facca, C., Pellegrino, N., Ceoldo, S., Tibaldo, M. & Sfriso, A. (2011). 'Trophic conditions in the waters of the Venice lagoon (Northern Adriatic Sea, Italy).' *The Open Oceanography Journal, 5*, pp.1-13.

Field, T., Brodribb, T., Iglesias, A., Chatelet, D., Baresch, A., Upchruch, G.R., Gomez, B., Mohr, B.A.R., Coiffard, C., Kvacek, J. & Jaramillo, C. (2011). 'Fossil evidence for cretaceous escalation in angiosperm leaf vein evolution,' *Proceedings of the National Academy of Sciences, 108*(20), pp.8363-8366.

Field, R.J., Körös, E. & Noyes, R. (1974). 'Oscillations in chemical systems, IV, Limit cycle behaviour in a model of a real chemical reaction.' *Journal of Chemical Physics, 60*, pp.1877–1884.

Fletcher, C., da Mosto, J. & Spencer, T. (2005). *Introduction: Ecological processes and environmental quality*. Cambridge: Cambridge University Press.

Fraunhofer. (20 December 2012). 'Clean air: New paints break down nitrogen oxides, Press Release.' [online] Available at: http://www.fraunhofer.de/en/press/research-news/2012/december/clean-air-new-paints-break-down-nitrogen-oxides.html. [Accessed 17 April 2014].

Frazer, J. (1995). *An Evolutionary Architecture*. London: Architectural Association.

Freud, S. (2003). *The Uncanny*. London: Penguin Classics.

Frichot, H. (2011). 'When things get wild: An introduction of the work of R&Sie(n).' In: C. Daly, S. Hicks, A. Keene & R.R. Ricardo (Eds.), *Paradigms of Nature: Post natural futures. Kerb 19. Journal of Landscape Architecture* (pp.56–59). Melbourne: Melbourne Books.

Friedlander, D. (20 June 2009). 'No impact man documentary film is low on carbon, high on awareness.' *Treehugger*. [online] Available at: http://www.treehugger.com/culture/no-impact-man-documentary-film-is-low-on-carbon-high-on-awareness.html. [Accessed 25 May 2014].

Frosch, R.A. & Gallopoulos, N.E. (1989). 'Strategies for manufacturing.' *Scientific American, 189*(3), p.152.

Fuller, R.B. (1969). *Operating Manual for Spaceship Earth*. Carbondale: Southern Illinois University Press/Simon & Schuster.

Fuller, S. (2011). *Humanity 2.0*. Basingstoke: Palgrave Macmillan.

Funtowicz, S.O. & Ravetz, J.R. (1992). 'Three types of risk assessment and the emergence of post-normal science.' In: S. Krimsky & D. Golding (Eds.), *Social Theories of Risk* (pp.251–273). Westport: Praeger.

Future Laboratory. (27 October 2010). 'A life distilled.' [online] Available at: http://www.thefuturelaboratory.com/blog/2011/10. [Accessed April 2014].

Gage, S. (2012). 'The bespoke is a way of working, not a style.' In: B. Sheil (Ed.), *Manufacturing the Bespoke: Making and prototyping architecture AD reader* (pp.29–41). London: John Wiley & Sons.

Gallagher, D. (26 November 2001). 'Le Corbusier.' Open Learn. The Open University. [online] Available at: http://www.open.edu/openlearn/history-the-arts/history/heritage/le-corbusier. [Accessed 27 April 2014].

Gambino, M. (21 February 2013). 'With biodesign, life is not only the subject but the medium too.' Collage of arts and sciences. Smithsonian,com. [online] Available at: http://blogs.smithsonianmag.com/artscience/2013/02/with-biodesign-life-is-not-only-the-subject-of-art-but-the-medium-too/. [Accessed 18 April 2014].

Gánti, T. (2003). *The Principles of Life*. New York: Oxford University Press.

García-Ruiz, J.M., Melero-García, E. & Hyde, S.T. (2009). 'Morphogenesis of self-assembled nanocrystalline materials of barium carbonate and silica. '*Science, 323*(5912), pp.362–365.

Gardner, M. (1970). 'The fantastic combinations of John Conway's new solitaire game 'life'.' *Scientific American, 223*, pp.120–123.

Gibson, D.G., Glass, J.I., Lartigue, C., Noskov, V.N., Chuang, R.Y., Algire, M.A., Benders, G.A., Montague, M.G., Ma, L., Moodie, M.M., Merryman, C., Vashee, S., Krishnakumar, R., Assad-Garcia, N., Andrews-Pfannkoch, C., Denisova, E.A., Young, L., Qi, Z.Q., Segall-Shapiro, T.H., Calvey, C.H., Parmar, P.P., Hutchison, C.A.3rd., Smith, H.O. & Venter, J.C. (2010). 'Creation of a bacterial cell controlled by a chemically synthesized genome.' *Science, 329*(5987), pp.52–56.

Gilster, P. (3 January 2014). 'Living at a time of post-natural ecologies.' *Centauri Dreams*. [online] Available at: http://www.centauri-dreams.org/?p=29757. [Accessed 2 November 2014].

Gissen, D. (2011). 'Thoughts on a heap of rubble.' In: C. Daly, S. Hicks, A. Keene & R.R. Ricardo (Eds.), *Paradigms of nature: Post natural futures. Kerb 19, Journal of Landscape Architecture* (pp.50–52). Melbourne: RMIT University.

Giulietti, M., Seckler, M.M., Derenzo, S. & Valarelli, J.V. (1996). 'Changes in copper sulfate crystal habit during cooling crystallization.' *Journal of Crystal Growth, 166*(1–4), pp.1089–1093.

Glansdorff, P. & Prigogine, I. (1971). *TheTheory of Structure, Stability and Fluctuations*. New York: Wiley.

Glauber, J.R. (1651). *Furni Novi Philosophici: Sive, Descriptio artis destillatoriae novae, nec non spirituum, oleorum, florum, aliorumque medicamentorum illius beneficio, facilimâ quâdam [et] peculiari viâ è vegetabilibus, animalibus [et] mineralibus conficiendorum [et] quidem magno cum lucro; agens quoque de illorum usu, tàm chymico, quàm medico, edita [et] publicata in gratiam veritatis studiosorum*. Amsterdam: Prostant apud J. Janssonium.

Gleick, J. (2008). *Chaos: Making a new science*. Revised edition. New York: Penguin.

Godfrey-Cook, M. (2011). *The Zero Carbon House*. Wilshire: The Crowood Press.

Goodall, J. (1969). *My Friends the Wild Chimpanzees*. Washington: National Geographic Society.

Gould, S.J. (1997). 'Kropotkin was no crackpot.' *Natural History, 106*, pp.12–21.

Gould, S.J. (1994). 'The evolution of life on the Earth.' *Scientific American, 271*(4), p.84.

Gould, S.J. & Lewontin, R.C. (1979). 'The spandrels of San Marco and the Panglossian paradigm: A critique of the adaptionist programme.' *Proceedings of the Royal Society of London, Series B, 205*(1161), pp.581–598.

Guardian. (2013). 'Sustainable living hub, Sustainable business.' [online] Available at: http://www.guardian.co.uk/sustainable-business/sustainable-living. [Accessed 25 April 2014].

Guardian. (14 May 2012). 'Prof Alice Roberts: 'Public engagement should be part of academic life'.' [video] Available at: http://www.guardian.co.uk/commentisfree/video/2012/may/14/alice-roberts-public-engagement-academic-video. [Accessed 18 April 2014].

Gudrais, E. (March – April 2004). 'The way of trout: Up a choppy river.' *Harvard Magazine*. [online] Available at: http://harvardmagazine.com/2004/03/the-way-of-trout.html. [Accessed 19 July 2014].

Györgyi, L. & Field, R. (1992). 'A three-variable model of deterministic chaos in the Belousov–Zhabotinsky reaction.' *Nature, 355*, pp.808–810.

Gyr, C. (29 April 2009). 'Chemical inchworms, the latest in robotic technology?' *Popsci*. [online] Available at: http://www.popsci.com/scitech/article/2009-04/chemical-inchworms-latest-robotic-technology. [Accessed 19 April 2014].

Haldane, J.B.S. (1926). 'On being the right size.' [online] Available at: http://www.phys.ufl.edu/courses/phy3221/spring10/HaldaneRightSize.pdf. [Accessed 24 May 2014].

Hamilton, C. (26 May 2013a). 'Geoengineering: Our last hope, or a false promise?' *The New York Times*. [online] Available at: http://www.nytimes.com/2013/05/27/opinion/geoengineering-our-last-hope-or-a-false-promise.html?pagewanted=all&_r=0 [Accessed 7 June 2014].

Hamilton, C. (22 March 2013b). 'Why geoengineering has immediate appeal to China.' Environment, *Guardian*. [online] Available at: http://www.guardian.co.uk/environment/2013/mar/22/geoengineering-china-climate-change. [Accessed 16 May 2014].

Hanczyc, M.M. (2011a). 'Structure and the synthesis of life.' *Architectural Design, 81*(2), pp.26–33.

Hanczyc, M.M. (2011b). 'Metabolism and motility in prebiotic structures.' *Philosophical Transactions of the Royal Society B, 366*, pp.2885–2893.

Hanczyc, M.M. & Ikegami, T. (2009). 'Protocells as smart agents for architectural design.' *Technoetic Arts, 7*(2), pp.117–120.

Hanczyc M.M. (2008). 'The Early History of Protocells: The search for the recipe of life. 'In: S. Rasmussen, M.A. Bedau, L. Chen, D. Deamer, D. C. Krakauer, N.H. Packard & P.F. Stadler, (Eds.), *Protocells: Bridging non-living and living matter* (pp.3–18). Cambridge: MIT Press.

Hanczyc, M.M., Toyota, T., Ikegami, T., Packard, N.H. & Sugawara, T. (2007). 'Fatty acid chemistry at the oil-water interface: Self-propelled oil droplets.' *Journal of the American Chemical Society, 129*(30), pp.9386–9391.

Hanczyc, M.M., Fujikawa, S.M. & Szostak, J.W. (2003). 'Experimental models of primitive cellular compartments: encapsulation, growth and division.' *Science, 302*, pp.618–622.

Harvard School of Engineering and Applied Sciences. (16 May 2013). 'Beautiful 'flowers' self-assemble in a beaker.' [online] Available at: http://www.seas.harvard.edu/news-events/press-releases/beautiful-flowers-self-assemble-in-a-beaker. [Accessed 16 May 2014].

Haeckel, E. (1866). *Generelle Morphologie der Organismen. Allgemeine Grundzüge der organischen Formen-Wissenschaft, mechanisch begründet durch die von Charles Darwin reformirte Descendenztheorie*. Berlin: George Reimer. Available at: http://archive.org/details/generellemorphol01haec. [Accessed 18 April 2014].

Harvey, D. (1996). *Justice, Nature and the Geography of Difference*. Cambridge: Blackwell.

Hauser, H. (2013). 'Morphological computation and soft robotics, a perfect match?' *Soft Robotics and Morphological Computing*, 2013 International Workshop, Ascona, Switzerland 14–19 July, 2013. Workshop Digest. p.K-12.

Hays, K.M. (2000). *Architecture Theory Since 1968* (Columbia Books of Architecture). New edition. Cambridge (MA): The MIT Press.

Heaney, P.J. & Davis, A.M. 1995. 'Observation and origin of self-organised textures in agates.' *Science, 269*, pp.1562–1565.

Heidegger, M. (1993). 'The question concerning technology.' In: William Lovitt & David Farrell Krell (Eds.), *Martin Heidegger: Basic writings*. Revised and expanded edition (pp.287–311). New York: HarperCollins.

Henderson, H. (29 June 2012). 'Transitioning to the Solar Age,' Ethical Markets. [online] Available at: http://www.ethicalmarkets.com/2012/06/29/transitioning-to-the-solar-age/. [Accessed 17 April 2014].

Hewitt, J. (2013). 'Myelin exploits phase transitions to drive its assembly.' Phys.Org. [online] Available at: http://medicalxpress.com/news/2013-07-myelin-exploits-phase-transitions.html [Accessed 19 July 2014].

Hickman, L. (2 November 2011). 'Morocco to host first solar farm in €400bn renewables network.' *The Guardian*. [online] Available at: http://www.guardian.co.uk/environment/2011/nov/02/morocco-solar-farm-renewables. [Accessed 17 May 2014].

Hillier, B. & Hanson, J. (1984). *The Social Logic of Space*. Cambridge: Cambridge University Press.

Hird, M. (2009). *The Origins of Sociable Life: Evolution after science studies*. London: Palgrave Macmillan.

Hodgkin, J., Horvitz, H.R., Jasny, B.R. & Kimble, J. (2011). 'C. elegans: Sequence to biology.' *Science, 282*(5396), p.2011.

Horibe, N., Hanczyc, M.M. & Ikegami, T. (2011). Mode switching and collective behaviour in chemical oil droplets. *Entropy, 13*, pp.709–719.

Howard, C. (January/February 2011). "Mother trees' use fungal networks to feed the forest.' Ecology, *Canadian Geographic*. [online] Available at: https://www.canadiangeographic.ca/magazine/jf11/fungal_systems.asp. [Accessed 19 April 2014].

HGNC. (Not dated). HUGO Gene Nomenclature Committee Home Page. [online] Available at: http://www.genenames.org. [Accessed 28 May 2014].

Hulme, M. (14 March 2007). 'The appliance of science.' Society, *The Guardian*. [online] Available at: http://www.guardian.co.uk/society/2007/mar/14/scienceofclimatechange.climatechange. [Accessed 19 April 2014].

Hundertwasser, F. (Not dated). 'Mould manifesto against rationalism in architecture.' *Hundertwasser Manifestos and Texts*. [online] Available at: http://www.hundertwasser.com/english/texts/philo_verschimmelungsmanifest.php. [Accessed 24 May 2014].

Hundertwasser, F. (1976). 'Hundertwasser: Mould manifesto against rationalist architecture. 'In: J. Conrads (Ed.), *Programs and manifestos on 20th century architecture* (pp.157-161). Cambridge: MIT Press.

Huxley. J. 1957. *New Bottles for New Wine*. London: Chatto and Windus.

Hsu, T.J., Mou, C.Y. & Lee, D.J. 1994. 'Effects of macromixing on the Oregonator model of the Belousov–Zhabotinsky reaction in a stirred reactor.' *Chemical Engineering Science, 49*, pp.5291-5305.

Ignite. (29 January 2013). 'ArtScience Prize and more...' [online] Available at: http://www.ignitefutures.org.uk/ignite-blog/artscience-prize-and-more/. [Accessed, 19 April 2014].

Ikegami, T. (5 August 2008). 'Alife XI Keynote Lecture Takashi Ikegami.' ECS-TV Archive. [video] Available at: http://archive.zepler.tv/194/. [Accessed 19 April 2014].

Imperial College London. (2013). 'Sustainable society network +, Digital economy lab.' [online] Available at: http://www3.imperial.ac.uk/digital-economy-lab/partnernetworks/sustainablesocietynetwork. [Accessed 25 April 2014].

Jablonka, E. & Lamb, M. (2006). *Evolution in Four Dimensions*. Cambridge (MA): The MIT Press.

Jacobs, F. (3 October 2011). '534 – The eggs of Price: An ovo-urban analogy.' *Big Think*. [online] Available at: http://bigthink.com/strange-maps/534-the-eggs-of-price-an-ovo-urban-analogy. [Accessed 22 May 2014].

Jha, A. (20 November 2011). 'Is the end of the world really nigh?' Science, *The Observer*. [online] Available at: http://www.guardian.co.uk/science/2011/nov/20/doomsday-alok-jha-science-extinct. [Accessed 25 April 2014].

Jonkers H.M. (2007). 'Self healing concrete: A biological approach.' In: S. van der Zwaag (Ed.), *Self Healing Materials. An alternative approach to 20 centuries of materials science* (pp.195-204). The Netherlands: Springer.

Joy, B. 8 April (2000). 'Why the future doesn't need us.' *Wired*. [online] Available at: http:// www.wired.com/wired/archive/8.04/joy.html. [Accessed 19 April 2014].

Kalderon, M.E. (2005). *Fictionalism in Metaphysics*. Broadbridge: Clarendon Press.

Kanellos, M. (7 April 2009). 'Bacteria that eat plastic to make plastic to go commercial.' *Cleantech Investing*. [online] Available at: http://www.greentechmedia.com/green-light/post/bacteria-that-eat-plastic-to-make-plastic-going-commercial-1352. [Accessed 27 April 2014].

Kauffman, S.A. (11 May 2012). 'Evolving by magic. Sustainable to evolvable.' *Arup Thoughts*. [online] Available at: http://thoughts.arup.com/post/details/200/evolving-by-magic. [Accessed 26 April 2014].

Kauffman, S.A. (2011). 'Approaches to the origin of life on earth.' *Life, 1*(1), pp.34–48.

Kauffman, S.A. (2008). *Reinventing the Sacred: A new view of science, reason, and religion*. First trade paper edition. New York: Basic Books.

Kazan, C. (11 June 2008). 'Is the city an 'organism' Operating beyond the bounds of biology?' [online] Available at: http://www.dailygalaxy.com/my_weblog/2008/06/is-the-human-ci.html. [Accessed 19 April 2014].

Kellermeier, M., Melero-García, E., Glaab, F., Eiblmeier, J., Kienle, L., Rachel, R., Kunz, W. & García-Ruiz J.M. (2012). 'Growth behaviour and kinetics of self-assembled silica-carbonate biomorphs.' *Chemistry, 18*(8), pp.2272–2282.

Kelly, K. (2010). *What Technology Wants*. New York: Viking.

Khan, O. (2011). An architectural chemistry. *Architectural Design, 81*(2), pp.50–59.

Kimmelman, M. (21 February 1992). 'Just what is a drawing? Definitions, definitions.' *The New York Times*. [online] Available at: http://www.nytimes.com/1992/02/21/arts/review-art-just-what-is-a-drawing-definitions-definitions.html. [Accessed 1 June 2014].

Kingsnorth, P. (January/February 2013). 'Dark ecology.' *Orion Magazine*. [online] Available at: http://www.orionmagazine.org/index.php/articles/article/7277. [Accessed 17 April 2014].

Koch, S. (9 February 2013). 'When objects predict the future.' *Geneva Tribune*. [online] Available at: http://www.tdg.ch/high-tech/web/objets-mettent-predire-avenir/story/21837561. [Accessed 19 April 2014].

Koolhaas, R. (1995). 'Singapore Songlines: Portrait of a Potemkin metropolis or Thirty years of tabula rasa.' In: R. Koolhaas, B. Mau, H. Werlemann & J. Sigler (Eds.), Small, Medium, Large, Extra-Large (p.1011). New York: Monacelli Press.

Koolhaas, R. (Not dated). Sustainability: advancement vs. apocalypse. *OMA*. [online] Available at: http://oma.eu/lectures/sustainability-advancement-vs-apocalypse. [Accessed 28 May 2014].

Kroessler, J., Richard, F. & Finkelpearl, T. (2005). *Odd lots: Revisiting Gordon Matta-Clark's fake estates*. New York: Cabinet Books.

Kipnis, J. & Leeser, T. (1997). *Chora L Works: Jacques Derrida and Peter Eisenman*. New York: The Monacelli Press.

Lamarck, J.B. (September 1999). *Zoological Philosophy (1809)*. [online] Available at: http://records.viu.ca/~johnstoi/lamarck/tofc.htm. [Accessed 19 April 2014].

Langton, C. (1980). 'Computation at the edge of chaos: Phase Transitions and emergent computation.' *Physica D, 4*, pp.12–37.

Latour, B. (2013). *Once Out of Nature: Natural religion as a pleonasm*. Gifford Lecture, University of Edinburgh. [online] Available at: http://www.youtube.com/watch?v=MC3E6vdQEzk. [Accessed 7 July 2014].

Latour, B. (2005). *Reassembling the Social: An introduction to actor-network-theory*. Oxford: Oxford University Press.

Latour, B. (1996). 'On actor network theory: A few clarifications.' *Soziale Welt, 47*(4), pp.369–381.

Latour, B. (1993). *We Have Never Been Modern*. Cambridge: Harvard University Press.

Latour, B. & Woolgar, S. (1979). *Laboratory Life: The social construction of scientific facts*. Beverly Hills: Sage Publications.

Lauder, G.V., Lim, J., Shelton, R., Witt, C., Anderson, E. & Tangorra, J.T. (2011). 'Robotic models for studying undulatory locomotion in fish.' *Marine Technology Society Journal, 45*(4), pp.41–55.

Laylin, T. (25 February 2012). 'Trash to treasure. Six awesome buildings made of recycled materials.' *Inhabitat*. [online] Available at: http://inhabitat.com/trash-to-treasure-6-awesome-buildings-made-of-recycled-materials/. [Accessed 18 April 2014].

Le Corbusier (C.E. Jeanneret). (1923/reprint 2007). *Toward an Architecture*. Los Angeles: Getty Research Institute.

Lederman, L.M. & Teresi, D. (1993). *The God Particle: If the universe is the answer, what is the question?* New York: Bantam Press.

Leduc, S. (1911). *The Mechanism of Life*. Butcher, W.D. translation. London: William Heinemann.

Lee, J.J. (15 February 2013). 'Severe weather more likely thanks to climate change.' *National Geographic News*. [online] Available at: http://news.nationalgeographic.co.in/news/2013/13/130215-severe-storm-climate-change-weather-science/. [Accessed 18 April 2014].

Lee, M. (2011). *Oceanic Ontology and Problematic Thought*, NOOK Book/Barnes and Noble. [online] Available at: http://www.barnesandnoble.com/w/oceanic-ontology-and-problematic-thought-matt-lee/1105805765. [Accessed 19 April 2014].

Lee, R.C. (2006). 'Cell injury by electrical forces.' *Annals of the New York Academy of Sciences, 1066*, pp.85–89.

Lefebvre, H. (2004). *Rhythmanalysis: Space, time and everyday life*. New York: Continuum.

Lefebvre, H. (1991). *The Production of Space*. First English edition. Nicholson-Smith, D. translation. Oxford: Blackwell Publishers.

Lewis, C.S. (2001). *The Lion, the Witch and the Wardrobe* (The Chronicles of Narnia, 2). London: HarperCollins.

Liesegang, R.E. (1869). 'Ueber einige Eigenschaften von Gallerten,' *Naturwissenschaftliche Wochenschrift, 11*(30), pp.353–362.

Linnæus, C. (1735). 'Systema naturæ, sive regna tria naturæ systematice proposita per classes, ordines, genera et species.' Lugduni Batavorum. Haak. pp.1–12.

Lionni, L. (1977). *Parallel Botany*. London: Random House.

Liu, Y.Y., Slotine, J.J. & Barabási, A.L. (2011). 'Controllability of complex networks.' *Nature, 473*, pp.167–173.

Living Root Bridges. (5 August 2009). 'Atlas Obscura.' [online] Available at: http://rootbridges.blogspot.co.uk. [Accessed 18 April 2014].

Logan, W.B. (2007). *Dirt: The Ecstatic Skin of the Earth*. New York: W.W. Norton & Company.

Lovelock, J.E. (1979). *Gaia: A new look at life on earth*. Oxford: Oxford University Press.

Luebkeman, C. (19 April 2012). 'Designing the new normal.' Resources, *Arup Thoughts*. [online] Available at: http://thoughts.arup.com/post/details/185/designing-the-new-normal. [Accessed 30 April 2014].

Luisi, P.L. (2010). *The Emergence of Life: From chemical origins to synthetic biology*. Cambridge: Cambridge University Press.

Lynn, G. (2000). 'Greg Lynn: Embryological Houses.' *Architectural Design, 70*(3), pp.26–35.

MacLennan, B. (2011). 'Artificial morphogenesis as an example of embodied computation.' *International Journal of Unconventional Computing, 7*(12), pp.3–23.

MacLulich, D.A. (1937). *Fluctuations in the numbers of the varying hare (Lepus americanus)* (University of Toronto Studies, Biological Series, 43). Toronto: University of Toronto Press.

Malik, Z. (15 March 2010). 'Is Pakistan ready for water war?' *Pakistan Observer*. [online] Available at: http://pakobserver.net/detailnews.asp?id=20374. [Accessed 25 April 2014].

Malthus, T.R. & Gilbert, G. (2008). *An Essay on the Principle of Population* (Oxford World's Classics). New York: Oxford University Press.

Mandel, K. (14 February 2013). ''Bioconcrete' uses bacteria to heal itself.'*Green Futures Magazine*. [online] Available at: http://www.forumforthefuture.org/greenfutures/articles/bioconcrete-uses-bacteria-heal-itself. [Accessed 18 April 2014].

Matsumoto, K. Ueda, T. & Kobatake, Y. (1998). 'Reversal of thermotaxis with oscillatory stimulation in the plasmodium of Physarum polycephalum.' *Journal of Theoretical Biology, 131*, pp.175–182.

Matthäus, B. & Özcan, M.M. (2011). 'Determination of fatty acid, tocopherol, sterol contents and 1,2- and 1,3-diacylglycerols in four different virgin olive oils.' *Journal of Food Process Technology, 2*, p.4.

Maturana, H.R. & Varela, F.J. (1980). *Autopoieses and Cognition: The realization of the living*. Dordrecht: D. Reidel.

Max Planck Institute for Dynamics and Self-Organization. (2003–2013). 'Spatiotemporal chaos and pattern formation.' [online] Available at: http://www.ds.mpg.de/LFPB/chaos. [Accessed 22 April 2014].

McDonough, W. & Braungart, M. (2002). *Cradle to Cradle: Remaking the way we make things*. 1st edition. New York: North Point Press.

McGuirk, J. (21 November 2011). 'Philips microbial home takes kitchen design back to the future.' Art and Design, *The Guardian*. [online] Available at: http://www.guardian.co.uk/artanddesign/2011/nov/21/philips-kitchen-design-microbial-home. [Accessed 19 April 2014].

McLendon, R. (24 February 2010). 'What is the great Pacific Ocean garbage patch?' *Mother Nature Network*. [online] Available at: http://www.mnn.com/earth-matters/translating-uncle-sam/stories/what-is-the-great-pacific-ocean-garbage-patch. [Accessed 18 April 2014].

Meadows, D.H, Meadows, D.L. Randers, J and Behrens III, W.W. (1972). *The Limits to Growth*. New York: Universe Books.

Meier, A. (17 January 2013). 'Emerging: Artist Sterling Wells creates post natural landscapes.' *Inhabitat*. [online] Available at: http://www.artinfo.com/news/story/857655/emerging-miniature-landscapes-are-painting-subjects-and, [Accessed 18 April 2014].

Merkle, R. (1992). 'The risks of nanotechnology.' In: B. Crandall & J. Lewis (Eds.), *Nanotechnology research and perspectives* (pp.287–294). Cambridge: MIT Press.

Meta.morf. (2012a). Xandra van der Eijk. 'A matter of feeling.' [online] Available at: http://metamorf.no/2012/?p=263&lang=en. [Accessed 27 April 2013].

Meta.morf. (2012b). 'Presentation of speakers at *Meta.Morf* 2012.' [online] Available at: http://metamorf.no/2012/?page_id=1960&lang=en. [Accessed 29 April 2014].

Meyer, J. (2000). 'Using qualitative methods in health related action research.' *British Medical Journal, 320*, pp.178–181.

Microbial Home. (19 October 2011). 'Design probes,' Design Portfolio, Philips. [online] Available at: http://www.design.philips.com/about/design/designportfolio/design_futures/microbial_home.page. [Accessed 17 April 2014].

Mitsch, W.J. (1993). 'Ecological engineering.' *Environmental Science and Technology, 27*, pp.438–444.

Miyasaki, D. (2011). 'An ontological Interpretation of Nietzsche's 'will to power'.' Online research in philosophy. [online] Available at: http://philpapers.org/rec/MINNWO. [Accessed 19 April 2014].

Moran, L.A. (3 July 2008). Good Science Writers: Richard Lewontin. *Sandwalk.* [online]. Available at http://sandwalk.blogspot.co.uk/2008/07/good-science-writersrichard-lewontin.html. [Accessed 19 April 2014].

Morin, B. (5 February 2013). 'What is the maker movement and why should you care?' [online] Available at: http://www.huffingtonpost.com/brit-morin/what-is-the-maker-movemen_b_3201977.html. [Accessed 24 May 2014].

Morton, T. (2012). *The Ecological Thought.* Cambridge: Harvard University Press.

Morton, T. (2007). *Ecology without Nature: Rethinking environmental aesthetics.* Cambridge: Harvard University Press.

Moskowitz, C. (5 November 2012). 'Quantum mystery of light revealed by new experiment.' [online] Available at: http://www.livescience.com/24509-light-wave-particle-duality-experiment.html. [Accessed 26 April 2013].

Mother Shipton. (Not dated). [online] Available at: http://www.mothershipton.co.uk. [Accessed 18 April 2014].

Morris, M. (2011). 'Dream a little dream.' *Architectural Design, 81*(2), pp.44–49.

Morris, W. (Not dated). News from nowhere: or an epoch of rest being some chapters from a utopian romance. [online] Available at: http://www.marxists.org/archive/morris/works/1890/nowhere/nowhere.htm. [Accessed 20 April 2014].

Morris, S.C. & Gould, S.J. (1998). 'Showdown on the Burgess shale.' *Natural History, 107*(10), pp.48–55.

Mourmouras, A. (1991) 'Competitive equilibria and sustainable growth in a life-cycle model with natural resources.' *Scandanavian Journal of Economics, 93*(4), pp. 585–591.

Murphy, D. (23 January 2013). 'Prototyping architecture.' Review. *Building Design.*[online] Available at: http://www.bdonline.co.uk/culture/prototyping-architecture/5049098.article. [Accessed 11 August 2014].

Myers, W. & Antonelli, P. (2013). *Bio Design: Nature, Science, Creativity.* London: Thames & Hudson/New York: MOMA.

Næss, A. (1989). *Ecology, Community and Lifestyle: Outline of an ecosophy.* Cambridge: Cambridge University Press.

National Aeronautics and Space Administration. (Not dated). 'Air traffic management system.' *Virtual Skies.* [online] Available at: http://virtualskies.arc.nasa.gov/atm/9.html. [Accessed 19 April 2014].

National Science Foundation. (3 May 2010). 'Got silk?' Special Report, *Science Nation* [online] Available at: http://www.nsf.gov/news/special_reports/science_nation/spidersilk.jsp, [Accessed 18 April 2014].

National Research Council. (2007). *The Limits of Organic Life in Planetary Systems*. Committee on the Origins and Evolution of Life. National Research Council. Washington: The National Academies Press.

NBC News. (2012). 'Then and now: The 2011 Japan tsunami.' World News. [online] Available at: http://www.nbcnews.com/id/46430477/ns/world_news-asia_pacific/t/then-now-japan-tsunami/#.UXMMrKVt1SU. [Accessed 19 April 2014].

Nealson, K.H., Platt, T. & Hastings, J.W. (1970). 'Cellular control of the synthesis and activity of the bacterial luminescence system.' *Journal of Bacteriology, 104*, pp.313–322.

Negroponte, N. (1975). *Soft Architecture Machines*. Cambridge: MIT Press.

Nesbitt, K. (1996). *Theorizing a New Agenda for Architecture: An anthology of architectural theory 1965–1995*. New York: Princeton Architectural Press.

Newitz, A. (21 May 2013a). 'Paolo Soleri and the cities of the future.' *io9*. [online] Available at: http://io9.com/paolo-soleri-and-the-cities-of-the-future-509049258. [Accessed 22 May 2014].

Newitz, A. (4 April 2013b). 'Tomorrow's material world is being built at the molecular foundry.' *io9*. [online] Available at: http://io9.com/tomorrows-material-world-is-being-built-at-the-molecul-466082320. [Accessed 19 April 2014].

Nicholson, D. (2009). 'Is the organism really a machine?' *Proceedings of the Eleventh International Conference on the Simulation and Synthesis of Living Systems Abstracts*, p.48.

Nielsen, A.T. & Houlihan, W.J. (1968). 'The aldol condensation.' *Organic Reactions, 16*, pp.1–438.

Nielsen, J.R. (1976). *Niels Bohr Collected Works, Vol. 3: The correspondence principle (1918–1923)*. Amsterdam: North-Holland Publishing.

Nimmrichter, S. & Hornberger, K. (2013). 'Macroscopity of mechanical quantum superposition states.' *Physics Review Letters, 110*, pp.1–11. [online] Available at: http://arxiv.org/pdf/1205.3447v2.pdf [Accessed 26 April 2014].

Nobelprize.org. (26 April 2013). 'History and properties of liquid crystals.' [online] Available at: http://www.nobelprize.org/educational/physics/liquid_crystals/history/index.html. [Accessed 27 April 2014].

Novak, M. (1992). 'Liquid architectures in cyberspace.' In: M. Benedikt (Ed.), *Cyberspace: First Steps* (pp.225–251). Cambridge (MA): The MIT Press.

Nye, J. (1990). *Bound to Lead: The changing nature of American power*. New York: Basic Books.

Odum, H.T. (1971). *Environment, Power and Society*. New York: John Wiley and Sons.

Oksiuta, Z. (2006). *Zbigniew Oksiuta: Breeding spaces*. Cologne: Buchhandlung Walther Konig GmbH & Co. KG. Abt. Verlag.

Olive Oil Source. (1998–2013). 'Chemical characteristics.' [online] Available at: http://www.oliveoilsource.com/page/chemical-characteristics. [Accessed 18 April 2014].

Oltean, M. (2009). 'Unconventional computing: a short introduction'. *Studia Universitatis Babes-Bolyai Series Informatica, 54*(1), pp.85–96.

Otto, F. (1996). *Finding form: Towards an architecture of the minimal*. Stuttgart: Edition Axel Menges.

Open Culture. (24 February 2013). 'Captivating collaboration: Artist Hubert Duprat uses insects to create golden sculptures.' [online] Available at: http://www.openculture.com/2013/02/artist_hubert_duprat_uses_insects_to_create_golden_sculptures.html. [Accessed 4 June 2014].

Orwell, G. (1979). *Animal Farm*. London: Heinemann.

Oxman, N. (24 September 2012). 'Fabvolution: Advances in digital fabrication.' *Materialecology*. [online] Available at: http://materialecology.blogspot.co.uk/2012/09/fabvolution-advances-in-digital.html. [Accessed 17 April 2014].

Palacci, J., Sacanna, S., Steinberg, A.P., Pine, D.J. & Chaikin, P.M. (2013). 'Living crystals of light-activated colloidal surfers.' *Science, 339*(6119), p.1.

Palmer, J. (11 January 2010) 'Chemical computer that mimics neurons to be created.' *BBC News*. [online] Available at: http://news.bbc.co.uk/1/hi/8452196.stm. [Accessed 22 April 2014].

Pask, G. (1995). 'Foreword.' In: J. Frazer (Ed.), *An Evolutionary Architecture* (pp.6–8). London: Architectural Association.

Parker, J. (24 February 2011). 'The 9 billion-people question.' *The Economist*. [online] Available at: http://www.economist.com/node/18200618. [Accessed 25 April 2014].

Pasteur, L. (1866). *Études Sur le Vin: Ses maladies, causes qui les provoquent, procédés nouveaux pour le conserver et pour le vieillir*. Paris: A'Imprimerie Impériale.

Patel, R. (June 2011). 'Venetian binds.' *Icon Magazine*. [online] Available at: http://www.iconeye.com/read-previous-issues/icon-096-%7C-june-2011/venetian-binds. [Accessed 19 April 2014].

Paul, R. (9 September 2011). 'SPURSE and the BMW Guggenheim Lab explore NY's most astonishing post-natural landscape.' *Inhabitat*. [online] Available at: http://inhabitat.com/nyc/spurse-the-guggenheim-bmw-lab-explore-nys-most-astonishing-post-natural-landscape/2/. [Accessed 18 April 2014].

Paun G. (2005). 'Bio-inspired computing paradigms (natural computing) in unconventional programming paradigms.' *Lecture Notes in Computer Science, 3566*, pp.155–160.

Pawlowski, M., Paterek, T., Kaszlikowski, D., Scarani, V. Winter, A.J. & Zukowski, M. (2009). 'Information causality as a physical principle.' *Nature, 461*(7267), pp.1101–1104.

Pearce, M. & Spiller, N. (Eds.) (1995). *Architects in cyberspace*. Architectural Design. London: John Wiley & Sons.

Pfeifer, R. & Iida, F. (2005). 'Morphological computation: Connecting body, brain and environment.' *Japanese Scientific Monthly, 58*(2), pp.48–54.

Philip Beesley Architect. (2013a). 'Radiant Soil.' [online] Available at: http://www.philipbeesleyarchitect.com/sculptures/1218_Radiant-Soil_Paris/index.php. [Accessed 1 June 2014].

Philip Beesley Architect. (2013b). 'Protocell Mesh.' [online] Available at: http://www.philipbeesleyarchitect.com/sculptures/1208_Epiphyte-Grove/index.php. [Accessed 20 April 2014].

Philip Beesley Architect. (2012). 'Epiphyte Grove.' [online] Available at: http://www.philipbeesleyarchitect.com/sculptures/1208_Epiphyte-Grove/index.php. [Accessed 20 April 2014].

Philip Beesley Architect. (2011). 'Hylozoic Veil.' [online] Available at: http://www.philipbeesleyarchitect.com/sculptures/1016_The_Leonardo/index.php. [Accessed 20 April 2014].

Philippe Rahm Architects. (2002). 'Hormonorium.' [online] Available at: http://www.philipperahm.com/data/projects/hormonorium/. [Accessed 1 June 2014].

Phys.org. (26 April 2010). 'How shape-memory materials remember.' [online] Available at: http://phys.org/news191502654.html. [Accessed 27 April 2014].

'Physical Properties of Liquids.' (Not author-credited or dated). [online] Available at: http://www.trimen.pl/witek/ciecze/old_liquids.html. [Accessed 19 April 2014].

Pileni, M.P. (2006). 'Reverse micelles used as templates: a new understanding in nanocrystal growth.' *Journal of Experimental Nanoscience, 1*(1), pp.13–27.

Pohl, E.B. (20 October 2011). *KERB* 19 / 'Paradigms of Nature: Postnatural Futures,' [online] Available at: http://www.archdaily.com/177905/kerb-19-paradigms-of-nature-postnatural-futures/. [Accessed 2 November 2014].

Popescu, S. (2006). 'Quantum mechanics: Why isn't nature more non-local?' *Nature Physics, 2*, pp.507–508.

Popp, J. (2011). 'Bit.flow circular.' Vimeo. [online] Available at: http://vimeo.com/22390871. [Accessed 1 August 2014].

Popper, K. (1959). *The Logic of Scientific Discovery*. London: Hutchinson.

Prigogine, I. (1997). *The End of Certainty: Time, chaos and the new laws of nature*. 1st edition. New York: The Free Press.

Prigogine, I. & Stengers, I. (1984). *Order Out of Chaos: Man's new dialogue with nature*. London: Heinemann.

Pruned Blogspot. (23 January 2012). 'Gardens as crypto water computers. On landscape architecture and related fields.' [online] Available at: http://pruned.blogspot.co.uk/2012/01/gardens-as-crypto-water-computers.html. [Accessed 1 June 2014].

Quay, A. (9 August 2013). 'Japanese architects create floating cloud in a transparent room.' *Design Taxi*. [online] Available at: http://designtaxi.com/news/359909/Japanese-Architects-Create-Floating-Cloud-In-A-Transparent-Room/. [Accessed 9 August 2014].

Raconteur. (6 June 2012). 'Bacteria, active paints and aquariums.' [online] Available at: http://theraconteur.co.uk/bacteria-active-paints-and-aquariums/. [Accessed 17 April 2014].

Rasmussen, S., Bedau, M.A., Chen, L., Deamer, D., Krakauer, D.C., Packard, N.H. & Stadler, P.F. (2008). *Protocells: Bridging non-living and living matter*. Cambridge (MA): The MIT Press.

Rasmussen, S., Chen, L., Nilsson, M. & Abe, S. 2003. ;Bridging non-living and living matter.' *Artificial Life, 9*(3), pp.269–316.

Ravera, O. (2000). 'The lagoon of Venice: the result of both natural factors and human influence.' *Journal of Limnology, 59*(1), pp.19–30.

Ray, N. (2005). *Alvar Aalto*. New Haven: Yale University Press.

Resnick, L.B. (1989). Introduction. In: L.B. Resnick, (Ed.), *Knowing, Learning and Instruction: Essays in honor of Robert Glaser* (pp.1–24). Hillsdale: Erlbaum.

Reutgers. (21 October 2011). 'Professor 'going green' in New York City with lichen project.' [online] Available at: http://news.rutgers.edu/medrel/news-releases/2011/10/professor-going-gree-20111021. [Accessed 18 April 2014].

Robbins, J. (31 December 2012). 'Worms produce another kind of gold for farmers.' *The New York Times*. [online] Available at: http://www.nytimes.com/2013/01/01/science/worms-produce-another-kind-of-gold-for-farmers.html?_r=0. [Accessed 19 April 2014].

Roco, M.C. & Bainbridge, W.S. (2003). *Converging Technologies for Improving Human Performance, Nanotechnology, Biotechnology, Information Technology and Cognitive Science*. NSF/DOC-sponsored report. Dordrecht: Springer.

Rodriguez, C. (21 August 2012). 'From Bill Gates's toilets to Louvre aArt: A poop potpourri.' *Forbes*. [online] Available at: http://www.forbes.com/sites/ceciliarodriguez/2012/08/21/a-poop-potpourri-from-bill-gatess-toilets-to-a-louvre-exhibition/. [Accessed 19 April 2014].

Rosen, R. (1999). *Essays on Life Itself*. New York: Columbia University Press.

Rosen, R. (1996). 'On the limits of scientific knowledge.' In: J.L. Casti & A. Karlqvist (Eds.), *Boundaries and barriers: On the limits to scientific knowledge* (pp.199–214). Reading: Addison-Wesley.

Rowlinson, A. (2012). 'The woman regrowing the planet.' *Red Bulletin*, March 2012, pp.76–81.

Royal Society. (1 September 2009). 'Geoengineering the climate: science, governance and uncertainty.' [online] Available at: http://royalsociety.org/policy/publications/2009/geoengineering-climate/. [Accessed 17 May 2014].

Royal Society of Chemistry. (19 August 2011). 'Bacteria create missing 'atmospheric brooms' the sweep the sky clean.' [online] Available at: http://www.rsc.org/chemistryworld/News/2011/August/19081101.asp. [Accessed 19 April 2014].

Runge F.F. (1850). *Zur Farben-Chemie. Musterbilder für Freunde des Schönen und zum Gebrauch für Zeichner, Maler, verzierer und Zeugdrucker. 1. Lieferung. Dargestellt durch chemische Wechelwirkung*. Berlin: Verlage E. S. Mittler und Sohn.

Ruskin, J. (1989). *The seven lamps of architecture*. Mineola: Dover Publications.

Sadler, P. & Smart, V. (2012). Tomorrow's value, Tomorrow's value lecture series, Tomorrow's Company. [online] Available at: http://tomorrowscompany.com/tomorrows-value-lecture-series. [Accessed 17 April 2014].

Sagan, C. (2 April 2007). 'Pale blue dot.' YouTube. [video] Available at: http://www.youtube.com/watch?v=p86BPM1GV8M. [Accessed 18 April 2014].

Sample, I. (28 July 2009). 'Human activity is driving Earth's sixth great extinction event.' *The Guardian*. [online] Available at: http://www.guardian.co.uk/environment/2009/jul/28/species-extinction-hotspots-australia. [Accessed 19 April 2014].

Savory, A. (1998). *Holistic Resource Management*. Washington: Island Press.

Savory, A. (1991). 'Holistic resource management: a conceptual framework for ecologically sound economic modelling.' *Ecological Economics, 3*, pp.181–191.

Schrödinger, E. (1944). 'What is life? The physical aspect of the living cell. Based on lectures delivered under the auspices of the Dublin Institute for Advanced Studies at Trinity College, Dublin, in February 1943.' [online] Available at: http://whatislife.stanford.edu/LoCo_files/What-is-Life.pdf. [Accessed 16 April 2013].

Schumacher, P. (2009). 'A new global style for architecture and urban design.' *Architectural Design, 79*(4), pp.14–23.

Schuster, P. (2007). 'Self-organization: An old paradigm revisited: Nonlinear dynamics from physics to biology.' *Complexity, 14*(4), pp.9–11.

Schwartz, J. (13 May 2008). 'Museum kills live exhibit.' Science, *The New York Times*. [online] Available at: http://www.nytimes.com/2008/05/13/science/13coat.html?_r=4&partner=rssnyt&emc=rss&oref=slogin&. [Accessed 18 April 2014].

Science Daily. (8 April 2013). 'Spooky action at a distance' aboard the International Space Station'. Science News. [online] Available at: http://www.sciencedaily.com/releases/2013/04/130409095412.htm. [Accessed 27 April 2014].

Science Daily. (25 June 1999). 'Physicist finds out why "We are stardust..."' [online] Available at: http://www.sciencedaily.com/releases/1999/06/990625080416.htm. [Accessed 17 April 2014]. Searle, A. (4 September 2008). 'Don't forget your wellies.' *The Guardian*. [online] Available at: http://www.guardian.co.uk/artanddesign/2008/sep/04/art. [Accessed 30 May 2014].

Secret Cinema. (2 July 2012). 'Prometheus, are you seeing this?' Available at: http://www.secretcinema.org/newsletter/nl6.html. [Accessed 18 April 2014].

Seibt, J. (15 October 2012). 'Process philosophy.' *Stanford Encyclopedia of Philosophy*. [online] Available at: http://plato.stanford.edu/entries/process-philosophy/. [Accessed 16 April 2014].

Sellars, S. (19 December 2011). 'Living architecture.' *Australian Design Review*. [online] Available at: http://www.australiandesignreview.com/features/15151-interview-rachel-armstrong. [Accessed 19 April 2014].

Serb, J.M. & Eernisse, D.J. (2008). 'Charting evolution's trajectory: using molluscan eye diversity to understand parallel and convergent evolution.' *Evolution: Education and Outreach, 1*, pp.439–447.

Shark Bay World Heritage Area. (2009). 'Nature of Shark Bay.' [online] Available at: http://www.sharkbay.org/stromatolites.aspx. [Accessed 19 April 2014].

Siggraph. (Not dated). 'Hot pool. Driessen and Verstappen.' [online] Available at: http://www.siggraph.org/connect/digital-arts/siggraph2011/artists/driessenVerstappen.html. [Accessed 9 August 2014].

Simonite, T. (28 April 2009). 'Chemical caterpillar points towards electronics free robots.' *New Scientist*. [online] Available at: http://www.newscientist.com/article/dn16910-chemical-caterpillar-points-to-electronicsfree-robots.html, [Accessed 17 April 2014].

Sis, P. (2000). *The Starry Messenger: Galileo Galilei*. Zephyr-Cove: Sunburst.

Smithsonian Institution Archives. (2004). 'The history of PCR.' [online] Available at: http://siarchives.si.edu/research/videohistory_catalog9577.html [Accessed 18 April 2014].

Smolin, L. (1997). *The Life of the Cosmos*. Oxford: Oxford University Press.

Smuts, J.C. (1998). *Holism and Evolution*. Gouldsboro: Gestalt Journal Press.

Snow, C.P. (1959). *The Two Cultures*. London: Cambridge University Press.

Sontag, S. (1988). *Aids and its Mmetaphors*. London: Penguin.

Spanos, N. & Koutsoukos, P.G. (1998). 'Kinetics of precipitation of calcium carbonate in alkaline pH at constant supersaturation. Spontaneous and seeded growth.' *Journal of Physical Chemistry B, 102*(34), pp.6679–6684.

Spiller, N. (2011). 'Protocells: The universal solvent.' *Architectural Design, 81*(2), pp.60–67.

Spiller, N. & Armstrong, R. (2011). 'It's a brand new morning.' *Architectural Design, 81*(2), pp.14–23.

Spiller, N., Armstrong, R. & Hanczyc, M.M. (2011). 'A manifesto for protocell architecture: Against biological formalism.' *Architectural Design, 81*(2), pp.24–25.

Spiller, N. (2007). *Visionary Architecture: Blueprints of the modern imagination*. Reprint edition. London: Thames and Hudson.

Spuybroek, L. (2011). *The Sympathy of Things: Ruskin and the ecology of design*. Rotterdam: NAi010/V2.

Stamp, J. (23 November 2004). 'Bernard Tschumi's 'Advertisement for Architecture'.' [online] Available at: http://lifewithoutbuildings.net/2004/11/bernhard-tschumis-advertisement-for-architecture-1978.html. [Accessed 12 May 2014].

Stanford News. (19 July 2012). 'Stanford researchers produce first complete computer model of an organism.' [online] Available at: http://news.stanford.edu/news/2012/july/computer-model-organism-071812.html. [Accessed 19 April 2014].

Steadman, I. (16 April 2013). 'Hamburg unveils world's first algae-powered building.' *Wired*. [online] Available at: http://www.wired.com/design/2013/04/algae-powered-building/. [Accessed 18 April 2014].

Stengers, I. (2000). 'God's heart and the stuff of life.' *Pli, 9*, pp.86–118.

Stengers, I. (Not dated). 'Gilles Deleuze's last message.' [online] Available at: http://www.recalcitrance.com/deleuzelast.htm. [Accessed 19 April 2013].

Sterling, B. (4 February 2013). 'Architecture fiction: future perfect and Lisbon Architecture Triennale.' Beyond the Beyond. *Wired*. [online] Available at: http://www.wired.com/beyond_the_beyond/2013/02/architecture-fiction-future-perfect-and-lisbon-architecture-triennale/ [Accessed 18 April 2014].

Sterling, B. (3 January 2012). 'Design fiction: Growth assembly. Sascha Pohflepp and Daisy Ginsberg.' Beyond the Beyond. *Wired*. [online] Available at: http://www.wired.com/beyond_the_beyond/2012/01/design-fiction-sascha-pohflepp-daisy-ginsberg-growth-assembly/. [Accessed 18 April 2014].

Sterling, B. (21 December 2011). 'Architecture fiction: Rachel Armstrong.' Beyond the Beyond. *Wired*. [online] Available at: http://www.wired.com/beyond_the_beyond/2011/12/architecture-fiction-rachel-armstrong/. [Accessed 17 April 2014].

Su, H., Cheng, Y., Oswald, R., Behrendt, T., Trebs, I., Meixner, F.X., Andreae, M.O., Cheng, P., Zhang, Y. & Pöschl, U. (2011). 'Soil nitrite as a source of atmospheric HONO and OH radicals.' *Science, 333*(6049), pp.1616–1618.

Sustainable Now Technologies. (18 September 2012). 'Sustainable Now Technologies discusses the development of the 'energy conversion machine'.' *YouTube*. [online] Available at: http://www.youtube.com/watch?v=jbxlyDj7PX4. [Accessed 18 April 2014].

Swartz, J. (2 November 2010). 'Can profit and sustainability go together?' Guardian sustainable business. *The Guardian*. [online] Available at: http://www.guardian.co.uk/sustainable-business/blog/jeff-swartz-timberland-sustainability-profit. [Accessed 25 April 2014].

Synth-ethic. (2011). 'Synth-ethic: Art and Synthetic Biology Exhibition, Vienna, Austria.' Biofaction. [online] Available at: http://www.Biofaction.com/synth-ethic/. [Accessed 19 April 2014].

Szostak, J.W., Bartel, D.P. & Luisi, P.L. (2001). 'Synthesizing life.' *Nature, 409*, pp.387–390.

Tangen, U. & McCaskill, J.S. (2004-2008). *From Chemotaxis to Soft Robots*. PACE Report, Programmable Artificial Cell Evolution. [online] Available at: http://www.istpace.org/Web_Final_Report/Applications/Appl_potent_future_cells/descr_app_fut_cells/chemotaxis_soft_robots/index.html, [Accessed March 2014].

Teatini, P., Castelletto, N., Ferronato, M., Gambolati, G. & Tosi, L. (2011). 'A new hydrogeologic model to predict anthropogenic uplift of Venice.' *Water Resources Research, 47*(12), W12507.

TED.com. (February 2013a). 'Allan Savory. How to green the desert and reverse climate change.' [online] Available at: http://www.ted.com/talks/allan_savory_how_to_green_the_world_s_deserts_and_reverse_climate_change.html. [Accessed 19 April 2014].

TED.com. (February 2013b). 'Skylar Tibbits. The emergence of 4D printing.' [online] Available at: http://www.ted.com/talks/skylar_tibbits_the_emergence_of_4d_printing.html, [Accessed 19 April 2014].

TED.com. (March 2011). 'Aaron O'Connell: Making sense of a visible quantum object.' [online]. Available at: http://www.ted.com/talks/aaron_o_connell_making_sense_of_a_visible_quantum_object.html. [Accessed 25 May 2014].

Telegraph. (2013). 'Parts of Britain under water as torrential rain causes flash floods.' [online] Available at: http://www.telegraph.co.uk/topics/weather/9326993/Weather-parts-of-Britain-under-water-as-torrential-rain-causes-flash-floods.html. [Accessed 19 April 2014].

Textile Futures Research Centre. (Not dated). *En Vie/Alive*. University of the Arts London. [online] Available at: http://www.tfrc.org.uk/alive-carole-collet/. [Accessed 19 April 2014].

Thackara, J. (2 December 2012). Venice: From gated lagoon to bioregion. *Doors of Perception*. [online] Available at: http://www.doorsofperception.com/sustainability-design/venice-from-gated-lagoon-to-bioregion/#more-4170. [Accessed 19 April 2014].The Laboratory at Harvard. (2013). 'Artscience.' [online] Available at: http://thelaboratory.harvard.edu/concept/artscience/. [Accessed 18 April 2014].

'Theo Jansen's Strandbeest.' (Not author-credited or dated). [online] Available at: http://www.strandbeest.com. [Accessed 19 April 2014].

Thompson, W.D. (1917). *On Growth and Form*. Cambridge: Cambridge University Press.

'Thrilling Wonder Stories.' (Not author-credited or dated). Archive. [online] Available at: http://www.thrillingwonderstories.co.uk. [Accessed 11 August 2014].

Toyota, T., Maru, N., Hanczyc, M.M., Ikegami, T. & Sugawara, T. (2009). 'Self-propelled oil droplets consuming "fuel" surfactant.' *Journal of the American Chemical Society, 131*(14), pp.5012–5013.

Traube, M. (1867). 'Experimente zur Theorie der Zellbildung und Endomose.' *Archiv fur Anatomie, Physiologie und Wissenschaftliche Medicin*, 87, pp.129–165.

Treehugger. (25 July 2005). 'It slices, it dices: Pollution eating concrete.' [online] Available at: http://www.treehugger.com/green-architecture/it-slices-it-dices-pollution-eating-concrete.html. [Accessed 17 April 2014].

Tschumi, B. (2012). *Architecture Concepts: Red is not a colour*. New York: Rizzoli International Publications.

Turchin, P. (2009). 'Long-term population cycles in human societies: The year in ecology and conservation biology.' *Annals of the New York Academy of Sciences, 1162*, pp.1–17.

Turney, J. (2013). *Imagining Technology*. NESTA working paper. [online] Available at: http://www.nesta.org.uk/library/documents/Imagining_Technology.pdf. [Accessed 19 April 2014].

Turing, A.M. (1952). 'The chemical basis of morphogenesis.' *Philosophical Transactions of the Royal Society of London. Series B, Biological Sciences, 237*(641), pp.37–72.

Turing, A.M. (1936). 'On computable numbers, with an application to the Entscheidungs problem.' *Proceedings of the London Mathematical Society, 2*(42), pp.230–265.

Turrent, D. (2008). *Sustainable Architecture*. London: RIBA Publications.

Tyson, J.T. (1994). 'What everyone should know about the Belousov–Zhabotinsky reaction.' In S.A. Levin (Ed.), *Frontiers in mathematical biology* (pp.569–587). New York: Springer Verlag,

UC Davis. (9 April 2013). 'New green technology squeezes out building leaks. 'UC Davis News and Information. [online] Available at: http://news.ucdavis.edu/search/news_detail.lasso?id=10553. [Accessed 21 May 2014].

UK ArtScience Prize. (18 April 2013). 'Rachel Armstrong.' [video] Available at: http://www.youtube.com/watch?v=B8Hh6z1mt5E. [Accessed 22 April 2014].

Ulissi, Z.W., Strano, M.S. & Braatz, R.D. (2013). 'Control of nano and microchemical systems.' *Computers and Chemical Engineering, 51*, pp.149–156.

University of Central Florida. (26 April 2013). 'Computer scientists suggest new spin on origins of evolvability.' *UCF Today*. [online] Available at: http://today.ucf.edu/computer-scientists-suggest-new-spin-on-origins-of-evolvability/. [Accessed 27 April 2014].

Ungerleider, N. (Not dated). 'The toilet of the future will turn poop into power.' *Fast CoExist*. [online] Available at: http://www.fastcoexist.com/1678292/the-toilet-of-the-future-will-turn-poop-into-power. [Accessed 19 April 2014].

United Nations Office for Disaster Risk Reduction. (21 December 2012). 'US$6.7 billion floodgates to protect Venice in 2014.' [online] Available at: http://www.unisdr.org/archive/30174. [Accessed 19 April 2014].

University of Oxford. (5 April 2013). '3D printer can build synthetic tissues.' [online] Available at: http://www.ox.ac.uk/media/news_stories/2013/130405.html [Accessed, 18 April 2014].

University of Southampton. (Not dated). 'Artificial wet neuronal networks from compartmentalised excitable chemical media.' *Electronics and Computer Science*. [online] Available at: http://www.ecs.soton.ac.uk/research/projects/735. [Accessed 2 June 2014].

University of Southern Denmark. (23 August 2012). 'Initiative for science, society and policy.' [online] Available at: http://www.sdu.dk/en/Om_SDU/Institutter_centre/C_ISSP. [Accessed 17 April 2014].

US Department of Energy. (15 December 2011). 'Biomass basics.' [online] Available at: http://www1.eere.energy.gov/biomass/biobasics.html. [Accessed 25 April 2014].

US National Research Council. (2010). *Advancing the Science of Climate Change, America's climate choices*. Washington: National Academies Press. [online] Available at: http://dge.stanford.edu/labs/caldeiralab/Caldeira_research/pdf/ACC_Science_2010.pdf. [Accessed 18 April 2014].

Vaihinger, H. (1968). *The Philosophy of 'As If': A system of the theoretical, practical and religious fictions of mankind*. T.C.K. Ogden translation. New York: Barnes and Noble.

Van Der Eijk, X. (2013). 'Xandra van der Eijk.' [online] Available at: http://xandravandereijk.com. [Accessed 7 June 2014].

Van Mensvoort, K. & Grievink, H.J. (2012). *Next Nature: Nature changes along with us*. Barcelona: Actar.

Venter, J.C. (2007). *A Life Decoded: My genome, my life*. London: Allen Lane.

Vernadsky, V.I. (2007). *Geochemistry and the Biosphere*. Santa Fe: Synergetic Press.

Vernadsky, V.I. (1998). *The Biosphere*. Complete annotated edition. Göttingen: Copernicus.

Vernadsky, V.I. (1945). 'The biosphere and the noösphere.' *American Scientist, 33*(1), pp.1–12.

Villar, G., Graham, A.D. & Bayley, H. (2013). 'A tissue-like printed material.' *Science, 340*(6128), pp.48–52.

Viswanatha, D.S., Ghosh, T.K., Prasad, D.H.L., Dutt, N.V.K. & Rani, K.Y. 2007. *Viscosity of Liquids: Theory, estimation, experiment, and data*. Dordrecht: Springer.

Volk, T. (1995). *Metapatterns: Across space, time, and mind*. New York: Columbia University Press.

Von Bertalanffy, L. (1969). *General System Theory: Foundations, development, applications*. Revised edition. New York: George Braziller.

Von Bertalanffy, L. (1950). 'An outline for general systems theory.' *British Journal for the Philosophy of Science, 1*(2), pp.134–165.

Waddington, C.H. (1957). *The Strategy of the Genes*. London: George Allen and Unwin.

Warneke, B., Last, M., Liebowitz, B. & Pister, K.S.J. (2001). 'Smart dust: Communicating with a cubic millimeter.' *Computer, 34*, pp.44–51.

Warr, P. (4 March 2013). 'Bioengineers produce 'Google Map' of human metabolism.' *Wired UK*. [online] Available at: http://www.wired.co.uk/news/archive/2013-03/04/map-of-human-metabolism. [Accessed 19 April 2014].

Watson, J.D. & Crick, F.H.C. (1953). 'A structure for deoxyribose nucleic acid.' *Nature, 171*, pp.737–738.

Watts Towers. (2006–2013). 'The history of the Watts Towers by Sam Rodia.' Official site. [online] Available at: http://www.wattstowers.us/history.htm. [Accessed 25 May 2014].

Webster, G. (14 October 2011). 'Living buildings could inhale city carbon emissions.' CNN. [online] Available at: http://edition.cnn.com/2011/10/14/tech/innovation/living-buildings-carbon/index.html. [Accessed 18 April 2014].

Weintraub, A. & Hess, A. (2012). *Frank Lloyd Wright Natural Design: Lessons for building green from an American original*. New York: Rizzoli International Publications.

Weizmann Institute of Science News. (23 July 2012). 'Fools' gold found to regulate oxygen.' Environment. [online] Available at: http://wis-wander.weizmann.ac.il/fools-gold-found-to-regulate-oxygen#.UXKRyqVt1SU [Accessed 19 April 2014].

Wellcome Trust. (2009). 'Insight and exchange: An evaluation of the Wellcome Trust's Sciart programme.' [online] Available at: www.wellcome.ac.uk/sciartevaluation. [Accessed April 2014].

WETFab. (24–25 January 2011). Event booklet. School of Chemistry, University of Glasgow. [online] Available at: http://www.chem.gla.ac.uk/cronin/files/news/WETFAB.pdf. [Accessed 26 May 2014].

Whitehead, A.N. (1979). *Process and Reality* (Gifford Lectures). 2nd revised edition. New York: The Free Press.

Whitehead, A.N. (1970). *Science and the Modern World*. New York: Simon and Schuster.

Whitehead, A.N. (1968). *Modes of Thought*. New York: The Free Press.

Whitehouse, D. (13 August 1999). 'Once upon a time 3.5 billion years ago.' Sci/Tech. BBC News. [online] Available at: http://news.bbc.co.uk/1/hi/sci/tech/418540.stm. [Accessed 19 April 2014].

Whitelaw, M. (2003). 'Morphogenetics: generative processes in the work of Driessens and Verstappen.' *Digital Creativity, 14*(1), pp.43-53.

Whitelaw, S., Beattie, A., Balogh, R. & Watson, J. (2003). *A Review of the Nature of Action Research*. Cardiff: Welsh Assembly Government.

Whitesides Research Group. (2011). 'Soft robotics, research.' Harvard University Department of Chemistry and Chemical Biology. [online] Available at: http://gmwgroup.harvard.edu/research/index.php?page=23. [Accessed 18 April 2014].

Wickson, F. (2008). 'What is nature, if it's more than just a place without people?' *Nature, 456*(29), p.6.

Wiener, N. (1998). *The Human Use of Human Beings: Cybernetics and society* (Da Capo Paperback). New edition. Cambridge: DaCapo Press.

Wiener, N. (1948). *Cybernetics or Control and Communication in the Animal and the Machine*. Cambridge: MIT Press.

Williams, K. (26 February 2009). 'Examined Life, a film by Astra Taylor.' The sly fox film reviews. [online] Available at: http://www.kamwilliams.com/2009/02/examined-life.html. [Accessed 1 June 2014].

Willis, A.J. 1997. 'The ecosystem: An evolving concept viewed historically.' *Functional Ecology, 11*(2), pp. 268–271.

Wimberley, E.T. (2009). *Nested Ecology: The place of humans in the ecological hierarchy*. Baltimore: Johns Hopkins University Press.

Wöhler, F. (1828). 'Ueber künstliche Bildung des Harnstoffs.' *Annalen der Physik und Chemie, 88*(2), pp.253–256.

Woods, L. 27 (January 2012a). 'A tree is a tree is a ... building? Lebbeus Woods.' [online] Available at: http://lebbeuswoods.wordpress.com/2012/01/27/a-tree-is-a-tree-is-a-building/. [Accessed 22 May 2014].

Woods, L. (9 July 2012b). 'Inevitable architecture. Lebbeus Woods.'[online] Available at: http://lebbeuswoods.wordpress.com/2012/07/09/inevitable-architecture/ [Accessed 17 April 2014.].

Woods, L. (28 November 2011). 'Return of the manifesto. Lebbeus Woods.' [online] Available at: http://lebbeuswoods.wordpress.com/2011/11/28/return-of-the-manifesto/. [Accessed 10 August 2014].

Woody, T. (5 September 2009). 'Alcoa's self-cleaning, smog, eating, buildings.' *Forbes*. [online] Available at: http://www.forbes.com/sites/toddwoody/2011/05/09/alcoas-self-cleaning-smog-eating-buildings/. [Accessed 20 April 2014].

World Commission on Environment and Development. (1987). *Our Common Future*. Report of the World Commission on Environment and Development. Published as Annex to General Assembly document A/42/427. Available at: http://www.un-documents.net/our-common-future.pdf. [Accessed 25 April 2014].

WSP. (Not dated). 'Built ecology: We are WSP built ecology.' [online] Available at: http://www.wspgroup.com/en/Built-Ecology/homepage-BE/About-Us/. [Accessed 18 April 2014].

Yarkoni, O., Donlon, L. & Frankel, D. (2012). 'Creating a bio-hybrid signal transduction pathway: Opening a new channel of communication between cells and machines.' *Bioinspiration and Biomimetics, 7*(4), [046017].

YouTube. (27 May 2013). 'Giant swirling straws help a galaxy grow.' NASA Jet Propulsion Laboratory Space Science HD video. [video] Available at: http://www.guardian.co.uk/science/video/2013/may/28/galaxy-formation-universe-video. [Accessed 29 May 2014].

YouTube. (6 October 2011). 'A magazine is an iPad that doesn't work.' [video] Available at: http://www.youtube.com/watch?v=aXV-yaFmQNk. [Accessed 30 April 2014].

YouTube. (16 March 2009). *Extreme Sheep LED Art*. [video]. Available at: http://www.youtube.com/watch?v=D2FX9rviEhw. [Accessed 30 April 2014].

YouTube. (9 March 2008). *Brigadoon 1954*, Trailer. [video] Available at: http://www.youtube.com/watch?v=61MEKyAQSpg. [Accessed 27 April 2014].

Young, L. (2014). *Brave New Now*. Lisbon Architecture Triennale. Close, Closer ebook series. Kindle. [online] Available at: http://www.amazon.com/Brave-New-Lisbon-Architecture-Triennale-ebook/dp/B00KDOGRF0. [Accessed 2 June 2014].

Young, L. (2011). 'Specimens of an unnatural history: A near future bestiary.' In: C. Daly, S. Hicks, A. Keene & R.R. Ricardo (Eds.), *Paradigms of nature: Post natural futures. Kerb 19, Journal of Landscape Architecture* (pp.12–15). Melbourne: Melbourne Books.

Zarinabadi, S., Kharrat, R. & Yazdi, A.V. (2010). 'Estimate solubility of canola oil (oleic acid) in supercritical carbon dioxide: experimental and modeling.' *Journal of American Science, 6*, p.10.

Zhabotinsky, A.M. (1964). 'Periodic processes of malonic acid oxidation in a liquid phase.' *Biofizika, 9*, pp.306–311.

Zimmerman, C. & Gossel, P. (2007). *Mies van der Rohe. Less is more: finding perfection in purity* (Taschen Basic Architecture Series). Cologne: Taschen GmbH.

Žižek, S. (2011). *Living in the End Times*. London: Verso.

Zyga, L. (28 March 2013). 'Artificial muscle computer performs as a universal Turing machine.' *Phys. Org*. [online] Available at: http://phys.org/news/2013-03-artificial-muscle-universal-turing-machine.html, [Accessed 18 April 2014].

List of Figures

List of Movies

The following movies were taken during my period of laboratory research between July 2009 and December 2012 and uploaded to YouTube. I performed all the experiments and took the footage in these films except where I note otherwise. Many more videos of my laboratory experiments are archived and documented at TheGrayanat channel on YouTube.

has an immediate effect on the Bütschli system when added directly to the oil medium, causing the droplets to move quickly towards the source. The droplets become very agitated but this excitement is quickly quenched. Magnification 4× — **108**

Movie 6.23: YouTube. (14 December 2011). Multiple interfacing Bütschli droplets under the influence of butan-1-ol. [video] Available at: http://www.youtube.com/watch?v=r30w2ScbBHA. [Accessed 30 April 2014]. Bütschli droplets appear to form assemblages more readily in the presence of butan-1-ol. Magnification 4× — **110**

Movie 8.1: YouTube. (23 December 2011). Growth of Traube cell membrane, light microscopy 10×. [video] Available at: http://www.youtube.com/watch?v=vA_Y87DNkeo. [Accessed 28 May 2014]. Detail of Traube cell membrane undergoing almost instantaneous rupture and repair following the build-up of osmotic pressure inside a copper hexacyanoferrate semi-permeable membrane, which is produced when copper sulphate and potassium hexacyanoferrate react. The growth of the membrane is jerky and produces an undulating, seaweed-like structure. Magnification 40× — **178**

List of Tables

Index

Bei Fragen zur Produktsicherheit wenden Sie sich bitte an:
If you have any questions regarding product safety,
please contact:

Walter de Gruyter GmbH
Genthiner Straße 13
10785 Berlin
productsafety@degruyterbrill.com